Lecture Notes in Morphogenesis

Series editor

Alessandro Sarti, CREA/CNRS, Paris, France
e-mail: alessandro.sarti@polytechnique.edu

For further volumes:
http://www.springer.com/series/11247

Giuseppe Longo · Maël Montévil

Perspectives on Organisms

Biological Time, Symmetries and Singularities

Giuseppe Longo
Centre Interdisciplinaire Cavaillès
  (CIRPHLES)
CNRS and Ecole Normale Supérieure
Paris
France

Maël Montévil
Anatomy and Cell Biology
Tuft University
Boston
USA

ISSN 2195-1934   ISSN 2195-1942   (electronic)
ISBN 978-3-642-35937-8   ISBN 978-3-642-35938-5   (eBook)
DOI 10.1007/978-3-642-35938-5
Springer Heidelberg New York Dordrecht London

Library of Congress Control Number: 2013954680

© Springer-Verlag Berlin Heidelberg 2014

This work is subject to copyright. All rights are reserved by the Publisher, whether the whole or part of the material is concerned, specifically the rights of translation, reprinting, reuse of illustrations, recitation, broadcasting, reproduction on microfilms or in any other physical way, and transmission or information storage and retrieval, electronic adaptation, computer software, or by similar or dissimilar methodology now known or hereafter developed. Exempted from this legal reservation are brief excerpts in connection with reviews or scholarly analysis or material supplied specifically for the purpose of being entered and executed on a computer system, for exclusive use by the purchaser of the work. Duplication of this publication or parts thereof is permitted only under the provisions of the Copyright Law of the Publisher's location, in its current version, and permission for use must always be obtained from Springer. Permissions for use may be obtained through RightsLink at the Copyright Clearance Center. Violations are liable to prosecution under the respective Copyright Law.

The use of general descriptive names, registered names, trademarks, service marks, etc. in this publication does not imply, even in the absence of a specific statement, that such names are exempt from the relevant protective laws and regulations and therefore free for general use.

While the advice and information in this book are believed to be true and accurate at the date of publication, neither the authors nor the editors nor the publisher can accept any legal responsibility for any errors or omissions that may be made. The publisher makes no warranty, express or implied, with respect to the material contained herein.

Printed on acid-free paper

Springer is part of Springer Science+Business Media (www.springer.com)

To Francis Bailly,
 for his humanism in science, his discreet enthusiasm, his openness to others' ideas while staying firm in his principles, his driven commitment to understand the thinking of others, his trusting generosity in the common endeavour to knowledge, his critical thinking tailored to better advance beyond the mainstream.

Foreword

by *Denis Noble*

During most of the twentieth century experimental and theoretical biologists lived separate lives. As the authors of this book express it, "there was a belief that experimental and theoretical thinking could be decoupled." This was a strange divorce. No other science has experienced such a separation. It is inconceivable that physical experiments could be done without extensive mathematical theory being used to give quantitative and conceptual expression to the ideas that motivate the questions that experimentalists try to answer. It would be impossible for the physicists at the large hadron collider, for example, to search for what we call the Higgs boson without the theoretical background that can make sense of what the Higgs boson could be. The gigantic masses of data that come out of such experimentation would be an un-interpretable mass without the theory. Similarly, modern cosmology and the interpretation of the huge amounts of data obtained through new forms of telescopes would be inconceivable without the theoretical structure provided by Einstein's general theory of relativity. The phenomenon of gravitational lensing, for example, would be impossible to understand or even to discover. The physics of the smallest scales of the universe would also be impossible to manage without the theoretical structure of quantum mechanics.

So, how did experimental biology apparently manage for so many years without such theoretical structures? Actually, it didn't. The divorce was only apparent.

First, there was a general theoretical structure provided by evolutionary biology. Very little in biology makes much sense without the theory of evolution. But this theory does not make specific predictions in the way in which the Higgs boson or gravitational lensing were predicted for physicists. The idea of evolution is more that of a general framework within which biology is interpreted.

Second, there was theory in biology. In fact there were many theories, and in many different forms. Moreover, these theories were used by experimental biologists. They were the ideas in the minds of experimental biologists. No science can be done without theoretical constructs. The so-called Central Dogma of Molecular Biology, for example, was an expression of the background of ideas that were circulating during the early heydays of molecular biology: that causation was one way (genes to phenotypes), and that inheritance was entirely attributable to DNA, by which an organism could be completely defined. This was a theory, except that it was not formulated as such. It was presented as fact, a *fait accompli*. Meanwhile the pages of journals of theoretical and mathematical biology continued to be filled with fascinating and difficult papers to which experimentalists, by and large, paid little or no attention.

We can call the theories that experimentalists had in mind implicit theories. Often they were not even recognised as theory. When Richard Dawkins wrote his persuasive book *The Selfish Gene* in 1976 he was not only giving expression to many of these implicit theories, he also misinterpreted them through failing to understand the role of metaphor in biology. Indeed, he originally stated "that was no metaphor"! As Poincaré pointed out in his lovely book *Science and Hypothesis* (*La science et l'hypothèse*) the worst mistakes in science are made by those who proudly proclaim that they are not philosophers, as though philosophy had already completed its task and had been completely replaced by empirical science. The truth is very different. The advance of science itself creates new philosophical questions. Those who tackle such questions are philosophers, even if they do not acknowledge that name. That is particularly true of the kind of theory that could be described as meta-theory: the creation of the framework within which new theory can be developed. I see creating that framework as one of the challenges to which this book responds.

Just as physicists would not know what to do with the gigantic data pouring out of their colliders and telescopes without a structure of interpretative theory, biology has hit up against exactly the same problem. We also are now generating gigantic amounts of genomic, proteomic, metabolomic and physiomic data. We are swimming in data. The problem is that the theoretical structures within which to interpret it are underdeveloped or have been ignored and forgotten. The cracks are appearing everywhere. Even the central theory of biology, evolution, is undergoing reassessment in the light of discoveries showing that what the modern synthesis said was impossible, such as the inheritance of acquired characters, does in fact occur. There is an essential incompleteness in biological theory that calls out to be filled.

That brings me to the question how to characterise this book. It is ambitious. It aims at nothing less than filling that gap. It openly aims at bringing the rigour of theory in physics to bear on the role of theory in biology. It is a highly welcome challenge to theorists and experimentalists alike. My belief is that, as we progressively make sense of the masses of experimental data we will find ourselves developing the conceptual foundations of biology in rigorous mathematical forms. One day (who knows when?), biology will become more like physics in this respect: theory and experimental work will be inextricably intertwined.

However, it is important that readers should appreciate that such intertwining does not mean that biology becomes, or could be, reducible to physics. As the authors say, even if we wanted such a reduction, to what physics should the reduction occur? Physics is not a static structure from which biologists can, as it were, take things 'off the shelf'. Physics has undergone revolutionary change during the last century or so. There is no sign that we are at the end of this process. Nor would it be safe to assume that, even if it did seem to be true. It seemed true to early and mid-nineteenth century biologists, such as Jean-Baptiste Lamarck, Claude Bernard, and many others. They could assume, with Laplace, that the fundamental laws of nature were strictly deterministic. Today, we know both that the fundamental laws do not work in that way, and that stochasticity is also important in biology. The lesson of the history of science is that surprises turn up just when we think we have achieved or are approaching completeness.

The claim made in this book is that there is no current theory of biological organisation. The authors also explain the reason for that. It lies in the multi-level nature of biological interactions, with lower level molecular processes just as dependent on higher-level organisation and processes, as they in their turn are dependent on the molecular processes. The error of twentieth century biology was to assume far too readily that causation is one-way. As the authors say, "the molecular level does not accommodate phenomena that occur typically at other *levels of organisation.*" I encountered this insight in 1960 when I was interpreting experimental data on cardiac potassium channels using mathematical modelling to reconstruct heart rhythm. The rhythm simply does not exist at the molecular level. The process occurs only when the molecules are constrained by the whole cardiac cell to be controlled by causation running in the opposite direction: from the cell to the molecular components. This insight is general. Of course, cells form an extremely important level of organisation, without which organisms with tissues, organs and whole-body systems would be impossible. But the other levels are also important in their own ways. Ultimately, even the environment can influence gene expression levels. There is no *a priori* reason to privilege any one level in causation. This is the principle of biological relativity.

The principle does not mean that the various levels are in any sense equivalent. To quote the authors again: "In no way do we mean to negate that DNA and the molecular cascades that are related to it, play an important role, yet their investigations are far from *complete* regarding the description of life phenomena." Completeness is the key concept. That is true for biological inheritance as well as for phenotype-genotype relations. New experimental work is revealing that there is much more to inheritance than DNA.

The avoidance of engagement with theoretical work in biology was based largely on the assumption that analysis at the molecular level could be, and was in principle, complete. In contrast, the authros write, "these [molecular] cascades may causally depend on activities at different levels of analysis, which interact with them and also deserve proper insights." Those 'proper insights' must begin by identifying the entities and processes that can be said to exist at the higher levels: "finding ways to constitute theoretically biological objects and objectivise their behaviour." To achieve this we have to distance ourselves from the notion, prevalent in biology today, that the fundamental must be conceptually elementary. As the authors point out, this is not even true in physics. "Moreover, the proper elementary observable doesn't need to be "simple". "Elementary particles" are not conceptually/mathematically simple."

There is therefore a need for a general theory of biological objects and their dynamics. This book is a major step in achieving that aim. It points the way to some of the important principles, such as the principle of symmetry, that must form the basis of such a theory. It also treats biological time in an innovative way, it explores the concept of extended criticality and it introduces the idea of anti-entropy. If these terms are unfamiliar to you, this book will explain them and why they help us to conceptualize the results of experimental biology. They in turn will lead the way by which experimentalists can identify and characterize the new biological objects around which a fully theoretical biology could be constructed.

Oxford University, Denis Noble

June 2013

Preface

In this book, we propose original perspectives in theoretical biology. We refer extensively to physical methods of understanding phenomena but in an untraditional manner. At times, we directly employ methods from physics, but more importantly, we radically contrast physical ways of constructing knowledge with what, we claim, is required for conceptual constructions in biology.

One of the difficult aspects of biology, especially with respect to physical insights, is the understanding of organisms and by extension the implications of what it means for an object of knowledge to be a part of an organism. The question of which conceptual and technical frameworks are needed to achieve this understanding is remarkably open. One such framework we propose is extended criticality. Extended criticality, one of our main themes, ties together the structure of coherence that forms an organism and the variability and historicity that characterize it. We also note that this framework is not meant to be pertinent in understanding the inert.

We are aware that our theoretical proposals are of a kind of abstraction that is unfamiliar to most biologists. An epistemological remark can hopefully make this kind of abstract thinking less unearthly. At the core of mathematical abstractions, not unlike in biological experiments, lies the "gesture" made by the scientist. By gesture we mean bodily movements, real or imagined, such as rearranging a sequence of numbers in the abstract or seeding the same number of cells over several wells. Gestures may remain mostly virtual in mathematics, yet any mathematical proof is basically a series of acceptable gestures made by the mathematician — both the ones described by a given formalism and the ones performed at the level of more fundamental intuitions (which motivate the formalisms themselves). For example, symmetries refer to applying transformations (e.g. rotating) and order refers to sorting (eg: the well-ordering of integer numbers and the ordering of oriented time), both of which are gestures. Since Greek geometry until contemporary physics, symmetries (defining invariance) and order (as for optimality) have jointly laid the foundation of mathematics and theoretical physics within the human spaces of action and knowledge. In summary, the theoretician singles out conceptual contours and organizes the World similarly as the experimenter prepares and executes scientific experiments.

From this perspective, biological theory directly relates to the acceptable moves, both abstract and concrete, that can be performedwhile experimenting and reflectiong on biological organisms. Symmetries and their changes, order and its breaking will guide our approach in an interplay with physics — often a marked differentiation. Again, the question of building a theory of organisms is a remarkably open one. With this book, we hope to contribute in explicitly raising this question and providing some elements of answer.

Interactions are as fundamental in knowledge construction as they are in biological evolution and ontogenesis. We would like to acknowledge that this book is the result and the continuation of an intense collaboration of three people: the listed authors and our friend Francis Bailly. The ideas presented here are extensions of work initiated by/with Francis, who passed away in 2009. We are extremely grateful to have had the priveledge to work with him. His insights sparked the beginning of the second author's PhD thesis which was completed in 2011.

We are also appreciative for the exchanges within the team "Complexité et Information Morphologique" (see Longo's web page), who included Matteo Mossio, Nicole Perret, Arnaud Pocheville and Paul Villoutreix. We also extend gratitude to our main "interlocuteurs" Carlos Sonnenschein and Ana Soto, Marcello Buiatti, Nadine Peyreiras, Jean Lassègue and Paul-Antoine Miquel. Additionally, we are grateful to Denis Noble and Stuart Kauffman who not only encouraged our perspective but also wrote a motivating preface and inspired a joint paper, respectively. We would also like to thank Michael Sweeney and Christopher Talbot who helped us with the english grammar.

Paris, June 2013

Giuseppe Longo[1]
Maël Montévil[2]

---

[1] Centre Cavaillès, CIRPHLES, École Normale Supérieure and CNRS, Paris, France
http://www.di.ens.fr/users/longo/
giuseppe.longo@ens.fr

[2] Centre de recherche interdisciplinaire, université Paris V, and École Normale Supérieure, Paris, France.
Tufts University Medical School, Dept. of Anatomy and Cell Biology, Boston, USA
http://montevil.theobio.org/
mael.montevil@gmail.com

Contents

1 Introduction 1
  1.1 Towards Biology 2
  1.2 Objectivization and Theories 5
      1.2.1 A Critique of Common Philosophical Classifications 8
      1.2.2 The Elementary and the Simple 11
  1.3 A Short Synthesis of Our Approach to Biological Phenomena 13
  1.4 A More Detailed Account of Our Main Themes: Time Geometry, Extended Criticality, Symmetry Changes and Enablement, Anti-Entropy 15
      1.4.1 Biological Time 16
      1.4.2 Extended Criticality 17
      1.4.3 Symmetry Changes and Enablement 19
      1.4.4 Anti-entropy 19
  1.5 Map of This Book 21

2 Scaling and Scale Symmetries in Biological Systems 23
  2.1 Introduction 23
      2.1.1 Power Laws 24
  2.2 Allometry 26
      2.2.1 Principles 26
      2.2.2 Metabolism 28
      2.2.3 Rhythms and Rates 32
      2.2.4 Cell and Organ Allometry 34
      2.2.5 Conclusion 37
  2.3 Morphological Fractal-Like Structures 38
      2.3.1 Principles 38
      2.3.2 Cellular and Intracellular Membranes 44
      2.3.3 Branching Trees 45
      2.3.4 Some Other Morphological Fractal Analyses 50
      2.3.5 Conclusion 51

	2.4	Elementary Yet Complex Biological Dynamics	52
	2.4.1	Principles	52
	2.4.2	A Non-exhaustive List of Fractal-Like Biological Dynamics	57
	2.4.3	The Case of Cardiac Rhythm	59
	2.4.4	Conclusion	62
	2.5	Anomalous Diffusion	63
	2.5.1	Principle	63
	2.5.2	Examples from Cellular Biology	66
	2.5.3	Conclusion	67
	2.6	Networks	67
	2.6.1	Structures	67
	2.6.2	Dynamics	69
	2.6.3	Conclusion	71
	2.7	Conclusion	71
3	**A 2-Dimensional Geometry for Biological Time**	75
	3.1	Introduction	75
	3.1.1	Methodological Remarks	77
	3.2	An Abstract Schema for Biological Temporality	78
	3.2.1	Premise: Rhythms	78
	3.2.2	External and Internal Rhythms	78
	3.3	Mathematical Description	81
	3.3.1	Qualitative Drawings of Our Schemata	81
	3.3.2	Quantitative Scheme of Biological Time	84
	3.4	Analysis of the Model	85
	3.4.1	Physical Periodicity of Compactified Time	86
	3.4.2	Biological Irreversibility	86
	3.4.3	Allometry and Physical Rhythms	87
	3.4.4	Rate Variability	88
	3.5	More Discussion on the General Schema 3.1	92
	3.5.1	The Evolutionary Axis (τ), Its Angles with the Horizontal $\varphi(t)$ and Its Gradients $\tan(\varphi(t))$	92
	3.5.2	The "Helicoidal" Cylinder of Revolution $\mathscr{C}_e$: Its Thread p_e, Its Radius R_i	94
	3.5.3	The Circular Helix $\mathscr{C}_i$ on the Cylinder and Its Thread p_i	94
	3.5.4	On the Interpretation of the Ordinate t'	94
4	**Protention and Retention in Biological Systems**	99
	4.1	Introduction	99
	4.1.1	Methodological Remarks	101
	4.2	Characteristic Time and Correlation Lengths	102
	4.2.1	Critical States and Correlation Length	104
	4.3	Retention and Protention	104
	4.3.1	Principles	104

		4.3.2	Specifications	105
		4.3.3	Comments	107
		4.3.4	Global Protention	108
	4.4	Biological Inertia		110
		4.4.1	Analysis	111
	4.5	References and More Justifications for Biological Inertia		113
	4.6	Some Complementary Remarks		115
		4.6.1	Power Laws and Exponentials	115
		4.6.2	Causality and Analyticity	116
	4.7	Towards Human Cognition. From Trajectory to Space: The Continuity of the Cognitive Phenomena		117
5	**Symmetry and Symmetry Breakings in Physics**			121
	5.1	Introduction		122
	5.2	Symmetry and Objectivization in Physics		122
		5.2.1	Examples	122
		5.2.2	General Discussion	125
	5.3	Noether's Theorem		129
	5.4	Typology of Symmetry Breakings		131
		5.4.1	Goldstone Theorem	133
	5.5	Symmetries Breakings and Randomness		134
6	**Critical Phase Transitions**			137
	6.1	Symmetry Breakings and Criticality in Physics		137
	6.2	Renormalization and Scale Symmetry in Critical Transitions		141
		6.2.1	Landau Theory	141
		6.2.2	Some Aspect of Renormalization	150
		6.2.3	Critical Slowing-Down	155
		6.2.4	Self-tuned Criticality	158
	6.3	Conclusion		160
7	**From Physics to Biology by Extending Criticality and Symmetry Breakings**			161
	7.1	Introduction and Summary		161
		7.1.1	Hidden Variables in Biology?	163
	7.2	Biological Systems "Poised" at Criticality		165
		7.2.1	Principle	165
		7.2.2	Other Forms of Criticality	169
		7.2.3	Conclusion	171
	7.3	Extended Criticality: The Biological Object and Symmetry Breakings		172
	7.4	Additional Characteristics of Extended Criticality		177
		7.4.1	Remarks on Randomness and Time Irreversibility	179
	7.5	Compactified Time and Autonomy		180
		7.5.1	Simple Harmonic Oscillators in Physics	181

		7.5.2 Biological Oscillators: Symmetries and Compactified Time 183
		7.5.3 Conclusion 184
	7.6	Conclusion 184

8 Biological Phase Spaces and Enablement 187
8.1 Introduction 187
8.2 Phase Spaces and Symmetries in Physics 190
    8.2.1 More Lessons from Quantum and Statistical Mechanics ... 192
    8.2.2 Criticality and Symmetries 193
8.3 Non-ergodicity and Quantum/Classical Randomness in Biology .. 195
8.4 Randomness and Phase Spaces in Biology 199
    8.4.1 Non-optimality 202
8.5 A Non-conservation Principle 203
8.6 Causes and Enablement 205
8.7 Structural Stability, Autonomy and Constraints 209
8.8 Conclusion 210

9 Biological Order as a Consequence of Randomness: Anti-entropy and Symmetry Changes 215
9.1 Introduction 215
9.2 Preliminary Remarks on Entropy in Ontogenesis 217
9.3 Randomness and Complexification in Evolution 220
9.4 (Anti-)Entropy in Evolution 223
    9.4.1 The Diffusion of Bio-mass over Complexity 223
9.5 Regeneration of Anti-entropy 231
    9.5.1 A Tentative Analysis of the Biological Dynamics of Entropy and Anti-entropy 233
9.6 Interpretation of Anti-entropy as a Measure of Symmetry Changes 238
9.7 Theoretical Consequences of This Interpretation 243

10 A Philosophical Survey on How We Moved from Physics to Biology 249
10.1 Introduction 249
10.2 Physical Aspects 250
    10.2.1 The Exclusively Physical 250
    10.2.2 Physical Properties of the "Transition" towards the Living State of Matter 251
10.3 Biological Aspects 251
    10.3.1 The Maintenance of Biological Organization 252
    10.3.2 The Relationship to the Environment 253
    10.3.3 Passage to Analyses of the Organism 253

	10.4 A Definition of Life?... 254
	10.4.1 Interfaces of Incompleteness........................ 256
	10.5 Conclusion 257

A   **Mathematical Appendix** 259
    A.1 Scale Symmetries 259
    A.2 Noether's Theorem 260
        A.2.1 Classical Mechanics Version (Lagrangian) 260
        A.2.2 Field Theoretic Point of View 264

References 267

Chapter 1
Introduction

The historical dynamic of knowledge is a permanent search for "meaning" and "objectivity". In order to make natural phenomena intelligible, we *single out* objects and processes, by an active knowledge construction, within our always enriched historical experience. Yet, the scientific relevance of our endeavors towards knowledge may be analyzed and compared by making explicit the principles on which our conceptual, possibly mathematical, constructions are based.

For example, one may say that the Copernican understanding of the Solar system is the "true" or "good" one, when compared to the Ptolemaic. Yet, the Ptolemaic system is perfectly legitimate, if one takes the Earth as origin of the reference system, and there are good metaphysical reasons for doing so. However, an internal analysis of the two approaches may help for a scientific comparison in terms of the *principles* used. Typically, the Copernican system presents more "symmetries" in the description of the solar system, when compared to the "ad hoc" constructions of the Ptolemaic system: the later requires the very complex description of epicycles over epicycles, planet by planet.... On the opposite, by Newton's universal laws, a unified and synthetic understanding of the planets' Keplerian trajectories and even of falling apples was made possible. Later on, Hamilton's work and Noether's theorems (see chapter 5) further unified physics by giving a key role to optimality (Hamilton's approach to the "geodetic principle", often mentioned below) and to symmetries (at the core of our approach). And Newton's equations could be derived from Hamilton's approach. Since then, the geodetic principle and symmetries as conservation principles are fundamental "principles of intelligibility" that allow to understand at once physical phenomena. These principles provide objectivity and even define the objects of knowledge, by organizing the world around us. As we will extensively discuss, symmetries conceptually unified the physical universe, far away from the ad hoc construction of epicycles on top of epicycles.

Physical theorizing will guide our attempts in biology, without reductions to the "objects" of physics, but by a permanent reference, even by local reductions, to the *methodology* of physics. We are aware of the historical contingency of this method, yet by making explicit its working principles, we aim at its strongest possible conceptual stability and *adaptability*: "perturbing" our principles and even our methods may allow further progress in knowledge construction.

1.1 Towards Biology

Current biology is a discipline where most, and actually almost all, research activities are — highly dextrous — experimentations. For a natural science, this situation may not seem to be an issue. However, we fear that it is associated to a belief that experiments and theoretical thinking could be decoupled, and that experiments could actually be performed independently from theories. Yet, "concrete" experimentations cannot be conceived as autonomous with respect to theoretical considerations, which may have abstract means but also have very practical implications. In the field of molecular biology, for example, research is related to the finding of hypothesized molecules and molecular manipulations that would allow to understand biological phenomena and solve medical or other socially relevant problems. This experimental work can be carried on almost forever as biological molecular diversity is abundant. However, the understanding of the actual phenomena, beyond the differences induced by local molecular transformations is limited, precisely because such an understanding requires a theory, relating, in this case, the molecular level to the phenotype and the organism. In some cases, the argued theoretical frame is provided by the reference to an unspecified "information theoretical encoding", used as a metaphor more than as an actual scientific notion, [Fox Keller, 1995, Longo et al., 2012a]. This metaphor is used to legitimate observed correlations between molecular differential manipulations and phenotype changes, but it does so by putting aside considerable aspects of the phenomena under study. For example, there is a gap between a gene that is experimentally necessary to obtain a given shape in a strain and actually entailing this shape. In order to justify this "entailment", genes are understood as a "code", that is a one-dimensional discrete structure, meanwhile shapes are the result of a constitutive history in space and in time: the explanatory gap between the two is enormous. In our opinion, the absence or even the avoidance of theoretical thinking leads to the acceptance of the naive or common sense theory, possibly based on unspecified metaphors, which is generally insufficient for satisfactory explanations or even false — when it is well defined enough as to be proven false.

We can then informally describe the reasons for the need of new theoretical perspectives in biology as follows. First, there are empirical, theoretical and conceptual *instabilities* in current biological knowledge. This can be exemplified by the notion of the gene and its various and changing meanings [Fox Keller, 2002], or the unstable historical dynamics of research fields in molecular biology [Lazebnik, 2002]. In both cases, the reliability and the meaning of research results is at risk. Another issue is that the molecular level does not accommodate phenomena that occur typically at other *levels of organization*. We will take many examples in this book, but let's quote as for now the work on microtubules [Karsenti, 2008], on cancer at the level of tissues [Sonnenschein & Soto, 2000], or on cardiac functions at its different levels [Noble, 2010]. Some authors also emphasize the historical and conceptual shifts that have led to the current methodological and theoretical situation of molecular biology, which is, therefore, subject to ever changing interpretations [Amzallag, 2002, Stewart, 2004]. In general, when considering the molecular level, the

1.1 Towards Biology

problem of the composition of a great variety of molecular phenomena arises. Single molecule phenomena may be biologically irrelevant *per se*: they need to be related to other levels of organization (tissue, organ, organism, ...) in order to understand their possible biological significance.

In no way do we mean to negate that DNA and the molecular cascades related to it play a fundamental role, yet their investigations are far from *complete* regarding the description of life phenomena. Indeed, these cascades may causally depend on activities at different level of analysis, which interact with them and deserve proper insights.

Thus, it seems that, with respect to explicit theoretical frames in biology, the situation is not particularly satisfying, and this can be explained by the complexity of the phenomena of life. Theoretical approaches in biology are numerous and extremely diverse in comparison, say, with the situation in theoretical physics. In the latter field, theorizing has a deep methodological unity, even when there exists no unified theory between different classes of phenomena — typically, the Relativistic and Quantum Fields are not (yet) unified, [Weinberg, 1995, Bailly & Longo, 2011]. A key component of this methodological unity, in physics, is given by the role of "symmetries", which we will extensively stress. Biological theories instead range from conceptual frameworks to highly mathematized physical approaches, the latter mostly dealing with *local* properties of biological systems (e. g. organ formation). The most prominent conceptual theories are Darwin's approach to evolution — its principles, "descent with modification" and "selection", shed a major light on the dynamics of phylogenesis, the theory of common descent — all current organisms are the descendants of one or a few simple organisms, and cell theory — all organisms have a single cell life stage and are cells, or are composed of cells. It would be too long to quote work in the biophysical category: they mostly deal with the dynamics of forms of organs (morphogenesis), cellular networks of all sorts, dynamics of populations ... when needed, we will refer to specific analyses. Very often, this relevant mathematical work is identified as "theoretical biology", while we care for a distinction, in biology, between "theory" and "mathematics" analogous to the one in physics between theoretical physics and mathematical physics: the latter mostly or more completely formalizes and technically solves problems (equations, typically), as set up within or by theoretical proposals or directly derived from empirical data.

In our view, there is currently no satisfactory *theory* of biological organization as such, and in particular, in spite of many attempts, there is no theory of the organism. Darwin's theory, and neo-Darwinian approaches even more so, basically avoid as much as possible the problem raised by the organism. Darwin uses the duality between life and death as selection to understand why, between given biological forms, some are observed and others are not. That is, he gave us a remarkable theoretical frame for phylogenesis, without confronting the issue of what a theory of organisms could be. In the modern synthesis, since [Fisher, 1930], the properties of organisms and phenotypes, fitness in particular, are predetermined and defined, in principle, by genetics (hints to this view may be found already in Spencer's approach to evolution [Stiegler, 2001]). In modern terms, "(potential) fitness is already encoded in genes".

Thus, the "structure of determination" of organisms is understood as theoretically unnecessary and is not approached[1].

In physiology or developmental biology the question of the structure of determination of the system is often approached on qualitative grounds and the mathematical descriptions are usually limited to specific aspects of organs or tissues. Major examples are provided by the well established and relevant work in morphogenesis, since Turing, Thom and many others (see [Jean, 1994] for phillotaxis and [Fleury, 2009] for recent work on organogenesis), in a biophysical perspective. In cellular biology, the equivalent situation leads to (bio-)physical approaches to specific biological structures such as membranes, microtubules, ..., as hinted above. On the contrary, the tentative, possibly mathematical, approaches that aim to understand the proper structure of determination of organisms as a whole, are mostly based on ideas such as autonomy and autopoiesis, see for example [Rosen, 2005, Varela, 1979, Moreno & Mossio, 2013]. These ideas are philosophically very relevant and help to understand the structure of the organization of biological entities. However, they usually do not have a clear connection with experimental biology, and some of them mostly focus on the question of the definition of life and, possibly, of its origin, which is not our aim. Moreover, their relationship with the aforementioned biophysical and mathematical approaches is generally not made explicit. In a sense, our specific "perspectives" on the organism as a whole (time, criticality, anti-entropy, the main themes of this book) may be used to fill the gap, as on one side we try to ground them on some empirical work, on the other they may provide a theoretical frame relating the global analysis of organisms as autopoietic entities and the local analysis developed in biophysics.

In this context, physiology and developmental biology (and the study of related pathological aspects) are in a particularly interesting situation. These fields are directly confronted with empirical work and with the complexity of biological phenomena; recent methodological changes have been proposed and are usually described as "systems biology". These changes consist, briefly, in focusing on the systemic properties of biological objects instead of trying to understand their components, see [Noble, 2006, 2011, Sonnenschein & Soto, 1999] and, in particular, [Noble, 2008]. In the latter, it is acknowledged that, as for theories in systems biology:

> There are many more to be discovered; a genuine "theory of biology" does not yet exist. *[Noble, 2008]*

Systems biology has been recently and extensively developed, but it also corresponds to a long tradition. The aim of this book can be understood as a theoretical contribution to this research program. That is, we aim at a preliminary, yet possibly general theory of biological objects and their dynamics, by focusing on "perspectives" that shed some light on the unity of organisms from a specific point of view.

---

[1] By the general notion of structure of determination we refer to the theoretical determination of a conceptual frame, in more or less formalized terms. In physics, this determination is generally expressed by systems of equations or by functions describing the dynamics.

In this project, there are numerous pitfalls that should be avoided. In particular, the relation with the powerful physical theories is a recurring issue. In order to clarify the relationships between physics, mathematics and biology, a critical approach to the very foundations of physical theories and, more generally, to the relation between mathematized theories and natural phenomena is most helpful and we think even necessary. This analysis is at the core of [Bailly & Longo, 2011] and, in the rest of this introduction, we just review some of the key points in that book. By this, we provide below a brief account of the philosophical background and of the methodology that we follow in the rest of this book. We also discuss some elements of comparison with other theoretical approaches and then summarize some of the key ideas presented in this book.

1.2 Objectivization and Theories

As already stressed, theories are conceptual and — in physics — largely mathematized frameworks that frame the intelligibility of natural phenomena. We first briefly hint to a philosophical history of the understanding of what theories are.

The strength of theoretical accounts, especially in classical mechanics, and their cultural, including religious, background has led scientists to understand them as an intrinsic description of the very essence of nature. Galileo's remark that "the book of nature is written in the language of mathematics" (of Euclidean geometry, to be precise) is well known. It is a secular re-understanding of the "sacred book" of revealed religions. Similarly, Descartes writes:

> Par la nature considérée en général, je n'entends maintenant autre chose que Dieu même, ou bien l'ordre et la disposition que Dieu a établie dans les choses crées. [By nature considered in general, I mean nothing else but God himself, or the order and tendencies that God established in the created things.] *[Descartes, 1724]*

Besides, in [Descartes, 1724], the existence of God and its attributes legitimate, *in fine*, the theoretical accounts of the world: observations and clear thinking are truthful, as He should not be deceitful. In this context, the theory is thus an account of the "thing in itself" (das Ding an sich, in Kant's vocabulary). The validity and the existence of such an account are understood mainly by the mediation of a deity, in relation with the perfection encountered in mathematics — a direct emanation of God, of which we know just a finite fragment, but an identical fragment to God's infinite knowledge (Galileo).

Kant, however, introduced another approach [Kant, 1781]. In Kant's philosophy, the notion of "transcendental" describes the focus on the *a priori* (before experience) conditions of possibility of knowledge. For example, objects cannot be represented outside space, which is, therefore, the *a priori* condition of possibility for their representation. By this methodology, the thing in itself is no longer knowable, and the accounts on phenomena are given, in particular, through the *a priori* form of the sensibility that are space and time. Following this line, mathematics is understood as *a priori* synthetic judgments: it is a form of knowledge that does not depend on experience, as it is only based on the conditions of possibility for experience, but neither is it based on the simple analysis of concepts. For example, $2 + 3 = 5$ is neither in the concept of 2 nor in the concept of 3 for Kant: it requires a synthesis, which is based on *a priori* concepts.

The transcendental approach of Kant has, however, strong limitations, highlighted, among others, by Hegel and later by Nietzsche. Hegel insists on the status of the knowledge of these *a priori* conditions, which he aims to understand dialectically, by the historicity of Reason and more precisely by the unfolding of its contradictions. Similarly, with a different background, Nietzsche criticizes also the validity of this transcendental knowledge.

> Wie sind synthetische Urtheile *a priori* möglich? fragte sich Kant, — und was antwortete er eigentlich? Vermöge eines Vermögens [...]. [How are *a priori* synthetic judgments possible?" Kant asks himself — and what is really his answer? By means of a means (faculty) [...]] *[Nietzsche, 1886]*

For Nietzsche, it is essential, in particular, to understand the genesis of such "faculties", or behaviors, by their roots in the body and therefore by the embodied subject [Stiegler, 2001]. One should also quote Merleau-Ponty and Patocka as for the epistemological role of our intercorporeal "being in the world" and for reflections on biological phenomena (for recent work and references on both these authors in one text, see [Marratto, 2012, Thompson, 2007, Pagni, 2012]).

In short, for us, the analysis of a genesis, of concepts in particular, is a fundamental component of an epistemological analysis. This does not mean fixing an origin, but providing an attempted explicitation of a constitutive paths. Any epistemology is also a critical history of ideas, including an investigation of that fragment of "history" which refers to our active and bodily presence in the world. And this, by making explicit, as much as it is possible, the purposes of our knowledge construction. Yet, Kant provided an early approach to a fundamental component of the systems biology we aim at, that is to the autonomy and unity of the living entities (the organisms as "Kantian wholes", quoted by many) and the acknowledgment of the peculiar needs of the biological theorizing with respect to the physical one[2].

One of the most difficult tasks is to insert this autonomy in the unavoidable ecosystem, both internal and external: life is variability *and* constraints, and neither make sense without the other. In this sense, the recent exploration in [Moreno & Mossio, 2013] relates constraints and autonomy in an original way and complements our effort. Both this "perspective" and ours are only possible when accessing living organisms in their unity and by taking this "wholeness" as a "condition of possibility" for the construction of biological knowledge. However, we do not discuss here this unity *per se*, nor directly analyze its auto-organizing structural stability. In this sense, these two complementary approaches may enrich each other and produce, by future work, a novel integrated framework.

As for the interplay with physics, our account particularly emphasize the *praxis* underlying scientific theorizing, including mathematical reasoning, as well as the

---

[2] For a recent synthetic view on Kantian frames, and many references to this very broad topic, in particular as for the transcendental role of "teleology" in biological investigations, one should consult [Perret, 2013].

1.2 Objectivization and Theories

cognitive resources mobilized and refined in the process of knowledge construction. From this perspective, mathematics and mathematized theories, in particular, are the result of human activities, in our historical space of humanity, [Husserl, 1970]. Yet, they are the most stable and conceptually invariant knowledge constructions we have ever produced. This singles them out from the other forms of knowledge. In particular, they are grounded on the constituted *invariants of our action*, gestures and language, and on the *transformations* that preserve them: the concept of number is an invariant of counting and ordering; symmetries are fundamental cognitive invariants and transformations of action and vision — made concepts by language, through history, [Dehaene, 1997, Longo & Viarouge, 2010]. More precisely, both ordering (the result of an action in space) and symmetries may be viewed as "principles of conceptual construction" and result from core cognitive activities, shared by all humans, well before language, yet spelled out in language. Thus, jointly to the "principles of (formal) proof", that is to (formalized) deductive methods, the principle of construction ground mathematics at the conjunction of action and language. And this is so beginning with the constructions by rotations and translations in Euclid's geometry (which are symmetries) and the axiomatic-deductive structure of Euclid's proofs (with their proof principles).

This distinction, construction principles vs. proof principles, is at the core of the analysis in [Bailly & Longo, 2011], which begins by comparing the situation in mathematics with the foundations of physics. The observation is that mathematics and physics share the same construction principles, which were largely co-constituted, at least since Galileo and Newton up to Noether and Weyl, in the XXth century[3]. One may formalize the role of symmetries and orders by the key notion of group. Mathematical groups correspond to symmetries, while semi-groups correspond to various forms of ordering. Groups and semi-groups provide, by this, the mathematical counterpart of some fundamental cognitive grounds for our conceptual constructions, shared by mathematics and physics: the active gestures which organize the world in space and time, by symmetries and orders.

Yet, mathematics and physics differ as for the principles of proof: these are the (possibly formalized) principles of deduction in mathematics, while proofs need to be grounded on experiments and empirical verification, in physics. What can we say as for biology? On one side, "empirical evidence" is at the core of its proofs, as in any science of nature, yet mathematical invariance and its transformations do not seem to be sufficiently robust and general as to construct biological knowledge, at least not at the level of organisms and their dynamics, where variability is one of the major "invariant". So, biology and physics share the principles of proofs, in a broad sense, while we claim that the principles of conceptual constructions cannot be transferred as such. The aim of this book is to highlight and apply some cases where this can be done, by some major changes though, and other cases where

---

[3] Archimedes should be quoted as well: why a balance with equal weights is at equilibrium? for symmetry reasons, says he. This is how physicists still argue now: why is there that particle? for symmetry reasons — see the case of anti-matter and the negative solution of Dirac's equations, [Dirac, 1928].

one needs radically different insights, from those proper to the so beautifully and extensively mathematized theories of the inert.

It should be clear by now, that our foundational perspective concerns in priority the methodology (and the practice) that allows establishment of scientific objectivity in our theories of nature. As a matter of fact, in our views, the constitution of theoretical thinking is at the same time a process of objectivization. That is, this very process co-constitutes, jointly to the empirical friction on the world, the object of study in a way that simultaneously allows its intelligibility. The case of quantum mechanics is paradigmatic for us, as a quanton (and even its reference system) is the result of active measurement and its practical and theoretical preparation. In this perspective, then, the objects are defined by measuring and theorizing that simultaneously give their intelligibility, while the validity of the theory (the proofs, in a sense) is given by further experiments. Thus, in quantum physics, measurement has a particular status, since it is not only the access to an object that would be there beyond and before measurement, but it contributes to the constitution of the very object measured. More generally, in natural sciences, measurement deals with the questions: where to look, how to measure, where to set borders to objects and phenomena, which correlations to check and even propose .... This co-constitution can be intrinsic to some theories such as quantum mechanics, but a discussion seems crucial to us also in biology, see [Montévil, 2013].

Following this line of reasoning, the research program we follow towards a theory of organism aims at finding ways to constitute theoretically biological objects and objectivize their behavior. Differences and analogies, by conceptual continuities or dualities with physics will be at the core of our method (as for dualities, see, for example, our understanding of "genericity vs. specificity" in physics vs. biology in chapter 7), while the correlations with other theories can, perhaps, be understood later[4]. In this context, thus, a certain number of problems in the philosophy of biology are not methodological barriers; on the contrary, they may provide new links between remote theorizing such as physical and social ones, which would not be based on the transfer of already constructed mathematical models.

1.2.1 A Critique of Common Philosophical Classifications

As a side issue to our approach, we briefly discuss some common wording of philosophical perspectives in the philosophy of biology — the list pretends no depth nor completeness and its main purpose is to prevent some "easy" objections.

PHYSICALISM   In the epistemic sense (i.e. with respect to knowledge), physicalism can be crudely stated as follows:

---

[4] The "adjacent" fields are, following [Bailly, 1991], physical theories in one direction and social sciences in another. The notion of "extended criticality", say, in chapter 7, may prove to be useful in economics, since we seem to be always in a permanent, extended, crisis or critical transition, very far from economic equilibria.

1.2 Objectivization and Theories

> the majority of scientists [recognize] that life can be explained on the basis of the existing laws of Physics .
> *[Perutz, 1987]*

The most surprising word in this statement is "existing". Fortunately, Galileo and Newton, Einstein and the founders of quantum mechanics, did not rely on *existing* laws of physics to give us modern science. Note that Galileo, Copernicus and Newton where not even facing new phenomena, as anybody could let two different stones fall or look at the planets, yet, following different *perspectives* on familiar phenomena, they proposed radically new theories and "laws"[5].

There is no doubt that a wide range of isolated biological phenomena can be accommodated in the main existing physical theories, such as classical mechanics, thermodynamics, statistical mechanics, hydrodynamics, quantum mechanics, general relativity, ..., unfortunately, some of these physical theories are not unified, and, *a fortiori*, one cannot reduce one to the other nor provide by them a unified biological understanding. However, as soon as the phenomena we want to understand differs radically or are seen from a different perspective (the view of the organism), new theoretical approaches may be required, as it happened along the history of physics. There is little doubt that an organism may be seen as a bunch of molecules, yet we, the living objects, are rather funny bunches of molecules and the issue is: which *theory* may provide a sound perspective and account of these physically singular bunches of molecules? For us, this is an epistemic, a knowledge issue, not an ontological one.

Such lines are common within physics as well, in particular in areas that are directly relevant for our approach. For example, the understanding of critical transitions requires the introduction of a new structure of determination, as classes of parameterized models and the focusing on new observables, such as the critical exponents, see chapter 6. Similarly, going from macrophysics (classical mechanics) to microscopic phenomena (quanta) necessitates the loss of determinism, while the understanding of gravity in terms of quantum fields leads to a radical transformation of the classical and relativistic structure of space-time (e. g. by non-commutative geometry, [Connes, 1994]) or radically new objects (string theory, [Green et al., 1988]). It happens that these audacious new accounts of quantum mechanics, which aim to unify it with general relativity, are not compatible with each other. Moving backwards in time, another example is the link between heat and motion, which required the invention of thermodynamics and the introduction of a new quantity (entropy). The latter allowed to describe, in particular, the irreversibility of time, which is incompatible with a finite combination of Newtonian trajectories. Notice, though, that the current physical understanding of systems far from thermodynamical equilibrium is seriously limited because there is no general theory of them, see for example [Vilar & Rubí, 2001].

---
[5] What an unsatisfactory word, borrowed from religious tables of laws and/or the writing of social links — we will avoid it. Physical theories are better understood as the explicitation of (relative) reference systems, of measures on them and of the corresponding fundamental symmetries, see [Weyl, 1983, Van Fraassen, 1989, Bailly & Longo, 2011].

And biological entities, if considered as physical systems, would most probably fall at least in this category.

VITALISM. For similar reasons, the question and the debates around the notion of vitalism lead to a flawed approach to biological systems. We exclude, by principle, the various sorts of intrinsic teleologism (evolution leading to our human perfection), internal living forces, encoded homunculi in DNA or alike. From our theoretical point of view, what matters is to find ways to objectivize the phenomena we want to study, similarly as what has been done along the history of physics. However, the fear of negatively connoted vitalist interpretations leads to blind spots in the understanding of biological phenomena, since it hinders original approaches, strictly pertinent to the object of observation. If the search for an adequate theory for the living state of matter, in an autonomous interplay of differences and analogies with theories of the inert, is vitalism, then the researchers in hydrodynamics may be shamefully accused to be "hydrodynamicists" as, so far, there is no way to reduce to (nor to understand in terms of) elementary particles that compose fluids, of quantum mechanics say, the incompressibility and fluidity in continua at the core of their science. Those are understood in terms of new or different symmetries from the one founding the theory of particles (quanta): the suitable symmetries yield radically different and irreducible equations and mathematically objectivize the otherwise vague notions of fluidity and incompressibility in a continuum. Our colleagues in hydrodynamics are not "dualist" for this, nor they believe in a "soul" of fluids, against the vulgar matter of particles. Similarly, in thermodynamics, the founding fathers invented new observable quantities (entropy) and original phase spaces (P, V, T, pressure, temperature and volume) for thermodynamic trajectories (the thermodynamic cycle). By this, they disregarded the particles out of which gases are made. Later, Boltzmann did not reduce thermodynamics to Newton-Laplace trajectories of particles. He assumed molecular chaos and the random exploration of the entire intended physical space (ergodicity, see chapter 8), which are far away from the Newton-Laplace mathematical frame of an entailed trajectory in the momentum / position phase space. The new unit of analysis is the volume of each microstate in the phase space. He then unified asymptotically the molecular approach and the second principle of thermodynamics: given his hypotheses, in the thermodynamic integral, an infinite sum, the ratio of particles over a volume stabilizes only at the infinite limit of both. In short, the asymptotic hypothesis and treatment allowed Boltzmann to ignore the entailed Newtonian trajectory of individual particles and to give statistical account of thermodynamics.

The unity of science is a beautiful project, such as today's search for a theory unifying relativistic and quantum fields, yet unity cannot be imposed by a philosophical prejudice. It is instead the result of hard work and autonomous theorizing, followed, perhaps and if possible, by unification. And, if we do not have different theories, as for different phenomenal frames, there is nothing to unify.

REDUCTIONISM (SCALE). The methodological assumption that we should understand phenomena beginning at the small scales is, again, at odds with the history of physics. Thermodynamics started at macroscopic scales, as we said. As for gravitation and quantum fields, once more, in spite of almost one century of research, macroscopic and microscopic are not (yet) understood in a unified framework. And Galileo's and Einstein's theories remain fundamental even though they do not deal with the elementary.

The hope for "theory of everything" aims to overcome, first, this major difficulty, while there is no *a priori* reason why it would help, for example, in the understanding of non-equilibrium thermodynamics (except possibly in the case of black holes thermodynamics, [Rovelli, 1996], a remote issue from ours). Non-equilibrium thermodynamics remains mainly under theoretical construction and seems instead particularly relevant for life sciences. Moreover, and this point is crucial for this critique of reductionism, the current understanding of microscopic interactions, in the standard model, does not involve a fundamental, small scale; on the contrary it "hangs" between scales (by renormalization methods):

> QFT [Quantum Field Theory] is not required to be physically consistent at very short distance where it is no longer a valid approximation and where it can be rendered finite by a modification that is, to a large extent, arbitrary.   *[Zinn-Justin, 2007]*

Another example is the question of (scale) reductionism, which is approached by [Soto et al., 2008]. In the latter, the key role of time, with respect to biological levels of organization, is evidenced. We will approach this question in a complementary way, on smaller time scales — yet with a proper biological time — an "operator", we shall say in biology, both in a mathematical sense and by the role of the historical formation of biological entities.

Finally, scale reductionism is in contrast with the modern analysis of renormalization in critical transitions, see [Longo et al., 2012c], where scales are treated by cascades of mathematical models with no privileged level of observation. Critical transitions will be extensively discussed in this book.

The conclusion of this section is that we understand biological theorizing as a process of constitution of objectivity and, in particular, of organisms as *theoretical objects*. Science is not the progressive occupation of reality by more or less familiar conceptual and technical tools, but the permanent construction of new objects of knowledge, new perspectives and tools for their organization and understanding, yet grounded also on historically constructed knowledge and empirical friction.

1.2.2 The Elementary and the Simple

We mentioned that the points we made above are not philosophical prerequisites for a genuine intelligibility of biological phenomena, however, the technical aspects we hinted to in our critique will help us to provide both, we hope, philosophical and scientific insights. This is our aim as for the notion of "the physical singularity of life phenomena" developed in [Bailly & Longo, 2011], which we recall and further develop here. The "singularity" stems both from the technical notion of extended criticality below and from the historical specificity of living objects. Critical transitions are mathematical singularities in physics, yet they are non-extended as they are described by point-wise transitions, see chapter 6.

Biological objects are "singular" also in the sense of "being individual", that is, the result of a unique history. One may better say that they are specific (see the duality in chapter 7 with respect to physics).

In other words, we will widely use insights from physical theories, but these insights will mainly be a methodological and conceptual reference, and will not be rooted in an epistemic physicalism. Indeed, our approach may lead almost to the opposite: we will use the examples from physical theorizing as tools on the way to construct objectivity, and this will lead us, in some cases, to oppose biological theorizing to the very foundations of physical theories — typically, by the different role played by theoretical symmetries (in chapter 7 in particular). Moreover, we will recall the genericity of the inert objects, as invariant with respect the theory and the experiments, and the specificity of their trajectories (uniquely determined by the geodetic principle). And we will oppose them to the specificity (historical nature) of the living entities and the genericity of their phylogenetic trajectories, as possible or compatible ones in a co-determined ecosystem, see chapter 7. Yet, the very idea of this (mathematical) distinction, generic vs. specific, is borrowed from physical theorizing.

Further relations with physical theories will be developed progressively in our text, when needed for our theoretical developments in biology.

Before specifying further our approach to biological objects, we have to further challenge the Cartesian and Laplacian view that the fundamental is always elementary and that the elementary is always simple. According to this view, in biology only the molecular analysis would be fundamental.

As we mentioned, Galileo and Einstein proposed fundamental theories of gravitation and inertia, with no references to Democritus' atoms nor quanta composing their falling bodies or planets. Then, Einstein, and still now physicists, struggle for *unification*, not reduction of the relativistic field to the quantum one. Boltzmann did not reduce thermodynamics to the Newtonian trajectories of particles, but assumed the original principles recalled above and *unified* at the asymptotic limit the two intended theories, thermodynamics and particles' trajectories.

Thus, there is no reason in biology to claim that the fundamental must be conceptually elementary (molecular), as this is false also in physics. Moreover, the proper elementary observable doesn't need to be "simple". "Elementary particles" are not conceptually / mathematically simple, in quantum field theories nor in string theory. In biology, the elementary living component, the cell, is (very) complex, a further anti-Cartesian stand at the core of our proposal: a cell should already be seen as a Kantian whole.

In an organism, no reduction to the parts allows the understanding of the whole, because the relevant degrees of freedom of the parts, as associated to the whole, are *functional* and this defines their compatibility within the whole and of the whole in the ecosystem. In other terms, they are definable as components of the causal consequences of properties of the parts. Thus, only the microscopic degrees of freedom of the parts can be understood as physical. Further, because of the non-ergodicity of the universe above the level of atoms, inasmuch at ergodicity is well defined in this context (see chapter 8), most macromolecules and organs will never exist. Note also that ergodicity would prevent selection since it would mean that a negatively selected phenotype would "come back" in the long run, anyway.

As mentioned above and further discussed below, the theoretical frame establishes the pertinent observables and parameters, i.e. the ever changing and unprestatable phase space of evolution. Note that, in biology, we consider the observable and parameters that are derived from or relative to Darwinian evolution and this is fundamental for our approach. Their very definition depends on the intended organism and its integration in and regulation by an ecosystem. Selection, acting at the level of the evolving organism in its environment, selects organisms on functions (thus on and by organs in an organism) as interacting with an ecosystem. The phenotype, in this sense constitutes the observables we focus on.

1.3 A Short Synthesis of Our Approach to Biological Phenomena

A methodological point that we first want to emphasize is that we will focus on "current" organisms, as a result an in the process of biological evolution. Indeed, numerous theoretical researches are performed on the question of the origin of life. Most of these analyses use physical or almost physical theories as such, that is they try to analyze how, from a mix of (existing) physical theories, one can obtain "organic" or evolutive systems. We will not work at the (interesting, per se) problem of the origin of life, as the transition from the inert to the living state of matter, but we will work at the transition from *theories* of the inert to *theories* of living objects. In a sense this may contribute also to the "origin" problem, as a sound theory of organisms, if any, may help to specify what the transition from the inert leads to, and therefore what it requires.

More precisely, the method of mathematical biology and biophysical modeling quoted above is usually the transformation of *a part* of an organism (more generally, of a living system) into a physical system, in general separated from the organism and from the biological context it belongs to. This methodology often allows an understanding of some biological phenomena, from morphogenesis (phyllotaxis, formation of some organs ...) to cellular networks and more, see above. For example, the modeling of microtubules allows to approach their self-organization properties [Karsenti, 2008], but it corresponds to a theoretical (and experimental) *in vitro* situation, and their relation with the cell is not understood by the physical approach alone. The understanding of the system in the cell requires an approach external to the structure of determination at play in the purely physical modeling. Thus, to this technically difficult work ranging from morphogenesis and phyllotaxis to cellular networks, one should add an insufficiently analyzed issue: these organs or nets, whose shape and dynamics are investigated by physical tools, are generally part of an organism. That is, they are regulated and integrated in and by the organism and never develop like isolated or generic (completely defined by invariant rules) crystals or physical forms. It is instead this integration and regulation in the coherent

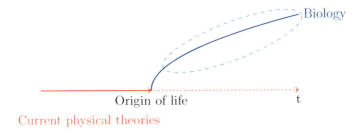

Fig. 1.1 *A scheme of the relation between physics and biology, from a diachronic point of view.* Theoretical approaches that focus on the origin of life usually follow the physical line (stay within existing physical theories) and try to approach the "bifurcation" point. The latter is not well defined since we don't have a proper theory for the biological entities that are supposed to emerge. Usually, the necessary ingredients for Darwinian evolution are used as goals. From our perspective, a proper understanding of biological phenomena need to focus directly, at least as a first (huge) step, on the properly biological domain, where the Darwinian tools soundly apply, but also where organisms are constituted. It may then be easier to fill the gap.

structure of an organism that contributes in making the biologically relevant situations, which is mostly non-generic, [Lesne & Victor, 2006].

The general strategy we use, is to approach the biological phenomena from different perspectives, each of them focusing on different *aspects* of biological organization, not on different *parts* such as organs or cellular nets in tissues .... The aim is to propose a basis for a partially mathematized theoretical understanding. This strategy allows us to obtain relatively autonomous progresses on the corresponding aspects of living systems. An essential difficulty is that, *in fine*, these concepts are fully meaningful only in the interaction with each other, that is to say in a unified framework that we are contributing to establish. In this sense, then, we are making progresses by revolving around this not yet existing framework, proposing and browsing these different perspectives in the process. However, this allows a stronger relation to empirical work, in contrast to theories of biological autonomy, without losing the sense of the biological unity of an organism.

The method we follow in order to progress in each of these specific aspects of life can mostly be understood as taking different points of view on organisms: we look at them from the point of view of time and rhythms, of the interplay of global stability vs. instability, of the formation and maintenance of organization through changes .... As a result, we will combine in this book a few of these theoretical perspectives, the principal common organizing concepts will be biological time, on one side, and extended criticality on the other. More specifically, the main conceptual frames that we will either follow directly or that will make recurrent appearance in this text are the following:

BIOLOGICAL TEMPORAL ORGANIZATION. The idea is that, more than space or energy, biological time is a crucial leverage to understand biological organization. This does not mean that space or energy are irrelevant, but they have a different role from the one they play in physics. The reason for this will be explained progressively throughout the book. The approach in terms of symmetry changes that we develop in chapter 7 provides a radical argument for this point of view. Intuitively, the idea is that what matters in biological theorizing is the notion of "organization" and the way it is constructed along and, we dare to say, *by* time, since biological time will be an operator for us, in a precise mathematical sense. In contrast to this, the energetic level (say, between mammals of different sizes) is relatively contingent, as we will argue on the grounds of the allometric relations, in chapter 2, where energy or mass appear as a parameter. Some preliminary arguments from physics are provided by the role of time (entropy production) in dissipative structures [Nicolis & Prigogine, 1977] and by the non-ergodicity of the molecular phase space, discussed in [Kauffman, 2002, Longo et al., 2012b].

EXTENDED CRITICAL TRANSITIONS. A large part of our work will use the notion of extended critical transition [Bailly, 1991, Bailly & Longo, 2008, 2011] to understand biological systems. This notion is relatively complex, in particular because of its physical prerequisites, and we will introduce it progressively in this book. Notice that it provides a precise meaning to the idea of the physical singularity of life phenomena.

ENABLEMENT. Biologists working on evolution often refer to a contingent state of the ecosystem as "enabling" a given form of life. A niche, typically, enables a, possibly new, organism; yet, a niche may be also constructed by an organism. In [Longo et al., 2012b] et [Longo & Montévil, 2013] an attempt is made to frame this informal notion in a rigorous context. We borrow here from that work to correlate enablement to the role of symmetry changes and we provide by this a further conceptual transition from physics to biology.

ANTI-ENTROPY. This aims to quantify the "amount of biological organization" of an organism [Bailly & Longo, 2003, 2009] as a non-reducible opposite of entropy. It also determines some temporal aspects of biological organization. This aspect of our investigation gives a major role to randomness. The notion of randomness is related to entropy and to the irreversibility of time in thermodynamics and statistical mechanics. As a result, we consider a proper notion of biological randomness as related to anti-entropy, to be added on top of the many (at least three) forms of randomness present in physical theories (classical, thermodynamical, quantum).

1.4 A More Detailed Account of Our Main Themes: Time Geometry, Extended Criticality, Symmetry Changes and Enablement, Anti-Entropy

The purpose of this book is to focus on some biological phenomenalities, which seem particularly preeminent, and try to approach them in a conceptually robust manner. The four points below briefly outline the basic ideas developed and are meant to provide the reader with the core ideas of our approach, whose precise meaning, however, can only be clarified by the technical details to which this book is dedicated.

1.4.1 Biological Time

The analysis of biological rhythms does not seem to have an adequate counterpart in mathematical formalization of physical clocks, which are based on *frequencies* along the usual, possibly oriented, time. Following [Bailly et al., 2011], we present a two-dimensional manifold as a mathematical frame for accommodating autonomous biological rhythms: the second dimension is compactified, that is, it is a circular fiber orthogonal to the oriented representation of physical time. Life is temporally paced by both external (physical) rhythms (circadian, typically), which are frequencies, and internal ones (metabolism, respiration, cardiac rhythms). The addition of a new (compactified) dimension for biological time is justified by the peculiar dimensional status of *internal* biological rhythms. These are pure numbers, not frequencies: they become average frequencies and produce the time of life span, when used as coefficient in scaling laws, see chapter 2. These rhythms have also singular behaviors (multi-scale variations) with respect to the physical time, which can be visualized in our framework. In contradiction with physical situations, the scaling, however, does not seem to be associated to a stable exponent. These two peculiar features (pure numbers and fractal-like time series) are the main evidences of the mathematical autonomy of our compactified time with respect to the physical time. Thus, the usual physical (linear) representation of time may be conveniently enriched, in our view, for the understanding of some phenomena of life: we will do it by adding one dimension to the ordinary physical representation of time.

Besides rhythms, an extended form of present is more adequate for the understanding of memory or elementary retention, since this is an essential component of learning, for the purposes of future action, even in some unicellular organisms. Learning is based on both memory and "protention", as pre-conscious expectation. Now, while memory, as retention, is treated by some physical theories (relaxation phenomena), protention seems outside the scope of physics. We then suggest some simple functional representation of biological retention and protention.

The two new aspects of biological time allow to introduce the abstract notion of "biological inertia", as a component of the conceptual time analysis of organisms. Our approach to protention and retention focuses on local aspects of biological time, yet it may provide a basis to accommodate the long range correlations observed experimentally, see [Grigolini et al., 2009]. Indeed, this kind of correlations is relevant for both aspects of our approach to biological time, and fits in the conceptual framework of extended criticality below.

Another aspect of biological time, discussed in chapter 7, is the time constituted by the cascade of symmetry changes which takes place in extended critical transitions. In other terms, this time is defined by the ubiquitous organizational transformations occurring in biological matter. Here, time corresponds to the *historicity* of biological objects and to the process of biological individuation, both

1.4 A More Detailed Account of Our Main Themes

ontogenetic and phylogenetic. Indeed, time is no longer the parameter of trajectories in the phase, space since the latter is unstable (chapter 8); therefore we will stress that temporality, defined by the changes of phase space, requires an original insight, in biology.

1.4.2 Extended Criticality

The biological relevance of physical theories of criticality is due first to the fact that, in physics, *critical phase transitions* are processes of changes of state where, by a sudden change (a singularity w. r. to a control parameter), the global structure of the system is involved in the behavior of its elements: the local situation depends upon (is correlated to) the global situation. Mathematically, this may be expressed by the fact that the correlation length formally tends towards infinity (e. g. in second order transitions, such as the para-/ferromagnetic transition). Physically, this means that the determination is global and not local. In other words, a critical transition is related to a change of phase and to the appearing of critical behaviors of some observable — magnetization, density, for example — or of some of its particular characteristics — such as correlation lengths. It is likely to appear at equilibrium (null fluxes) or far from equilibrium (non-null fluxes). In the first case, the physico-mathematical aspects are rather well-understood (renormalization as for the mathematics [Binney et al., 1992], thermodynamics for the bridge between microscopic and macroscopic description), while, in the second case, we are far from having theories as satisfactory. We present physical critical transitions in chapter 6.

Some specific cases, without particular emphasis on the far from equilibrium situation, have been extensively developed and publicized by Bak, Kauffman and others (see [Bak et al., 1988, Kauffman, 1993, Nykter et al., 2008a]). The sand pile, whose criticality reduces to the angle of formation of avalanches in all scales, percolation (see [Bak et al., 1988, Laguës & Lesne, 2003]) or even the formation of a snowflake are interesting examples. The perspective assumed is, in part, complementary to Prigogine's: it is not fluctuations within a weakly ordered situation that matter in the formation of coherence structures, but the "order that stems from chaos" [Kauffman, 1993]. Yet, in both cases potential correlations are suddenly made possible by a change in one or more control parameter for a specific (point-wise) value of this parameter. For example, the forces attracting water molecules towards each other, as ice, are there: the passage below a precise temperature, as decreasing Brownian motion, at a certain value of pressure and humidity, allows these forces to dominate the situation and, thus, the formation of a snow flake, typically.

Critical transitions should also be understood as sudden symmetry changes (symmetry breaking and formation of new symmetries), and a transition between two different macroscopic physical objects (two different states of matter, in the language of condensed matter physics), with a conservation of the symmetries of the components. The specific, local and global, symmetry breakings give the variety and unpredictability of organized forms and their regularities (the new symmetries) as these transitions are constituted by the fluctuations in the vicinity of criticality.

In physics, the point-wise nature of the "critical point" of the control parameter is an essential mathematical issue, as for the treatment by the relevant mathematics of "renormalization" in theories of criticality, see chapter 6 and [Binney et al., 1992].

Along the lines of the physical approaches to criticality, but within the frame of far from equilibrium thermodynamics, we consider living systems as "coherent structures" in a continual (extended) critical transition. The permanent state of transition is maintained, at each level of organization, by the integration/regulation activities of the organism, that is by its global coherent structure.

In short, following recent work [Bailly & Longo, 2008, Longo & Montévil, 2011a], but also on the grounds of early ideas in [Bailly, 1991], we propose to analyze the organization of living matter as "extended critical transitions". These transitions are extended in space-time and with respect to all pertinent control parameters (pressure, temperature etc.), their unity being ensured through global causal relations between levels of organization (through integration and regulation). More precisely, our main theoretical paradigm is provided by the analysis of critical phase transitions, as this peculiar form of critical states presents some particularly interesting aspects for the biological frame: the formation of extended (mathematically diverging) correlation lengths and coherence structures, the divergence of some observables with respect to the control parameter(s) and the change of symmetries associated to potentially swift organizational changes. However, the "coherent critical structures" which are the main focus of our work cannot be reduced to existing physical approaches, since phase transitions, in physics, are treated as "singular events", corresponding to a specific well-defined value of the control parameter, just one (critical!) point as we said. Whereas our claim is that in the case of living systems, these coherent critical transitions are "extended" and maintained in such a way that they persist in the many dimensional space of analysis, while preserving all the physical properties mentioned above (diverging correlation lengths, new coherence structures, symmetry changes ...). In other words, the critical transitions we look at are to be analyzed as taking place through an interval, not just a point, with respect to each control parameter. Thus, a living object is understood not only as a dynamic or a process, in the various possible senses analyzed by physical theories, but it is a *permanent critical transition*: it is always going through changes, of symmetries changes in particular, as analyzed below. We then have an extended, permanently reconstructed and changing *global* organization constituted by an interaction between local and global structures, since the global/local interplay is proper to critical transitions. We consider this perspective as a conceptual tools for understanding diversity and adaptivity.

Our analysis of extended criticality is largely conceptual, because of the loss of the mathematics of renormalization, which applies to point-wise phase transitions. Moreover, there seems to be little known Mathematical Physics that applies to physically singular, far from equilibrium critical transitions, *a fortiori* when the transition is extended. The other major conceptual and technical difficulty is also due to the instability of the symmetries involved. The issue we will focus on then, is how to objectivize biological phenomena, since, in contradiction with the physical cases, they do not not seem to be theoretically determinable within a specific, pre-given phase space and this because of the key biological role of symmetry changes.

1.4.3 Symmetry Changes and Enablement

As a fundamental conceptual transition between theories of the inert and of the living, we extensively focus on the different role of symmetry changes. Symmetry changes correspond, in physics, to the transition to a new state of the matter, or, even, in some cases, to a radical change of theory (recall the transition from theories of particles to hydrodynamics). In biology, instead, we will focus in chapter 7 on their constitutive role: the analysis of symmetry changes provide a key tool for constructing a coherent biological knowledge. As mentioned above, extended criticality is based on symmetry breakings and (re)constructions; our understanding of randomness, variability, adaptivity and diversity of life will largely rely on them. Moreover, in the passage from physics to biology, we will use these permanent dynamics to justify the introduction of "enablement" in [Longo et al., 2012b] and [Longo & Montévil, 2013], see chapter 8. Life and ecosystemic changes allow (enable) new life. (Changing) niches enable novelty produced by "descent with modification", a fundamental principle of Darwin's, while new phenotypes produce or co-constitute new niches. In our view, enablement is a fundamental notion, often used in the language of evolution, that we try to frame here in a coherent theoretical perspective. In contrast to the inert, whose default state is, of course, inertia, organisms interact with the surrounding world by acting (reproduction with modification and motility), use enabling conditions (are enabled by the environment), while producing new enabling conditions for further forms of life.

The analysis of enablement will lead us to the final main theme of this book: an understanding of the increasing complexity of phenotypes, through evolution. Often by sudden transitions, or by "explosions" as for richness of news phenotypes (Eldredge's and Gould's punctuated equilibria, see [Eldredge & Gould, 1972]), organisms complexify as for the anatomical structure through evolution. Our aim is to objectivize this intuition and the paleontological facts supporting it, by a sound mathematical understanding: anti-entropy will provide a possible quantification of phenotypic complexity and of its unbiased diffusion towards increasing values. It only makes sense in presence of continual symmetry changes and enablement.

1.4.4 Anti-entropy

In chapter 9, we develop our systemic perspective for biological complexity, both in phylogenesis and ontogenesis, by an analysis of organization in terms of "anti-entropy", a notion which conceptually differs from the common use of "negative entropy". Note that both the formation and maintenance of organization, as a permanent reconstruction of the organism's coherent structure, go in the opposite direction of entropy increase. This is also Schrödinger's concern in the second part of his 1944 book. He considers the possible decrease of entropy by the construction of "order from order", that he informally calls negative entropy. In our approach, anti-entropy is mathematically presented as a new observable, as it is not just entropy with a negative sign (negative entropy, as more rigorously presented in Shannon and in [Brillouin, 1956]). Typically, when summed up, equal entropy and negative entropy give 0. In our approach, entropy and anti-entropy are found simultaneously only in the non-null critical interval of the living state of matter. A purely conceptual analogy may be done with anti-matter in Quantum Physics: this is a new observable, relative to new particles, whose properties (charge, energy) have opposite sign. Along our wild analogy, matter and anti-matter never give 0, but a new energy state: the double energy production as gamma rays, when they encounter in a (mathematically point-wise!) singularity. Analogously, entropy and anti-entropy coexist in an organism, as a peculiar "singularity": an extended zone (interval) of criticality.

To this purpose, we introduced two principles (existence and maintenance of anti-entropy), in addition to the thermodynamic ones. These principles are (mathematically) compatible with the classical thermodynamic ones, but do not need to have meaning with regard to inert matter. The idea is that anti-entropy represents the key property of an organism, even a unicellular one, to be describable by several levels of organization (also a eukaryotic cell possesses organelles, say), regulating, integrating each other — they are parts that functionally integrate into a whole, and the whole regulates them. This corresponds to the formation and maintenance of a global coherence structure, in correspondence to its extended criticality: organization increases, along embryogenesis say, and is maintained, by contrasting the ongoing entropy production due to all irreversible processes. No extended criticality nor its key property of coherence would be possible without anti-entropy production, since always renewed organization expresses and allows the maintenance of the extended critical transition.

Following [Bailly & Longo, 2009], we apply the notion of anti-entropy to an analysis of Gould's work on the complexification of life along evolution in [Gould, 1997]. We thus extend a traditional balance equation for the metabolism to the new notion as specified by the principles above. This equation is inspired by Gibbs' analysis of free energy, which is hinted as a possible tool for the analysis of biological organization in a footnote in [Schrödinger, 2000]. We will examine far from equilibrium systems and focus in particular on the production of global entropy associated to the irreversible character of the processes. In [Bailly & Longo, 2009], a close analysis of anti-entropy has been performed from the perspective of a diffusion equation of biomass over phenotypic complexity along evolution. That is, we could reconstruct, on the grounds of general principles, Gould's complexity curve of biomass over complexity in evolution [Gould, 1997]. We will summarize and update some of the key ideas of that work. Once more, Quantum Mechanics indirectly inspired our mathematical approach: we borrow Schrödinger's operatorial approach in his famous equation but in a classical framework. Classically, that equation may be understood as a diffusion equation. As a key difference, which stresses the "analogical" frame, we use real coefficients instead of complex ones. Thus we are outside of the mathematical framework of quantum mechanics and just use the operatorial approach in a dual way, for a peculiar diffusion equation: the diffusion of bio-mass over phenotypic complexity.

1.5 Map of This Book

It should be clear by now that this book is at the crossroads of (theoretical) physics and biology. As a consequence, certain passages will use mathematical techniques that can seem of some difficulty for the non-mathematically trained reader. However, the main mathematical tools used in this book are very simple and we will try to explain them both conceptually and intuitively in the text. Similarly, we will refer to numerous physical ideas that we will explain qualitatively (and for a few of them, quantitatively). The prevalence of physical concepts will be especially marked in the chapter 5 and 6, however these concepts will be gradually introduced. In any cases, the the more technical parts of the book may be skipped at first reading, as suggested on place, and the qualitative explanations should be sufficient to proceed to our biological proposals. In general, we do not think at all that, in scientific disciplines, there is "as much scientific knowledge as there is mathematics". For example, the notions of extended criticality and enablement are represented only at a conceptual level. Mathematics is used here just when it helps to better specify concepts, if possible and needed, typically and more broadly to focus on invariance and symmetries. it is also used when it has some "generative" role, i. e. when it suggests how to go further by entailed consequences within or beyond proposed frameworks: the case of "biological inertia" in chapter 4 is a simple example of the latter form of entailment.

It is worth mentioning that despite conceptual and formal links between the chapters, most chapters retain a certain level of autonomy and can be read independently.

As for the references we will make to empirical evidences, we will start from some broadly accepted forms of "scaling". In chapter 2, we will review them in various contexts, where our choice of results is motivated by their relative robustness and by the theoretical role that they will play later. We will in particular try to assess their experimental reliability and the variability that is observed. This step is important since we will use these observations (including variability) both technically and conceptually, as examples, in the rest of this book.

Since biological rhythms are associated to relatively robust symmetries, we will consider the question of biological temporal organization directly, first by analyzing rhythms, in chapter 3, then by an analysis of the "non-linear" organization of biological time. More precisely, we first propose a bidimensional reference system for accommodating biological rhythms, by which we may take scaling behaviors of different nature into account. Then, in chapter 4, we will approach the local structure of biological time, through the notions of protention and retention, thus providing an elementary mathematical approach of the notion of "extended present".

Chapter 5 provides a conceptual (and light technical) introduction on the role of symmetries in physical theories. This chapter provides some background and examples to set the subsequent developments. The next chapter, chapter 6, will provide an elementary introduction to physical critical transitions. Both chapters are intended to introduce the notions required for the following.

In chapter 7, we will approach the structures of determination of biological phenomena by the notion of theoretical symmetries. This will allow us to contrast the status of biological objects with the status of the physical objects. As a matter of fact, for the latter, the theoretical symmetries are stable, while we will characterize biological processes as undergoing ubiquitous symmetry changes. This will allow us to provide a proper notion of variability and of biological historicity (as a cascade of symmetry changes).

Since this perspective yields a fundamental instability of biological objects, our theoretical proposal "destabilizes" the physical approach to objectivization, for biological objects. Chapter 8 explores the consequence of this approach on the notion of phase space in biology (that is on the space of the theoretical determination). Namely, in this context, the relevant space of description is changing and unpredictable. The notion of "enablement" provides an understanding of biological dynamics by adding on top of causality a novel theoretical insight on how the active default state of living entities continually constructs and occupies new niches and ecosystems.

In chapter 9, we revisit the quantified approach to biological complexity, as "anti-entropy", introduced in [Bailly & Longo, 2009]. By this, we will develop an analysis of that notion in terms of symmetry and symmetry changes, on one side, and analyze some regenerative aspects of biological organization on the other side. We will also discuss the issue of the associated notion of randomness.

We conclude by philosophical reflection on how we moved from physics to biology, chapter 10.

Chapter 2
Scaling and Scale Symmetries in Biological Systems

> Observations always involve theory.
>
> E. Hubble

Abstract. This chapter reviews experimental results showing *scaling*, as a fundamental form of "theoretical symmetry" in biology. Allometry and scaling are the transformations of quantitative biological observables engendered by considering organisms of different sizes and at different scales, respectively. We then analyze anatomical fractal-like structures, the latter being ubiquitous in organs' shape, yet with a fair amount of variability. We also discuss some observed temporal fractal-like structures in biological time series. In the final part, we will provide some examples of space-time and of network configurations and dynamics.

The few concepts and mathematics needed to understand allometry and scaling are progressively introduced, always accompanied by a discussion of the main experimental findings, either through special cases or more general results. We focus in particular on the robustness of these empirical observations and the corresponding variability.

Keywords: Scaling, variability, allometry, fractals, regulation, criticality.

2.1 Introduction

We propose, in this chapter, a unified picture of the empirical findings on allometry and scaling in organisms and cells. Although these findings mainly revolve around the notion of scale symmetry, they can take various forms. Therefore, we will provide brief accounts of the conceptual and mathematical basis leading to these empirical inquiries. These short introductions are also needed because they define the quantities that are tentatively constituted to be robust and biologically relevant.

We want first to emphasize the difference between the allometric relations and the scaling relationships *inside* a definite system.

ALLOMETRY. In the allometric methodology, the idea is *to compare properties of organisms with different sizes*. More precisely the core idea of allometry is that we can highlight fundamental aspects of biological organization by looking how quantities, such as various rates, sizes of components..., change with respect to a degree of freedom (the organism mass usually) and more precisely its scale.

This degree of freedom is not in general *per se* relevant (as a degree of freedom) to the organism. That is, usually we cannot change, in a relevant way, the mass of an adult organism (except by changing its organization, usually by obesity).

SCALING. This methodology aims at finding scaling relationships as *a property of a system observed at different scales* (especially spatial and temporal scales). Hence, this second approach aims at describing complex geometrical organization, usually by introducing a dependence of observed quantities on the resolution of observation. This can bring out significant results by looking how objects change as a function on this resolution, instead of "looking at the whole object at once" or on the average.

An extensive review of the second methodology, in the case of neuronal structures and activities, has been given in [Werner, 2010], see also [Werner, 2007, Ribeiro et al., 2010]. Since scaling and criticality, another major issue of this book, are already well reviewed for neuronal activities, we will, as for now, refer to those texts as for neuronal examples; more will be said in forthcoming chapters. For sake of generality, we focus here on the basic physiological properties of cells and organisms.

We are going first to look at allometric properties in biological systems. Then, we will consider the morphological fractal-like properties, their validity and basic properties. These properties concern the space organization of biological systems. Next, we will look at the temporal structure of organisms, by the observation of biological time series. Then, we will discuss anomalous diffusion as well as biological networks architectures and dynamics.

Before entering this discussion, we will present the basic mathematical forms that describe the simplest case of scaling, namely the renowned power laws, and the reason why they mathematically model scale symmetries. More complex definitions of scaling usually have these forms as mathematical building blocks.

This chapter deals with a few but fundamental approximate invariants of life, often highly debated. It requires, as we said, some (simple) mathematics and numerous references to empirical evidence, which is at the basis of our theorizing. However, the philosophically oriented reader may skip, at first reading, the technical details and move to more theoretical chapters of this book. The conclusion of each section will summarize some relevant aspects of our presentation.

2.1.1 Power Laws

As we said, power laws, that is laws described by functions f such as $f(x) = kx^{\alpha}$, are met almost everywhere when scaling is discussed. We will show in this section that there is a straightforward mathematical reason that justifies this situation. We will first present it informally, and a discussion with more technical details and generality can be found in annex A.1. The mathematics are very simple, it is more a matter of digesting some notation that true mathematical theorems and, at first reading, the reader may just try to assimilate it without going into the details of the (very simple) proofs.

2.1 Introduction

Let us consider a real function f of one variable. Suppose that f has a property of scale invariance, by which we mean that f is the same, modulo a coefficient when the variable is observed at different scales. Formally, this condition can be written as $f(\lambda x) = g(\lambda) f(x)$, where λ characterize the change of scale.

This condition can also be seen in terms of *symmetry*, as an *invariance by dilatation*: $f \circ \mathfrak{D}_\lambda = \tilde{\mathfrak{D}}_{g(\lambda)} \circ f$.

We will now show semi-informally that under this conditions, and the assumption that f is continuous, we have $f(x) = f(1) x^\alpha$.

- In this special case, we have $f(x1) = f(1) g(x) = g(x) f(1)$, meaning that f and g are proportional. We will in the following take an arbitrary value, a, not necessarily 1, as the starting point of the transformation. The aim is now to relate quantities $f(x)$, for any value of the variable x, to $f(a)$, provided that both $f(1)$ and $f(a)$ are different from 0. This assumption allows to define α as the unique value such that $f(a) = a^\alpha$ (i.e. $\alpha = \frac{\ln(f(a))}{\ln(a)}$).
- Considering an integer n, we have $f(a^n) = g(a^n) f(1)$ and $f(a^n) = f(aa^{n-1}) = g(a) f(a^{n-1}) = \cdots = g(a)^n f(1)$. Hence, $g(a^n) = g(a)^n$.
- If p and q are integers, $f(a^{qp/q}) = f(a^p) = g(a)^p f(1)$ and $f(a^{qp/q}) = f\left((a^{p/q})^q \right) = g(a^{p/q})^q f(1)$. So we conclude that $f(a^{p/q}) = g(a)^{p/q} f(1)$ and $g(a^{p/q}) = g(a)^{p/q}$.
- This is not sufficient to obtain the value $f(x) = f(a^y)$ for every *real* number y, because there is no *finite* algebraic transformation from the rationals to all real numbers. Nevertheless, since every *real number* is the limit of rational numbers and since we assumed that f is continuous, we can conclude that $f(a^y) = g(a)^y f(1)$.
- Writing $g(a) = a^\alpha$ (i.e. $\alpha = \frac{\ln(g(a))}{\ln(a)}$) we have then $f(x) = f(a^y) = g(a)^y f(1) = a^{\alpha y} f(1) = x^\alpha f(1)$.

We understand now that scale symmetry leads directly to power laws. It is noteworthy that such functions are described by two kinds of quantities: a coefficient, here $f(1)$, and one or more exponents, α here. This two kinds of quantities will be the at the core of the empirical results we discuss in this chapter. Notice, moreover that the exponent is a dimensionless quantity associated to the changing of scale itself, it determines the dilatation $\tilde{\mathfrak{D}}_{\lambda^\alpha}$ of the observable, associated to the dilatation $\mathfrak{D}_\lambda$ of the variable. The coefficient $f(1)$ is associated to the choice of units (of what 1 is, for the variables and the observable). Another aspect of this coefficient, in a given choice of units, is that it can be different for different sets of objects that still share the same symmetry by dilatation. In this case, they are not on the same orbit for these transformations: they have the same symmetry but are not symmetric for the intended transformation (they cannot be changed into each other by this symmetry, see below for an elementary example in the case of allometry).

By contrast with the situation of scale symmetry, the paradigmatic situation involving a characteristic scale is of the form $exp(-x/x_0)$. Since $exp(-\lambda x/x_0) = exp(-x/x_0)^{\lambda-1} exp(-x/x_0)$: the analog to g depends then crucially on x. From a more conceptual point of view, the nature of the exponential is to describe the decay (or growth) of a quantity which changes in proportion to its magnitude for a

2.2 Allometry

A foundational approach to allometry can be found in Galileo's work, [Galileo, 1638]. In his early studies on Dante's Inferno ("Two lectures on the size of the Hell", of the early 1580s) Galileo developed a close analysis of the structure of Hell and its covering — a major architectural problem as the Hell is a cone, centered at the center of the Earth with a base angle of 60 degrees. These technical studies were very much in fashion at the time and, most likely in order to improve his bibliometrics indices, the young Galileo felt obliged to be involved in them. When he obtained a tenured position at the University of Pisa, in 1589, he stopped this work and started his best known commitments (see [Lévy-Leblond, 2007]). In his "Two Lectures on the size of the Hell", Galileo mentioned, but did not discuss and apparently accepted as possible, the peculiar allometric properties of the Devil, who is 1,200 meters tall, but with the same proportions as a human. These properties may have later puzzled Galileo, as there are written traces that he refused to send out reprints of these lectures, when asked.

It is possible that this mistake enhanced his founding writings on allometry and relative resistance of materials in [Galileo, 1638], which we briefly summarize next[1].

2.2.1 *Principles*

As previously stated, the idea of allometry is to look at quantitative properties of objects of the same "nature" but of different sizes. Consider, for example, a cube of side length l, and a dilatation of all axes by a factor λ. If L is the length of the edges, S its surface and V its volume, we have:

$$\begin{array}{c} L = l \\ S = l^2 \\ V = l^3 \end{array} \longrightarrow \begin{array}{c} \lambda L \\ \lambda^2 S \\ \lambda^3 V \end{array}$$

Therefore, we have obtained quantities that have the form al^α, where α characterizes the nature of the quantity observed (a length, a surface or a volume). As we said before, the exponent α defines the scale symmetry associated to the considered quantity, whilst the coefficient a determines a class of objects which are symmetric, with respect to the observed quantity. For example, the set $\mathscr{C}$ of cubes of side length

---

[1] As soundly pointed out recently by J. M. Lévy-Leblond, Galileo's early oversight may have been motivated by the fact that the Devil's belly button coincides with the center of the Earth, so the gravitational effects are rather minor on Him (Her?). We also share this thesis, namely that the widely accepted "Devil's violation of allometric equations" is a *compatible fact* with the discussion below on these equations.

2.2 Allometry

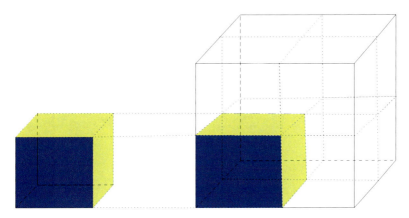

Fig. 2.1 *Allometry for a cube:* Doubling the lengths leads to a multiplication by 2^2 of the surfaces and a multiplication by 2^3 of the volume. The corresponding exponents, 2 and 3, characterize in a very general way the geometrical nature of the objects and property considered.

l and the set of spheres $\mathscr{S}$ of radii l share the same symmetries by dilatation for the surface (with exponent 2) and the volume (with exponent 3), but they have different coefficients (fully determined by the value at a given scale, say for $l = 1$).

More generally, the idea of allometry is to postulate that a class of objects, understood as parameterized in particular by their size (for example their mass, W), share a quantitative property $A(W, \ldots)$, and that this objects are similar in their constitution. The latter is formulated by saying that A verifies, at least approximately, a symmetry by dilatation with respect to the parameter W. As a result, the expected relationship, when everything else remains constant, is $A(W) \simeq aW^\alpha$. The latter relation describes the scale symmetry associated to the considered objects, then understood as different versions of the same organization for different sizes.

We should notice that this kind of dependency is geometrically fundamental and for "simple" systems does not lead to a plethora of situations. For example, in classical thermodynamic, with mass as a parameter, there are two kinds of quantities: the intensive quantities, with $\alpha = 0$ (for example temperature, pressure, concentrations, ...) and the extensive quantities, with $\alpha = 1$ (volume, energy, entropy, ...)[2].

In order to show allometry experimentally, the simplest methodology consists in representing the observable as a function of its parameter in log-log coordinates. In such a graph, the allometric relation correspond to a straight line, because of the relationship $\log A(W) \simeq \log(a) + \alpha \log(W)$. This relation also shows that the exponent appears as the slope of this line. In order to test such a relationship, a broad set of orders of magnitude is required, because multiplying $\log W$ by n requires measuring a weight W^n (from another point of view, the derivative of log tends to 0 when the

---

[2] The non-extensive thermodynamic approaches try to construct a framework where the situation is richer, starting from this point of view, see [Tsallis & Tirnakli, 2010] for a review.

variable tends to infinity). There are mainly two situations in biology that meet this criteria and which correspond to the two main types of allometry: the *intraspecific* or ontogenetic allometry (we observe organisms during their development) and the *interspecific* allometry (we observe adults of species encompassing a wide range of adult masses). From a biological point of view, an essential difficulty is to define A in such a way that A can be expected to be similar for organisms with various phylogenetic and ontogenetic histories and widely different sizes[3].

2.2.2 Metabolism

From the point of view of allometry, metabolism is a very interesting quantity since it is clearly a global quantity, involving the activity level of the whole organism. Besides, it is a quantity that can be measured for a wide class of organisms, in a similar way. Since the work of [Rubner, 1883] and [Kleiber, 1932], a considerable amount of studies has been performed focusing on metabolic allometry, also for historical reasons (see below).

In order to compare the metabolism of different organisms, it is necessary to consider them at comparable levels of activity. As a result, the levels of activity considered must be broadly defined, and should not exhibit too much specific effects, due to ontogenetic or phylogenetic individuation. The main levels of activity used are:

BASAL METABOLIC RATE (BMR). The BMR is the most used level of activity for allometry. It corresponds intuitively to a situation of an animal doing nothing in particular but being awaken. More precisely, the animal is at rest in a conscious post-absorptive state in a thermo-neutral environment and in the inactive phase of the circadian cycle. We should notice here that this can imply some difficulties. For example, ruminants are almost always in an absorptive state. This level is defined for homeothermic animals, and its analog for poikilothermic animals is the Standard Metabolic Rate. However, the distinction between the two terms is sometimes fuzzy in the literature.

MAXIMUM METABOLIC RATE (MMR). The MMR corresponds to a situation at the maximum level of sustainable exercise.

FIELD METABOLIC RATE (FMR). The FMR is measured on animals in the wild. As a result it needs specific measurement methodology, in order to not hinder animal's activity. Notice that, by definition, it may depend on the field considered, and on animal's habits.

In order to observe quantitatively metabolic activity, the simplest method is the measurement of the O_2 consumption rate, provided that for aerobic organisms this rate is assumed to be proportional to respiration, and the latter to metabolism. For plants, where the metabolic activity consumes and produces oxygen, the observed quantity

---

[3] Going back to the founder of these issues, Galileo applied his remarks to biological organisms, by observing that the diameter of bones had to grow like the cube, not linearly, since the wheigth grows like a volume, [Galileo, 1638].

is the increase of mass (dry matter). A less frequently used methodology consists in a calorimetric approach, allowing to directly observe the energy dispersed by organisms as heat. [Glazier, 2005] is an extensive survey on metabolic allometry.

2.2.2.1 Interspecific Metabolic Allometry

Allometry of the BMR and metabolic allometry in general has been first tested experimentally in biology by [Rubner, 1883]; and this experimental approach was initially motivated by a theoretical insight. The idea Rubner followed is that the metabolism, R, (measured by O_2 consumption rate) should be proportional to heat transfers with the environment. The latter is then assumed to be proportional to the surface of the body, leading to $R \propto S \propto l^2$, where l is a typical length. The mass is also assumed to be proportional to the volume, so we have $W \propto V \propto l^3$ and $l \propto W^{1/3}$. As a result, the expected relationship between metabolism and mass was $R \propto \left(W^{1/3}\right)^2 = W^{2/3}$. This relationship has been tested experimentally and seemed confirmed on dogs.

Nevertheless, the work of [Kleiber, 1932] lead to a largely different BMR allometric relationship for adult mammals: an exponent of $3/4$. This relation seems to hold for interspecific allometry, as for the observation of O_2 consumption rate among animals [Kleiber, 1961, Schmidt-Nielsen, 1984, Savage et al., 2004], but with different coefficients R_0 for mammals, birds or reptiles (where $R \approx R_0 W^{3/4}$). For example, this coefficient is roughly 10 times bigger for mammals than for reptiles, which comes as no surprise considering the deep metabolic differences between the two phylogenetic groups. Yet, the remarkable fact is the finding of similar allometric exponents. In the case of plants, for the observation of dry matter production, the same relation was also found in [Niklas & Enquist, 2001], with the same coefficient over 20 order of magnitude[4]. Nevertheless, notice that, in all cases, the variability remains high, in spite of the high regularizing effect of the allometric form. For example, in [Niklas & Enquist, 2001] the 95% confidence interval has a width of a factor 10 (for the metabolism).

Part of these results, however, are challenged by some authors. To understand the roots of the controversy, recall than any statistical analysis is based on an assumption of independence or controlled dependence between points, which lead to the weight given for various measurement. [Savage et al., 2004] takes averages on a *per species* basis, then gives equal statistical weights for equal $\log(W)$ intervals. The underlying idea is to give equal weight to all masses, which is the straightforward way to follow the form of allometric equation. This allows to deal with the fact that there are more small species than large species. It also indirectly compensates for phylogenetic correlations, since related species tends to have similar masses[5]. [White & Seymour, 2003], on the contrary, focuses on the question of phylogenetic dependence, but does not take into account the small size over-representation, and even

---

[4] This article is especially seen as compelling since one of its author was skeptical about the wide range validity of such a relationship.

[5] This approach is related to the theoretical justification of $3/4$ BMR allometric exponent proposed in [West et al., 1997, 1999], which emphasizes the scale symmetry aspects as a broad tendency.

accentuates it by ruling out species, usually big, where the post absorptive state was not clearly achieved[6]. This lead the two studies to find different, statistically incompatible exponents ... even though they are based on the same data set, see [Agutter & Wheatley, 2004] for a fairly balanced review on this issue.

One should note that temperature plays a key role for the metabolic rate. A rule, derived by theoretical arguments, based on thermodynamics, has been proposed and tested by [Gillooly et al., 2001]: $R \propto \exp(\frac{E}{kT})$. This rule is confirmed by empirical data, but seems far more scattered than the mass dependence, see §2.2.2.3.

Maximum metabolic rate, however, yields a higher exponent. For mammals, the exponent found is 0.87 in [Weibel et al., 2004] and of 0.83 in [Savage et al., 2004]; among mammals and birds, [Bishop, 1999] finds an exponent of 0.88. Even if the exponent is not stable in the literature, the fact that it is higher that the BMR exponent is well established. We emphasize that it is obvious that the coefficient R_0 for MMR is higher than that of BMR, because in the case where the mass is one, $R_{Max} = R_{0,Max} > R_{Basal} = R_{0,Basal}$; however, the two situations could *a priori* lead to similar exponents, like in the comparison of reptiles and mammals. There is, though, a simple mathematical reason, which is not discussed in the literature, why the exponent for MMR cannot be smaller than that of BMR: if this was the case, small animals would have a higher BMR than the MMR. A more biological argument is that the MMR is influenced by the inertial mass. Indeed the organism needs to move and it is fair to assume that the energy required for that is proportional to the inertial mass, which has an allometric exponent of 1.

The field metabolic rate depends on the habits of animals and on their ecological situation, which can be somewhat different in different habitats. As a result, the possible scale symmetry is significantly looser than for other metabolism values, see [Nagy, 2005]. The exponent found is roughly 0.8, but the importance of the temperature of the environment is emphasized (which comes as no surprise, since it leads to different contributions of the external surface/volume ratio).

In the special case of hibernation, the exponent seems to be 1, see [Singer et al., 1995]. If this result is confirmed — which may prove to be difficult because of the limited number of hibernating species —, it means that the metabolic rate is an extensive quantity in this very peculiar situation. Hence, we can hypothesize that in this case the organism behave mostly like a sum of its part. Indeed this global response is what is expected for a classical (i.e. not complex), physical system. We can accordingly hypothesize that the corresponding biological integration is weak, as far as the metabolism is concerned.

2.2.2.2 Intraspecific Metabolic Allometry

We want to emphasize first that intraspecific allometry actually means ontogenetic allometry, since the size variability among adults is not sufficient to perform allometric studies. Some pathological cases, such as obesity, dwarfism, ..., could also provide mass variability, but associated to very specific changes of organization.

---

[6] As we mentioned earlier, it is indeed extremely difficult, if ever possible, to put animals like ruminants in a post-absorptive state.

2.2 Allometry

Intraspecific allometry seems far less stable than interspecific allometry. A number of study has been performed, especially among fishes, where the intraspecific allometric exponents were found to vary, depending on the species, see for example [Bokma, 2004]. We have to mention, however, that the latter study does not distinguish field metabolic rate and standard metabolic rate, and use highly heterogeneous data.

The observation of intraspecific allometry allows to obtain results for single species[7], and, as a consequence, to examine the potential changes of organization in development through a shift of allometric exponent, or a curve in log-log space. A study of [Giguere et al., 1988], for example, finds that the routine metabolic rate of Atlantic mackerel, *Somber scombrus*, scales isometrically (exponent 1) for fishes at a larval stage, whilst the juveniles and adults have an exponent of 0.8. Another study [Moran & Wells, 2007] on the yellowtail kingfish, *Seriola lalandi*, leads to an exponent of 0.90. According to the authors, the situation is possibly better described by a quadratic curve in the log-log space (instead of a linear one), comforting the idea of possible changes of exponent during ontogeny. A general study, among five different phyla, [Glazier, 2006], found in particular that the exponent is higher for pelagic animals in comparison with benthic ones — including a shift when the considered species has a larval, pelagic lifestyle followed by a benthic adult stage. Such a shift occurs in the case of mussels, *Mytilus edulis*, from an exponent 0.9 to 0.7.

The conclusions we can draw from these results is that aquatic intraspecific allometry is correlated to the ecological lifestyle. Beyond this correlation, we want to emphasize that the variability is dominant; and allows to have extreme variations in allometric exponents, including for animals with related ecological statuses.

For terrestrial animals too, the situation is not particularly stable. For example, exponents from 0.62 to 0.68 have been found for different snakes species [Dmi'el, 1970]. A more extensive study of ectotherms, [Glazier, 2009], found an average basal metabolic rate exponent of 0.83 ± 0.10, where the high scattering of this exponent distribution is especially noteworthy. Similarly to the interspecific situation, the active metabolic rate exponent is found to be higher 0.92. A collection of specific results can be found in [Glazier, 2005].

2.2.2.3 A Word on Temperature

Temperature has a direct effect on the rate of chemical reactions, and, as a result, can be assumed to affect directly the metabolism of organisms. This effect is expected to be large when the body temperatures can be significantly different, for poikilotherms[8]. [Gillooly et al., 2001, 2006] propose to consider that the effect of temperature on metabolism can be deduced, at least as a tendency, from thermodynamics. Arrhenius equation provides the following relation between the kinetic constant, k, of a chemical reaction, and its activation energy E_i. In particular, the

---

[7] As a matter of fact results for single individuals could also be obtained, but such studies have not been performed to our knowledge.

[8] As a matter of fact, one can also change importantly the temperature of homeotherms, but this leads to a major global organizational change, namely hypothermia (or hibernation).

latter is assumed to be constant by the authors (independent of temperature and constant in a given group of organisms) because of the stability of the elementary reactions involved in respiration:

$$k = A\exp\left(-\frac{E_i}{k_b T}\right) \tag{2.1}$$

where k_b is Boltzmann constant and T is the temperature in Kelvin. This leads to a completed metabolic equation:

$$R = R_0 W^{3/4} \exp\left(-\frac{E_i}{k_b T}\right) \tag{2.2}$$

and if we take a reference temperature T_0

$$R(W,T) = R(1,T_0) W^{3/4} \exp\left(\frac{E_i(T-T_0)}{k_b T T_0}\right) \tag{2.3}$$

The results obtained in [Gillooly et al., 2001] is $E_i = 0.41\,\mathrm{eV}$ to $0.74\,\mathrm{eV}$, with variations depending in particular on the organism group, with an overall mean of $0.62\,\mathrm{eV}$. However, the variability of the data is quite high, even in a given phylogenetic group. Still, in relation with the level of activity, most allometric studies have to take the effect of temperature variations into account, when one has to compare organisms used to different temperatures.

However, this account is based on the assumption of a symmetry of the molecular (thermodynamic) activities at different temperature, which is not necessarily met in biological situations. Actually, even the intraspecific exponents (for mass allometry) can depend on temperature. In [Glazier, 2005], for example, the Arthropoda *Asellus aquaticus* has an exponent of ~ 0.8 for $10\,°\mathrm{C}$ to $20\,°\mathrm{C}$ and of ~ 0.4 for $25\,°\mathrm{C}$ to $30\,°\mathrm{C}$, whilst *Euphausia pacifica* has an exponent that remains approximately constant. We can interpret the situation by saying that temperature changes can lead to organizational changes. Therefore, we have, here, a complete departure from the analysis above.

Notice also that the temperature of the body is not well defined for some species, meaning that it is far from homogeneous. For example, tuna have temperature differences of $12\,°\mathrm{C}$ to $15\,°\mathrm{C}$, see [Kay, 1998].

2.2.3 Rhythms and Rates

Rhythmic processes are another aspect of biological phenomena that allometry can help to describe. The idea here is to look at processes that occur in similar ways for a great variety of organisms, and to measure the time τ_i needed for these processes to take place. These time intervals are, for example, the beat-to-beat interval of the heart or the entire lifespan.

The results in the literature, see [Lindstedt & Calder III, 1981, Savage et al., 2004, Günther & Morgado, 2005], is that $\tau_i \propto W^{1/4}$ for interspecific allometry of mammals

2.2 Allometry

Table 2.1 *Allometry of biological rhythms in mammals*, this table is established using results from [Lindstedt & Calder III, 1981] and represents $\tau_i = \tau_0 W^\alpha$ where W is in kg and τ_0 in min.

	$\log \tau_0$	exponent α
Life span in captivity	6.8	0.20
98% growth time	5.8	0.26
Time for population doubling	5.5	0.26
Time to reproductive maturity	5.5	0.18
50% growth time	5.3	0.25
Gestation period	5.0	0.25
Time to metabolize fat stores (0.1% body mass)	2.2	0.26
Drug half life (methotrexate)	1.8	0.19
Plasma clearance, inulin	0.8	0.27
Blood circulation time	−0.5	0.21
Gut beat duration	−1.3	0.31
Respiratory cycle	−1.7	0.26
Cardiac Cycle	−2.4	0.25
Twitch contraction time, soleus	−3.0	0.39
Twitch contraction time, extensor digitorum longus	−3.5	0.21

and birds, where W is the mass. Note that we can deduce from this relation that the metabolic rate should scale approximately like $W^{3/4}$. Indeed, the metabolic rate has the physical dimensionality of an energy over a time and we can then approximately write that energy iss proportional to the mass, thus: $R \propto \frac{W}{\tau_i} \propto W^{3/4}$.

The metabolic rate is unique for an organism (in a given measure), whilst there is at the same time a broad variety of rhythms taking place in an organism. Moreover, characteristic times of non-rhythmic phenomena can also be observed. Hence, a variety of allometric exponents can be observed for the same class of organisms, which makes temporal allometry especially interesting, *a priori*. Their theoretical value is confirmed *a posteriori* by the approximate stability of the obtained allometric exponents, see table 2.1. A consequence of this stability is that the average ratio of two such times does not depend on the mass of the organisms, and then seems to be an invariant quantity. The average number of iterations of a given rhythm during the entire life span of an organism is then especially relevant, as a variety of them remain approximately constant among mammals, for example. The drawback of this kind of approach is that observables, such as the cardiac rhythm, become undefined for organisms ... without a heart (and in particular unicellular ones).

It is usually argued that the maximum spiking rate of neurons does not depend on the mass of organism, because it is supposed to be determined by the parameters of Hodgkin-Huxley's equation, and this parameters are mostly determined by ion channels which remains symmetric for animals of different masses, and depend on the cell. In [Hempleman et al., 2005], allometry is thus found in the neural spiking pattern for intrapulmonary chemoreceptors, with exponents of 0.22 to 0.26 as for birds.

Plants, to our knowledge, have less relevant rhythms than animals from the allometric perspective. Nevertheless, [Marbà et al., 2007] considered the demographic parameters on a wide collection of plants (from phytoplankton to trees) and found that the times characterizing the mortality rate has an allometric exponent of 0.22 and that the exponent for birth characteristic time is 0.27. As a result, both of them are close to the usual value of $1/4$. Partial results on fishes support also comparable exponents [Gerkema, 2002].

Interestingly, these aspects directly impact the molecular level. For example, [Gillooly et al., 2005] found that the nucleotides substitution rates are associated to exponents of 0.21 to 0.23, depending on the precise phenomenon observed. As a result, the "molecular clock", if any, depends directly on the body size (and temperature) of the organisms involved. Temporal allometry also have direct consequences on the effect of drugs on an organism, see [Boxenbaum & DiLea, 1995, Kirman et al., 2003] and the discussion in the conclusion of this section below.

2.2.4 Cell and Organ Allometry

If we consider constituents of an organism (cells or organs, typically), the analysis of the allometry of their properties, with respect to the mass of the organism, provides original and compelling insights on their biological meaning. Indeed, it gives directly a quantitative relation between the considered part and the whole. Here, we will review some anatomic and metabolic allometric relationship for parts of organisms.

2.2.4.1 Cell Allometry

Let us suppose that an organism of mass W is constituted by a number of cell $N(W)$ and has a metabolism of $R = R_0 W^\alpha$ (understood here as a respiration rate). We will first consider two opposite organizational possibilities, as described in [Savage et al., 2007]. This description will allow to clarify what the logic of the situation is. We will first remark that a principle of conservation (provided that there is no production of O_2 in the organism and that metabolism occurs in cells) leads to $R = \sum R_i$, where R_i is the metabolism of the cell i. In other words, the oxygen consumption of the organism is the sum of the oxygen consumption of the cells[9].

1. If we assume that the mean cell size does not depend of the size of the organism ($m_c \propto W^0$), then we have $N(W) = \frac{W}{m_c} \propto W$, where m_c is the mean of the cells masses and N is the number of cells. However, in this case, the mean metabolic rate per cell, R_c, is $R_c = \frac{R(W)}{N(W)} = R_0 m_c W^{\alpha-1}$. As a result, if the cells sizes are

---

[9] The hypothesis of conservation is crucial (and also does not seem very problematic): it is this hypothesis which leads to an additive structure for the oxygen consumption. The same argument is used for the mass.

2.2 Allometry

Table 2.2 *Allometry of cell types*, this table is established on the grounds of results from [Savage et al., 2007]. α_V is the allometric exponent associated to the cells volume and α_N is the exponent associated to the estimated cell number.

Cell type	α_V	α_N
Expected 1	0	1
Expected 2	$1-\alpha \simeq 0.25$	$\alpha \simeq 0.75$
Alveolar macrophages	0.08	0.96
Erythrocytes	0	
Fibroblasts	0	
Fibrocytes	0.05	
Glomerular epithelium	0.05	
Goblet cells	0.07	
Henle loop cells	0.01	
Hepatocytes	−0.03	
Lung endothelial cells	0	1.00
Lung interstitial cells	0.06	1.08
Lung type I cells	0.05	0.95
Lung type II cells	0	0.98
Proximal convoluted tubules	0.04	
Sebaceous gland cells	0.05	
Adipocytes (dorsal wall of abdomen)	0.13	0.80
Adipocytes (skin)	0.17	
Cerebellar granule neurons	0.14	
Cerebellar Purkinje neurons	0.18	
Superior cervical ganglion neurons		0.68

independent of the size of the organism, their metabolism gets lower when we consider organisms of greater sizes.

2. On the contrary, if we consider that the mean cell metabolic rate stays invariant, we obtain that $N(W) = \frac{R(W)}{R_c} = \frac{R_0}{R_c} W^\alpha \propto W^\alpha$. Then the mean mass of cells become $m_c = \frac{W}{N(W)} = \frac{R_c}{R_0} W^{1-\alpha} \propto W^{1-\alpha}$. As a result, the cells get bigger when we consider larger organisms.

These two simple, opposite organizational possibilities have been proposed and empirically explored by [Savage et al., 2007], in an interspecific comparison among mammals, and for the basal metabolic rate. We reproduce their findings in table 2.2, where m_c is approached by the cellular volume. These results show that the two theoretical organizational tendencies seem to be biologically relevant.

2.2.4.2 Organ Allometry

The interspecific allometry of organs' weight is not trivial; it involves different exponents for different organs, see table 2.3. As a result, the relative weight of

Table 2.3 *Allometry of organs,* intraspecific after [Trieb et al., 1976] and interspecific [Schmidt-Nielsen, 1984, Wang et al., 2001]

Organs	Male rat[a]	Female rat[a]	Interspecific exponent
Body weight	1	1	1
Prostate	2.13	ND	
Testes	1.02	ND	
Ovaries	ND	0.68	
Liver	0.73	0.48	0.87[c]
Heart	0.67	0.65	0.98[c]
Kidney	0.67	0.57	0.85[c]
Thyroid	0.55	0.35	
Adrenal	0.55	1.01	
Pituitary	0.53	0.53	
Lungs	0.47	0.40	0.99[c]
Spleen	0.47	0.47	
Brain	0.16	0.18	0.76[b]; 0.70[c]

[a] Result from [Trieb et al., 1976]
[b] Results from [Wang et al., 2001]
[c] Results from [Schmidt-Nielsen, 1984]

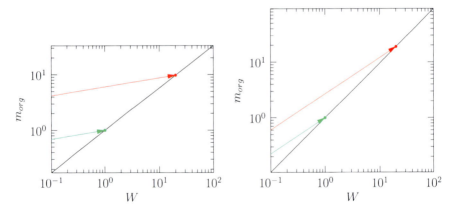

Fig. 2.2 *Illustration of the interspecific and intraspecific allometry of organs sizes.* Here, we consider the allometry of brain (LEFT) and heart sizes (RIGHT), see table 2.3. The black line corresponds to interspecific allometry while the colored line corresponds to developmental trajectories for two different species (with different adult masses). We assumed that they have the same allometric exponent; that is, qualitatively, we assumed that they have the same kind of developmental trajectory.

organs depends crucially on the adult mass of the organism considered. This approach provides a limited but general way to characterize organs' functions, through the analysis by size dependence [Schmidt-Nielsen, 1984].

The intraspecific approach provides a size-dependency of organ weights, which differs widely from interspecific allometry. The growth of an organism is indeed neither isometric (ie: preserving proportions) nor following interspecific relationships (a small dog is, anatomically, very different from an adult rat of the same weight). On the contrary, developmental organization leads usually to an early growth of vital organs, providing specific proportions to immature organisms, see table 2.3 and figure 2.2. Notice that these differences allow in particular the paleontological identification of juvenile and adults, when the adult size is *a priori* unknown and varies largely.

2.2.5 Conclusion

From a practical point of view, the question of allometry can be crucially found in the problem of pharmaceutical dosages, when no specific study has been performed before (in a veterinary context but also when going from animal models to humans). The naive approach of providing a dosage proportional to the mass of the organism can lead to catastrophic results. The accepted dosage depends, among others, of the nature of the toxicity involved, and on the mode of administration, the usual allometric exponents relevant are 0.75 and 1 [Boxenbaum & DiLea, 1995, Kirman et al., 2003].

Now, from a more theoretical perspective, we saw that *interspecific* allometry, and the corresponding scale symmetry, is supported by solid empirical data. The 3/4 exponent for BMR is not only supported by a study on plants over 20 orders of magnitude, but also by converging results from the study of biological rhythms and rates. However, this exponent can only be understood as a broad tendency and, even in the interspecific case, variability is relatively high.

For *intraspecific* allometry, however, the situation is far more scattered, phylogenetic individuation prevails; and specific exponents (or even exponent shifts) arise.

The differences between interspecific and intraspecific allometries are particularly compelling. They show clearly that a developing animal is not equivalent to an adult, even under these broad organizational tendencies that allometry allows to show. From this perspective, the organizational difference between adults and juveniles seems in particular characterized by oversized organs.

Notice also that the general consequence of the complex allometric relationships encountered in biology is that most biological quantities depend in a nontrivial way on the size of the organism considered. This is also relevant, even when no particular organizational changes are involved (for example, the mass distribution among organs depend on the adult size for mammals).

2.3 Morphological Fractal-Like Structures

In this section, we will focus on the geometry of anatomical structures, inasmuch scale symmetries are involved. Structures exhibiting this kind of symmetries, at least approximately, are widespread in biology and consequently we will limit ourselves to a few important examples. We will first introduce the elementary notion of fractal and fractal dimension; then we will mostly discuss the cases of cellular membranes and tree-like organs.

2.3.1 Principles

The term fractal was coined by Mandelbrot [Mandelbrot, 1983] for rugged geometric structures, usually exhibiting a scale symmetry. The fundamental idea of fractals come from an aspect of usual geometries (based on differential manifold typically) that has been found to be a limitation. These geometries are based on smooth structures, with possibly pointwise singularities (punctually undefined or infinite derivatives). As a result, when zooming, this structures invariably lead to a straight line (in arbitrary dimension, to a linear, flat, structure). However, many situations, let it be natural phenomena or mathematical constructs, do not seem to have such properties. On the contrary, when zooming this second type of structures display more and more details and does not converge towards a smooth linear object.

We should note that this kind of ideas originated first in pure mathematics, in particular in relation with the issues of conceptual instability that 19th century analysis encountered. It was in particular thought that any infinite sum of "usual" functions should be differentiable; however, Weierstrass constructed the counter-example of a "monstrous" nowhere differentiable function, constructed by summing cosine functions. Later, using the powerful and brand new set theoretic framework, Cantor constructed numerous strange functions and sets, such as a function of one variable filling the plane or the so called Cantor dust which, in one dimension, has the power of the continuum (a cardinality property) and is closed but is nowhere dense (see, for example, [Edgar, 1993] for a collection of historical papers on fractals).

We will now define some of the major kinds of scale symmetric sets that can be constructed, see, for example, [Le Méhauté et al., 1998, Falconer & Wiley, 2003] for complete presentations.

2.3.1.1 Scale Symmetry for Sets

To approach scale symmetric sets, we first have to define an *iterated function system* (IFS). Let D be an open subset of $\mathbb{R}^n$. A *contraction* is an application $S : D \to D$ such that there exists $c \in]0,1[$, with $\forall (x,y) \in D^2$, $\|S(x) - S(y)\| \leq c\|x-y\|$ (the case of equality defines the notion of *similarity*[10]). A finite family of contractions $(S_i)_{i \in [\![1,n]\!]}$

---

[10] In this case, the distance between points is conserved, modulo a constant factor, see also below.

2.3 Morphological Fractal-Like Structures

Fig. 2.3 *Example of a strictly self-similar fractal, obtained by a simple iterative procedure: the Sierpiński triangle.* Here, the fractal object is encountered at the limit, when the number of iterations tends to infinity. Notice that the figure become more pale when the number of iterations increases. This can be easily explained by the fact that the fractal structure itself has a null measure (for the natural measure of the embedding space, here a surface), so that the convergence to a fractal structure leads to the vanishing of color with iteration.

acts on a non-empty compact[11] set $F \subset D$ by

$$\mathscr{S}(F) = \bigcup_{i \in [\![1,n]\!]} S_i(F) \qquad (2.4)$$

With these definitions, there exists a unique F, which is called the *attractor* of the IFS and is a fixed point of $\mathscr{S}$, so that:

$$F = \mathscr{S}(F) \qquad (2.5)$$

Indeed, F can (and should) be seen as an attractive fixed point by iteration of $\mathscr{S}$ since F can be equivalently defined as:

$$F = \bigcap_{k=0}^{\infty} \mathscr{S}^k(E) \qquad (2.6)$$

Now, with these preliminary definitions, we can define sets exhibiting scale symmetry, by increasing order of generality (meaning that the symmetry we will consider will be less and less strict). These scale symmetric sets correspond to intuitive and for a part of them, constructive notions of fractal.

SELF-SIMILAR SETS. Strictly speaking, a self-similar set is the attractor of an IFS with contractions which are similarities. Similarities conserve all distances (modulo a global factor) and, as a result, are proportional to euclidean transformations, the transformation preserving usual geometric shape. The result of a similarity

---
[11] Here, the compact sets are bounded set stable for taking the limit of converging sequences.

is (here) a reduced version of the object (which can be translated, rotated and reflected). However, the name of self-similar fractal is sometimes loosely used for fractal defined by an IFS, whatever the nature of the contractions involved is.

SELF-AFFINE SETS. This case corresponds to attractors of an IFS composed of contractions which are affine transformations (the composition of a linear transformation and a translation). This means in particular that these transformations can deform proportions of objects (linearly).

QUASI-SELF-SIMILAR SETS. This notion describes situations where the above symmetries are not verified, but are approximately so. A set F is k-quasi-self-similar if there exists $r_0 > 0$ such that for any subset $U \subset F$, with $\|U\| = r \leq r_0$, there is an application $\phi : U \to F$, called a k-quasi-similarity such that:

$$\frac{1}{k}\|x-y\| \leq \frac{r}{r_0}\|\phi(x)-\phi(y)\| \leq k\|x-y\| \qquad (\forall x, y \in U) \qquad (2.7)$$

This definition means that a part of the set has approximately the same shape that the whole. We should notice that this kind of approximation concerns all subsets of F at the same time, with a definite level of approximation, and is thus different from usual analytic approximations, which tries to obtain an arbitrarily precise control on the object by considering increasingly small neighborhoods "successively".

STATISTICAL SELF-SIMILAR SETS. Such a set only has a statistical regularity between scales. This regularity is described by the existence of a fractal dimension (see next section).

MULTI-FRACTAL SETS. Such a set is similar to a statistical self-similar set; however, the fractal dimension is not well defined. Instead, different points have different local scaling behavior; and we have a spectrum of fractal dimensions. Notice also that the multi-fractal approach is more precise that the fractal one. For example, when one consider a fractal structure near a ball of the embedding space one only obtain the dimension of the embedding space, from a strictly fractal point of view.

We mentioned fractal dimensions in the above definitions; so we will now describe this notion with more details.

2.3.1.2 Fractal Dimension

The fractal dimension(s) is a way to evaluate how convoluted an object is, or more technically, how densely it occupies space. The classical dimensions are integers: 1 for a line, 2 for a surface 3 for a volume, 4 for a space-time block, ... However, fractal objects, like the Sierpiński triangle in figure 2.3, do not fit in this category.

2.3 Morphological Fractal-Like Structures

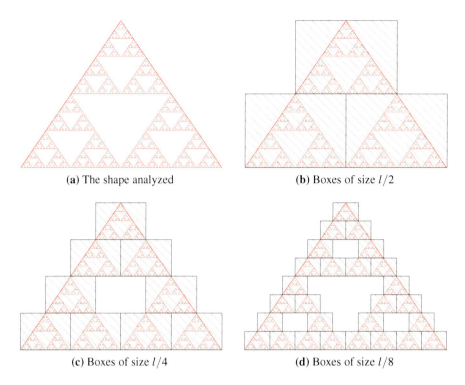

(a) The shape analyzed (b) Boxes of size $l/2$
(c) Boxes of size $l/4$ (d) Boxes of size $l/8$

Fig. 2.4 *Principle of the evaluation of the box-counting dimension illustrated on the Sierpiński triangle*. The finer the resolution becomes, the more holes due to fractality appear. As a result, the shape is not "full" enough to have the dimension of a surface.

If we measure the length of Sierpiński triangle's internal frontier, we get (modulo a factor $3l$ where l is the length of the side of the first triangle):

$$L = \frac{1}{2} \times 1 + \left(\frac{1}{2}\right)^2 \times 3 + \left(\frac{1}{2}\right)^3 \times 3^2 + \cdots \propto \left(\frac{3}{2}\right)^n \quad (2.8)$$

This length tends to infinity, in spite of the finite extent of the figure. However at each iteration we remove $1/4$ of the surface, so after n iteration the surface is $S_0 \left(\frac{3}{4}\right)^n$, thus the final surface is 0, as physically illustrated by the vanishing of the color's intensity in figure 2.3. We understand then, that both points of view, the 1-dimensional and 2-dimensional measures fail to provide an account of the situation; and that it is related to exponential drifts (either towards 0 or infinity).

Let us consider a cover of a line of length l with cubes of length r we need approximately $l/r \propto 1/r$ cubes. Now, if we want to cover a surface, each cube having a surface r^2, then we need approximately $l^2/r^2 \propto 1/r^2$ cubes. In the case of Sierpiński triangle, see also figure 2.4, we see that we need one cube for each triangle of the same size. After one iteration, we get 3 times more triangles with half size, so for cubes of size $r = l/2^n$ (which corresponds to $n = -\log(r)/\log(2)$) we need 3^n cubes; thus we obtain that the number of cubes is of the form $3^n = 3^{-\log(r)/\log(2)} = 1/r^{\log(3)/\log(2)}$. We can then intuitively identify the dimension of the fractal shape with $\log(3)/\log(2)$.

More generally the box-counting dimension, or Minkowski-Bouligand dimension of a subset F of a *metric* space is:

$$D(F) = \lim_{r \to 0} \frac{\log(N(F,r))}{\log(1/\varepsilon)} \tag{2.9}$$

where $N(F,r)$ is the number of balls of radius r needed to cover the set F.

This formula performs just the transformation needed to make the number appear, which we "manually" identified above. However, since it is defined as a limit, we see that this dimension has no particular reason to be well defined in all cases. When it is not well defined, that is when the sequence does not converge to a single value, we still have bounds for the asymptotic values it can take. As a result, the situation can be characterized by an infimum and a supremum. This defines a lower dimension (the infimum) and an upper dimension (the supremum). In such cases, the scaling properties of the objects are oscillating (not necessarily in a regular way) when we are zooming.

Fig. 2.5 *Illustration of topological dimension.* TOP, a line can be covered with open balls with points that are covered 1 or 2 times. BOTTOM, LEFT, there are missing points if we try to do the same thing in order to cover a surface. BOTTOM, RIGHT, to cover the surface some points are present in 3 balls.

There is a variety of different definitions of fractal dimensions [Falconer & Wiley, 2003, Le Méhauté et al., 1998], which generally coincide in the case of self-similar fractals, but do not in more complex cases. The box-counting dimension is not the best from a mathematical point of view. For example, the rationals have dimension

1 even though they form a countable set which therefore does not "really" occupies space. A better mathematical notion of dimension than the box-counting dimension is the Hausdorff dimension. It is based on a parameterized family of measures (in the sense of measure theory); then the Hausdorff dimension is the critical value of the parameter which yields neither infinite nor 0 weight to the the set under study. Notice that in this case, the fractal dimension is a notion associated to measure theory. In the case of rationals, the Hausdorff dimension is thus 0, which corresponds better to the intuition. However, the box-counting dimension is particularly useful since it allows to study empirical cases (with many possible tweaking). For a recent account of the methodology used to evaluate the properties of experimentally observed structures, including the various forms of fractal dimensions and the image analysis used, see [Lopes & Betrouni, 2009].

In estimating fractal dimension of natural phenomena, we generally have access to a limited range of scale. This range should be large enough to allow genuinely to claim the validity of a fractal description; however, the term fractal is also used by experimenters, even in physics, when this condition is not met, see [Avnir et al., 1998] for a critical discussion[12].

2.3.1.3 Definition of Fractals

So we have sets with scale symmetries and a notion of fractal dimension. What is the definition of a fractal then? The situation is not obvious. We could say that it is a set corresponding to one of the criteria for scale symmetry above. However, flat structures like a segment or a piece of a plane meet these criteria (the criteria are just overly complicated to be actually useful to describe them). Worse, they meet all of them (we see exactly the same thing when zooming in on a flat structure). Reciprocally, we could approach the notion of fractal by the notion of fractal dimension, and associate fractality with non-integer dimensions. However, there exists structures such as Peano curve that have a convoluted structure; but are nevertheless space-filling, so have an integer dimension (roughly, this criterion is already better, though).

We need a more precise criterion to grasp the concept of fractal. To formulate it, we need to introduce another concept: the concept of topological dimension. The definition of this notion is the following:

Definition 2.1 (Topological dimension (Lebesgue)). Let us consider a topological space X. The topological dimension is the integer n, such that for any finite *open* cover of X, a finite refinement of this cover can be found where each point of X is in at most $n+1$ open sets of the refined cover.

Behind the technical terminology, the idea behind this definition is relatively straightforward. We are considering open sets, which are intuitively sets without their boundaries, for example an interval $]a,b[$ without a and b. As a result, overlaps are needed to avoid gaps, see figure 2.5. However, if points are missing in the

---

[12] The case of critical transition, however, display sharp fractality, associated to a sound theoretical background.

structure of X (gaps), then there is no need for overlapping. As a result, the overlaps are engendered by the local, *gapless*, degrees of freedom on X. Of course, this dimension, in general, can be infinite in some cases (when there is no finite n which meets the criteria, but this can only happen in an infinite dimensional space, when a linear algebra dimension makes sense).

We understand then that there is a conceptual discrepancy between fractal dimension (Hausdorff dimension, a metrical notion) and topological dimension (a topological notion). Mandelbrot then defines fractals as structures where these notions do not coincide, meaning that they do not lead to the same dimensions:

Definition 2.2 (Fractals). A fractal is a set F with $\dim_H F \neq \dim_{topo} F$

Notice that the fractal dimension is not always defined. When it is defined, the fractal is also called monofractal to emphasize the uniqueness of the dimension (for a given definition) and to contrast the situation with more complex structures. When the dimension is not defined, as we said earlier, we nevertheless have an infimum and a supremum for it. This corresponds, for example, to an oscillation when looking at smaller scales. Alternatively, multifractals correspond to a situation where the fractal dimension is not the same when zooming to look at different points.

Since natural objects studied, when fractality is found to be relevant, are limited to a range of scales, they are usually called prefractal to emphasize this limitation. However, and anticipating on the following of the rest of this thesis, biological cases usually do not have the same level of invariance than physical cases, especially those where fractality is theoretically justified (which is far from being the case for all reported cases of fractality [Avnir et al., 1998]). As a result, here, we will prefer the terminology of *fractal-like* situation. By the later, we mean to qualify situations where fractality is relevant as a tendency, and certainly more relevant than smooth geometries, but where the theoretical validity of an actual scale symmetry is nevertheless dubious because of the variability inside and between subjects and possibly a range of scales that is too limited.

2.3.2 Cellular and Intracellular Membranes

The need to use fractal geometries in morphometry, instead of notions using euclidean geometry, historically derived from very practical considerations. A nice historical example of the difficulties that led to this paradigm shift is provided in [Weibel, 1994]. In the 1970's, two different teams tried to evaluate the surface density of the endoplasmic reticulum in the liver. However, these teams surprisingly reported largely different results: $5.7 \, m^2/cm^3$ and $10.9 \, m^2/cm^3$. This puzzling situation was clarified by the notion of fractal: if considered fractal, the evaluated surfaces depend on the scale of observation (and tends to infinity with increasing magnification, in principle). This approach has been tested [Paumgartner et al., 1981] and leads to an estimate of the fractal dimension of 2.7 for endoplasmic reticulum's surface and of 2.54 for mitochondrial inner membranes. This results where found for data over 2 order of magnitude (which is a limited range, but nevertheless substantial[13]).

Comparisons of fractal-like properties of cell membranes have been investigated to discriminate between different relevant biological situations. Such investigations have especially been done with respect to cancer cells, and in order to contribute to cell type or cells activity classifications. In [Smith et al., 1996], for example, the fractal dimension reported for the boundaries of Purkinje neurons is 1.89 (in 2d) whilst the dimension for Spinal cord neurons is 1.62, see also [Losa, 2006] for other examples. In [De Vico et al., 2009], fractal analysis of cellular (internal and external) membranes was performed to analyze feline oocytes. Statistical self-similarity has been found over 2 orders of magnitude. An interesting point in this study is that the methodology followed by the authors involved different image analysis (basically different thresholds) in order to focus on different aspects of cellular morphology, and lead to different fractal dimensions. The authors also remark that, in general, active states (including cancerous cells) display a higher fractal dimension than inactivated ones, with some exceptions. One of the exceptions is found in [Losa et al., 1992], where the fractal dimensions observed for the pericellular membranes of leukocytes if of $\propto 1.2$ in the normal case and $\propto 1.1$ in the case of lymphoblastic leukemia.

2.3.3 *Branching Trees*

A branching tree is constituted by a starting branch and an iteration pattern: a branch splits in n sub-branches[14]. The number of sub-branches can be fixed, in which case the *topology*[15] of the tree is uniquely determined. However, n can also be random, uniformly or depending on the number of iterations.

The biologically relevant situations are usually cases of branches having tubular structures, embedded in the usual 3-dimensional space. For some observational methodologies or some cases, the tree is looked at in 2 dimensions, for example when using classical microscopy or in special cases like the morphological observation of the vascular tree of retina. Biological structures have also *metrical* properties. The quantities typically considered are the lengths of branches, L, and their diameters, D. The angles between branches are also commonly studied.

The natural question which arises then is the evolution of these lengths with the branch order, z (the number of branching from the first branch to the considered branch). These quantities and the aim of morphological measurement will depend

---

[13] Whether it is sufficient to talk about fractality or not depends on authors.

[14] Leonardo da Vinci inaugurated this analysis. In particular he observed that the sum of the sections of the sub-branches, in a tree, is equal to the section before branching, since "the nurture in the branch is divided in the sub-branches" — "ogni biforcazione di rami insieme giunta ricompone la grossezza del ramo che con essa si congiunge ... e questo nasce perché l'umore del piu' grosso si divide secondo i rami". Trattato di Pittura, unaccomplished, collected by an anonymous author, 1550.

[15] Here, by topology we mean properties that are invariant under smooth transformation.

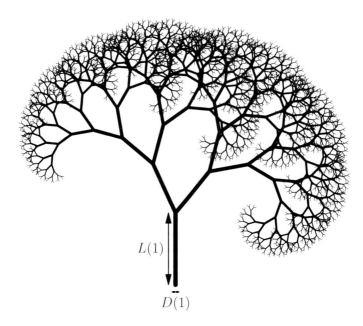

Fig. 2.6 *A binary tree.* Here, the branching order is a constant, 2, and the tree is generated deterministically by iterating the pattern encountered in the first branching. Note that the first branching alone can be insufficient to fully determine the structure, see [Le Méhauté et al., 1998] for more details on this issue.

in particular on the model used to understand the shapes encountered. In particular, [Gabrys et al., 2005] compares symmetrical and asymmetrical models of vascular morphology. This aspect is especially relevant because the scale range of analysis (and of potential fractality) is limited and because other structures, associated to scale dependent morphogenic principles, can exhibit similar features for a limited range of scales, leading to wrong estimations of fractality.

Generally speaking, there are different methodologies which enable to observe such structures. The main constraint on the methodologies used is that they need to be able to observe a geometrical shape in a similar way for a wide range of scales. The main methodologies are: working on silicone casts of the studied structure (for lower resolution studies), performing classical microscopy or photography observations (leading to 2-dimensional projection, but which can sometimes be performed *in vivo*), confocal microscopy (in 3D), tomography, ... The shape can be observed as a 2-dimensional projection or directly in 3-dimensions, as mentioned before. It is noteworthy that in general, the branching structure is embedded in the (smooth) geometrical structure of organs or tissues; for example the vascularization of the retina takes places approximately in a curved 2-dimensional structure (basically a piece of sphere) [Masters, 2004], and the curvature introduces a particular scale (which disappears in local analysis).

2.3 Morphological Fractal-Like Structures

2.3.3.1 Lungs

Lungs form highly convoluted structures, allowing fast bi-directional gas exchanges. Their structure vary among phylogenetic classes, but in the case of mammals, their topological shape is that of a branching tree[16]. This binary branching allows (for humans), to go from a single trachea to an observed number of alveoli of 274×10^6 to 790×10^6 [Ochs et al., 2004][17], meaning that the number of branching, from the trachea to an alveolus has a mean of approximately 29.

The early approaches to the understanding of the morphological properties of the lung assumed that a length S (diameter, length of the branch, ...) decreases exponentially with the branch order, i.e. $S(z+1) = qS(z)$ (as reported in [Bassingthwaighte et al., 1994]). This assumption fits the empirical results for approximately 10 branching levels; but the observed quantities depart clearly from this equation for smaller scales. In [West et al., 1986], it is argued that there is variability in the distribution of branch sizes of the same order z (instead of the basic factor q), and that this variability should be scale-free (at least for high values of z). Moreover, the authors argue that the branches should have sizes of the form q^n (see also [Bassingthwaighte et al., 1994]).

This approach leads to describe the mean evolution of a metric quantity S with the branch order z with the from: $S(z) \simeq z^{-\alpha} \left(A_0 + A_1 \cos \left(2\pi \frac{\ln(z)}{\ln(q)} + \phi \right) \right)$. This description coincides well with empirical results [Nelson et al., 1990]. Interestingly, for the four species studied (rat, dog, hamster and human), the exponents α_D (for the diameters) and α_L, are similar among species (as for the lengths of branches), even if the lung sizes are very different. However, the phase, ϕ, distinguishes human from the other species (among these four species). These findings correspond to a number of branching of ~ 22, spanning ~ 2 orders of magnitude for the lengths.

Interestingly, [Canals et al., 2000] argue that the fractal dimension of lungs is higher in the case of juveniles (1.626 ± 0.157 for males) than for adults (1.547 ± 0.012 for males) in *Rattus norvegicus*. This result, however, is based on a limited difference and would need confirmation.

Another study [Boser et al., 2005] shows a significantly reduced fractal dimension of lungs in the case of asthma; it is also noteworthy that this reduction is higher in the case of a lethal asthma condition.

It is interesting to mention that the alveolar perimeters also have fractal-like properties [Witten et al., 1997]. The fractal dimensions found are small; however, they are noteworthy in particular because they are correlated with the age of the subjects (and significant changes are found in diseases). Numerically, for the less than 16 year old subjects, the mean fractal dimension found is 1.047 ± 0.010, while for older subjects it is 1.093 ± 0.013. Note also that this supplementary structure is conceptually interesting because it shows the "concatenation" of the lungs tree structure and the alveolar rugged structure.

---

[16] Amphibian, for example, have simpler, more or less convoluted bag-like lungs.

[17] This study has been performed on 6 individuals, notice that the variability is high. However, the size of alveoli has been found to be stable in this study.

2.3.3.2 Vascular System

The vascular system is composed of structures with sizes ranging from the scale of cm to a few µm. It is composed of the arterial tree, going from the heart to organs and of the venous tree, from organs to the heart. It's function is to transport matter to organs (especially oxygen and nutrient) and from organs (especially carbon dioxide and wastes but also hormones, some cells, etc.). It forms a (mostly) closed system, where exchanges arise at the level of capillaries, between arteries and veins.

A considerable amount of research as been performed to evaluate the geometrical properties of the vascular system. Retina are particularly studied as they are easily, non-invasively accessible, mostly bi-dimensional, and medically relevant (for diabetes, typically). In [Masters, 2004], the methodological difficulties are discussed in the case of the retina, and some stability of the fractal dimension is found in this case around a value of 1.7. In table 2.4, we report other results compiled in [Lorthois & Cassot, 2010]. The crucial point is that low resolution measurements support a fractal-like structure with variable fractal dimension, but usually not very far from 2.7 (1.7 if projected). However, the studies looking at capillaries found a space filling structure, with dimension estimates comparable to the embedding space.

Table 2.4 *Evaluations of fractal dimensions for vascular networks.* The values reported here are from [Lorthois & Cassot, 2010]. Note that the sharp differences between the situations where capillaries are not considered and when they are considered (low and high resolutions). We do not report, here, the other methodological differences involved in these measurements (which come from different studies, in different laboratories). CAM stands for chorioallantoic membrane (this membrane plays a crucial exchange role during development for oxygen, calcium, ...); and ID stands for incubation days. The scale range of these studies is mostly around or below 2 order of magnitudes.

Resolution	System studied	evaluated fractal dimension
Low resolution	Subcutaneous AV	1.70 ± 0.03
	Developing CAM (ID 15)	$1.42 \pm 0.05 - 1.49 \pm 0.04$
	Developing CAM (ID 13–ID 18)	1.1–1.8
	Developing CAM (ID 13)	1.26 ± 0.03
	Developing CAM (ID 3–ID 6)	1.3–1.68
	Developing CAM (ID 6–ID 12)	$1.37 \pm 0.01 - 1.54 \pm 0.03$
	Placenta's arterial[a]	1.86
	Pial vasculature	1.31 ± 0.03
	Retinal vasculature	1.70 ± 0.02
High resolution	Subcutaneous capillary	1.99 ± 0.01
	Epifoveal vessels	2.00
	Developing CAM (ID 14)	1.86 ± 0.01
	Hepatic sinusoidal network	2.01 ± 0.01

[a] The dimension in 2d embedding space is estimated from a 3d result for comparative purposes.

2.3 Morphological Fractal-Like Structures

A study [Risser et al., 2006], which is important because of the technical breakthrough mobilized (3 orders of magnitude observed in 3 dimensions), leads to unexpected results. The vascular system has been found to be fractal-like at small (capillary) scales and homogeneous at bigger scale. These findings were highly unexpected because of the older results in table 2.4. However, the situation was clarified by [Lorthois & Cassot, 2010] by using considerations on the vascular development. The latter is indeed dominated by two different processes: the formation of a mesh-like structure (capillaries), homogeneous and space filling above a certain scale, followed by the growth of a fractal-like structure in this capillary mesh[18]. They showed that simulated models presenting these features can lead to the results of [Risser et al., 2006] and that an analysis of cortical vasculature adapted to this structure seems to confirm it.

In order to discuss the situation further, it is noteworthy to mention that the vascular structure has been hypothesized, for optimality reasons, to follow the Murray law:

$$d_0^x = d_1^x + d_2^x \tag{2.10}$$

where d_0 is the diameter of the mother branch and d_1 and d_2 are the diameters of the daughter branches. The initial formulation was for $x = 3$ and corresponded to an optimization principle taking into account the energy needed to transport the blood and the energy needed to confine it. However, the experimental results do not really back this relation up, especially since the data are highly scattered. As a result, it was proposed to extend this relation for values of x between 1 and 4. Even in these cases the variability remains high so that multi-fractal analysis should also be performed. This position is defended in [Zamir, 2001], in reference to results on the vascular structure of the heart, where multi-fractal analysis seems useful, as its criteria seems approximately met. This leads to an architectural variability of scaling in the vascular system, even when considering a part of it that is irrigating a single organ.

2.3.3.3 Further Aspects for These Organs

Following the properties above, we can raise the question of the possible geometrical relationship between different tree-shaped organs. In [Maina & van Gils, 2001], silicon casts have been obtained for the airway, venous and arterial systems of the lung, for a specimen of domestic pig, *Sus scropha*. This study focuses on the relation between these structures and shows that they have statistical correlations in their geometrical properties (diameter depending on the branching order for example). These structures obviously meet at the alveolar level (though this is not directly observed in silicon casts), but their geometrical relationship is not limited to this point (related to their function). The three structures, except the venous system in some regions, follow each other closely, both with respect to their branching patterns and

---

[18] More precisely, this growth is understood as equivalent to diffusion-limited aggregation, driven by the blood pressure field in the homogeneous capillary mesh.

spatial positions. Since a physical fractal growth usually requires some kind of randomness, according to its "singular" nature (that is, by its historical and irreversible paths and its specific result), these spatial and structural correlations at all scales indicate a correlated growth of these hierarchical structures, observable by abnormal similarities between them.

We therefore have a kind of "geometrical entanglement" that seems needed to understand the pulmonary structure. More results would be needed to formulate and characterize precisely the scope of this developmental "entanglement", let us note, however, that it seems to fall in the conceptual framework of [Soto et al., 2008], where the ontogenetic historicity is argued to be a crucial and irreducible part of biological structure of determination.

The question of the heritability of the properties of vascular structure has been studied in [Glenny et al., 2007] for armadillos[19]. The results, obtained by microsphere deposition, showed that correlations exists between animals originating from the same litter and the fractal dimension of blood flow distribution. However, the degree of variability in a litter was still very high, especially in the muscles. Interestingly, this study show that correlations are particularly high between heart and lung vasculatures.

One can also raise the question of the effects of activity levels on the vascular fractal dimension. A study, [Sinclair et al., 2000], using also microsphere deposition methods, establishes a change of the spatial structure of the blood flow in lungs of horses. More precisely, for the fractal dimension of this structure, they found high variability among individuals at rest and a change of the dimension between different levels of exercise. This change consisted in a reduction of the variability among individuals and a progressive reduction of the fractal dimension from trot to canter to gallop. This study was performed at low resolutions on four horses.

2.3.4 Some Other Morphological Fractal Analyses

The frontier of tumor growth has been reported as a tool for estimating the nature of the corresponding tumor, see [Baish & Jain, 2000]. It has also been shown that, at least in a specific case, a high fractal dimension for carcinoma was correlated to bad prognosis [Delides et al., 2005].

Another interesting kind of morphological fractal-like structure (even though a dynamical description is better) is at the level of proteins. The fractality by itself does not come as a surprise in this case, since a crude statistical mechanics description of them is a self-avoiding walk. However, proteins have variations with respect to their fractal dimension; and exhibit also specific, local, structures [Lewis & Rees, 1985]. As a result, fractality is biologically relevant in this context too.

Chromatin has also been reported to have a fractal-like structure [Bancaud et al., 2009], allowing to discriminate euchromatin, which has an estimated fractal

---

[19] The fact that armadillos have clonal offspring in a litter make it particularly useful for studies on clonal variability in complex metazoans.

2.3 Morphological Fractal-Like Structures

dimension of 2.2 and heterochromatin, with 2.6. We will discuss this situation further in the paragraph 2.5, on anomalous diffusion.

2.3.5 Conclusion

We have seen various aspects and cases of biological structures showing fractal-like properties in space. This review is, of course, by no means exhaustive; however, we can point out certain interesting and general aspects. First, the number of orders of magnitude involved rarely reaches 3, which means that the scope of the observed fractality is limited. As a result, one should be cautious when we are discussing it as a scale symmetry. Moreover, the determination of the fractal dimension is not fully stable: even with a constant methodology (and as [Masters, 2004] points out, even with the same experimenter) the evaluation of fractal dimensions leads to a high level of variability.

We can also now draw significant, positive conclusions. First, the space structure we have seen are clearly highly fractured. Moreover, they also display measure anomalies: lengths and surfaces should be considered scale dependent and do not seem to stabilize toward a finite value when observed with increasing precision. Thus, these hierarchical (non-homogeneous) and rugged structures are, in a loose, but nevertheless etymologically sound, sense fractals. In this sense, fractals are present in an organism from the scales of proteins to the scales of organs.

It is sometimes argued that the geometry of nature in general and of biological phenomena in particular is predominantly fractal. Putting aside the difficulties we discussed, there is an issue, in this point of view, which is that the fractal-like structures in an organism are widely heterogeneous in their (fractal) dimensions, geometrical structures, ... This point is sufficient to lead to a non-obvious situation (from a geometrical but also physical point of view). Indeed, most theoretical approaches depend on a unique parameter, roughly corresponding to a specific scaling symmetry (for example: Tsallis entropy, fractional derivatives and analysis, analysis on fractal structures, ...). Notice also that these physical and mathematical accounts are recent and remains somewhat unstable with respect to their physical applications.

In [Werner, 2010], examples are also provided from the nervous system. These examples include the neuronal dendritic trees, the white matter distribution, ...

Beyond these cases, we will see more fractal structures involving space in chapter 6, on critical transitions. Our global perspective will then allow to frame these dynamics and their variability, in terms of continual "symmetry changes", as an essential component of life structural stability. As for the iterations, which are proper to fractal formation, we may paraphrase Goethe's remark, which enables to understand the extensive presence of fractality: "life is the *never identical* iteration of a morphogenetic process" — we added the "never identical", which will turn out to correspond to symmetry changes. In all the examples above, the variability and irregularity of fractal structures are part of diversity and adaptability, which contribute to biological robustness in an essential way.

2.4 Elementary Yet Complex Biological Dynamics

Fractal or fractal-like structures are not only encountered in space, but also in the temporal behavior of biological systems. The specificity of fractal-like dynamics is that they lead to nonstandard statistical situations: instead of the usual reduction of variance with an increase in statistical sample, quantities like the statistical mean or the statistical variance diverge (they tend to infinity with the increase of samples). This situation is relatively counter-intuitive with respect to the classical notion of homeostasis, which is usually understood as a stabilization around a fixed mean. However, as we will see, fractal-like dynamics, are nevertheless ubiquitous in biology, as we said, in particular when we consider processes associated with physiological regulation. This paradigm shift, as to what biological regulation is, is discussed at length in [West, 2006].

In order to describe specific examples, we first provide some elementary methodological and mathematical background. Then, we will report examples of various natures. Finally, we will focus on the specific example of cardiac rhythms, and its peculiar properties, since this case constitutes half the available literature on biological fractal dynamics according to [Eke et al., 2002], and thus it allows a fine grained account of the variability and the statistical structure of biological dynamics.

2.4.1 Principles

The first point we should emphasize is that there are two elementary kinds of temporal structures observed, in comparison to physical time. The first one, commonly called time series, correspond to a quantity observed at different time points, usually regularly spaced. The second one is generated by a sequence of equivalent events, which thus leads to the observation of the time of their occurrence. Of course, the second category can be mathematically transformed into the first one by considering $s = \sum_i \delta_{t_i}$ (where δ_{t_i} is a pulse of magnitude 1 at time t_i). The time series, obtained by this procedure, is then $\int_{t_0}^{t} s(t) dt$. However, from a theoretical point of view there is still an associated peculiarity of the second situation, since such series can also be expressed as Δt_i with $t = t_0 + \sum_i \Delta t_i$. In other words, this form of time series is a purely temporal structure: there is no dimensional quantity involved in its description other than time.

In order to obtain an intuitive understanding of the situation, and before describing the analytic tools used to deal with experimental data, it is useful to look at the historical example of the Weierstrass function [Edgar, 1993].

$$f(x) = \sum_n a^n \cos(\pi b^n x) \qquad (2.11)$$

For $|a| < 1$, since $|\cos(\pi b^n x)| \leq 1$, the sum converges uniformly, thus f is continuous. However, if we look at the derivative of its terms we get $-\pi(ab)^n \sin(\pi b^n x)$, thus, if $ab > 1$, we see that the derivative corresponds to a divergent sum at all

2.4 Elementary Yet Complex Biological Dynamics

points (this is a heuristic argument not a rigorous proof, for the latter see for example [Edgar, 1993]). Even though this function has been introduced as a mathematical counter-example to the differentiability of continuous functions, it has a physical meaning: it is a sum of periodic functions with pulsations $\omega_n = 2b^{-n}$ and corresponding weights $\omega_n^{-\log(a)/\log(2b)}$. This corresponds to a power-law in the decomposition of f. As a result, such a process possesses a form of scale symmetry.

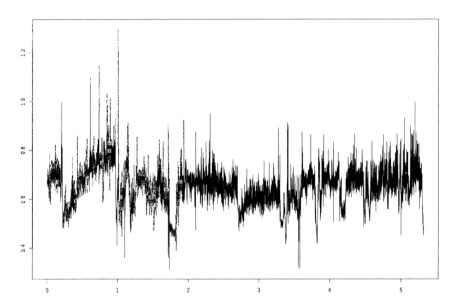

Fig. 2.7 *A sample of beat-to-beat intervals in human.* The beat-to-beat interval is given in minutes, and the time of each beat, in abscissa, is given in hours. Notice the significant variations in this rhythm, which are in particular characterized by its long-range temporal correlations. An important consequence of this situation is that evaluating the beat-to-beat interval at different times, even with a prolonged measurement, will not lead to a stable result. The data come from the Long-Term ST Database, [Goldberger et al., 2000], note that some unusually long beat-to-beat intervals can be erroneous (typically a heartbeat is not detected, which leads to an interval of approximately double length).

To understand another aspect of fractal dynamics, we can consider a classical example with infinite mean: the St. Petersburg game. In this game, the player tosses a coin as long as he obtains tails. When he finally gets heads he wins 2^n \$, where n is the number of tails obtained before. The mean gain is then $2^{-1}2^1 + \cdots + 2^{-n}2^n + \cdots = 1 + \cdots + 1 + \cdots = \infty$ [Bassingthwaighte et al., 1994]. Experimentally, it means that the more one plays such a game, the more the statistical means increases. A similar situation occurs when considering the classical Brownian motion, which has mean 0 but a variance which diverges (proportionally to n).

With more generality, [Eke et al., 2002] specify two forms of (fractal) processes. The *stationary* processes, on one side, for which the statistical description is stable along time evolution and the *non-stationary* processes, on the other side, which simply do not have this property. They propose the fractional Gaussian noise, as a paradigmatic stationary example, and the fractional Gaussian motion, as a non-stationary dynamics. The latter can be easily obtained by the summation over time of a fractional Gaussian noise. Reciprocally, the difference between two consecutive time points of a fractional Gaussian motion is a fractional Gaussian noise. This comes as no surprise because when one considers the simplest case of Brownian motion, the spatial displacement traveled at each time step has a constant probability distribution. In the notions of Gaussian noise and motion, the word Gaussian refers to the unstructured distribution of the values of the time series, when the temporal structure is no longer taken into account (that is to say when one considers properties that are invariant by data shuffling). We will not discuss these distributions closely since we are primarily interested in the temporal structures of some empirical time series. Their rather surprising fractal structure motivates our presentation.

We thus present and discuss some statistical quantities characterizing fractal-like time series. Remark that, in general, the validity of the approach of a process by such quantities depends of the stationarity or not of the process, see [Eke et al., 2002]. In the following, we consider a time series $f(t)$.

POWER SPECTRUM. This approach to temporal structures in biology is based on the classical method of Fourier analysis. It consists in describing the time series $f(t)$, originally given in the time domain, by its Fourier transform $\hat{f}(\omega)$, in the frequency domain. Then, the power spectrum S corresponds to the squared weight of each frequency.

$$S(\omega) = \left| \frac{1}{\sqrt{2\pi}} \int_{-\infty}^{\infty} f(t) e^{-\iota \omega t} dt \right|^2 = \frac{1}{2\pi} \hat{f}(\omega) \bar{\hat{f}}(\omega) \qquad (2.12)$$

The fractal situation is characterized by a behavior $S(\omega) \simeq b\omega^{-\beta}$ for large ω. It is then a behavior associated to scale-free contributions of the various frequencies. The auto-correlation function is the Fourier transform of the power spectrum. In the case of a power law for the power spectrum, the auto-correlation function decays like $C(0) - C(h) \simeq h^{\beta-1}$ (for small h). The fractal dimension of the graph of the signal is in this case $\frac{1}{2}(5-\beta)$ for $1 < \beta < 3$, see [Falconer & Wiley, 2003]. In practice, finite time series are used, and the power spectrum is found by FFT (fast Fourier transform). However, some care is needed in the use of this approach. In particular, a preliminary interpolation is generally required, and the high frequencies contributions should be neglected [Eke et al., 2002].

The more qualitative meaning of β is that it describes the distribution of "energy" among the various frequencies. When all frequencies have the same weight, we have $\beta = 0$ (white noise, see below). Notice that in finite cases, $\beta < 1$ is also possible and corresponds to an anti-correlation between different time points. $-1 < \beta < 1$ corresponds to a domination of high frequencies (the weight of high frequencies is infinite at the limit). This situation corresponds roughly to a case of stationarity because the large period behaviors are not relevant. Reciprocally, $1 < \beta < 3$ corresponds to low frequencies (large period) as the dominating contribution (and the infinite case converges), which leads also to a non-stationary situation. In general, a small value of β means a more noisy time series and a large β leads to a more regular time series.

DETRENDED FLUCTUATION ANALYSIS (DFA). This method has been introduced specifically for non-stationary biological series, initially to study DNA structure and heart rates variability [Peng et al., 1995]. In order to perform such an analysis, one first has to center the time series around its mean, and then to integrate it. The point of this procedure is to obtain an unbounded time series g with 0 average increase. Indeed, the local variations of rates (heart rate for example) are usually bounded by the viability domain; however, the accumulated distance from the mean[20] is not necessarily so. Then, the standard deviation of g with respect to local (linear) trends is computed at various resolutions, n. In other words, one evaluates the quadratic distance of g to g_n, the piecewise (best) linear approximation of g where the segments are of length n:

$$F(n) = \sqrt{\frac{1}{N} \sum_{k=1}^{N} [g(k) - g_n(k)]^2} \qquad (2.13)$$

Then, in scale-free cases $F(n) \simeq n^\alpha$. Moreover, we have the relation $\beta = 2\alpha - 1$ for infinite sequences. The interpretation of α can therefore be transferred from our discussion of the power spectrum and of its exponent β.

A crucial motivation for the introduction of this approach and its widespread use in biology is that it allows to study both stationary and non-stationary time series. Its drawback, however, is that it is usually less precise than more specialized approaches, dedicated to a specific structure of time series [Eke et al., 2002].

RELATIVE DISPERSIONAL ANALYSIS (RDA). Relative dispersion is the ratio of standard deviation by the mean, thus leading to a dimensionless quantity. In the context of fractal time series, RDA consists in evaluating relative dispersion as a function of the level of coarse graining applied to the time series. Note that it can only be applied to stationary processes.

The exponents α and β are usually noted in the literature with the same letters as in this text (with some exceptions).

There are three main kinds of random time series that are used for comparative purposes and they are named after colors. Let us recall that these designations correspond to the *temporal structure* of time series (and not to the unstructured distribution of their values).

WHITE NOISE. This case roughly corresponds to a situation where the time points are not related to each other. This can be defined by saying, for example, that the expected value of a process is 0, and the expected correlation between two different time points is also 0. White noise leads to $\beta = 0$, meaning that energy

---

[20] The accumulated distance from the mean is the distance of the number of heartbeat during a given time to the mean number of heartbeat during such a time.

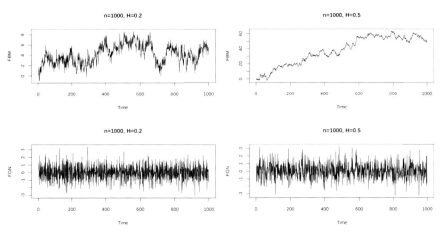

(a) Fractal Gaussian motion (UP) with anti-correlated increment (DOWN).
(b) Brown noise (UP) and white noise (DOWN).

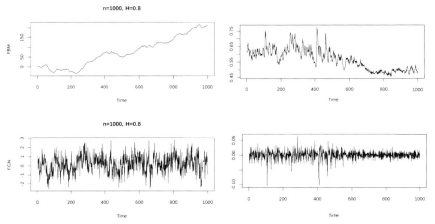

(c) Fractal Gaussian motion (UP) and noise (DOWN).
(d) Beat-to beat interval (UP) and its increment (DOWN).

Fig. 2.8 *Some fractal dynamics.* In each computer generated case, we represent the Gaussian motion up and the corresponding noise down. The last case corresponds to empirical data of the cardiac dynamic. Note that the latter increment does not seem completely stationary.

is equally distributed among all frequencies. This situation also leads to $\alpha = 0.5$. This case is illustrated in figure 2.8c, BOTTOM.

BROWN NOISE. This case is equivalent to Brownian motion, where $X_{t+\Delta t} = X_t + Y$, where Y has mean 0 (and is stationary). From another point of view, brown noise is obtained by integrating white noise. Brown noise leads to $\beta = 2$, so lower frequencies have more energy than higher frequency. This qualitative aspect is logical for a random motion, were the random contribution at each time point is added to the already existing displacement. We also have $\alpha = 1.5$. Figure 2.8c, TOP, corresponds to this case.

PINK NOISE. This case is intermediary between the preceding situations; quantitatively it corresponds to $\beta = 1$. As a result, pink noise corresponds to an equal distribution of energy per octave. We also have $\alpha = 1$ in this case. In a looser sense, situations with $0 < \beta < 2$ are also called pink noise.

The dynamics, which corresponds to scale-free behaviours with respect to one of these analyses, are usually called fractional dynamics or fractal dynamics.

2.4.2 A Non-exhaustive List of Fractal-Like Biological Dynamics

In this section, we will provide some examples of biological dynamics with long-range behaviors. This list is far from exhaustive; nevertheless, we took examples at various levels of organization, which also correspond to a wide variety of phenomena.

Blood Cell Number

In a study [Perazzo et al., 2000], the blood of two sheep has been sampled every day, for 1000 days, and the number of cells for different cell categories has been estimated. To the authors surprise, the observed time series are remarkably different from the main hypothesized structure, namely white noise fluctuations around a mean value. Indeed, they computed de-trended fluctuations and found a scale-free behavior of exponents α near 1, see table 2.5 (recall that 0.5 is expected for white noise and 1.5 for a random walk). A power spectrum analysis shows also that these long-range correlations take the form of anti-correlations ($\beta < 1$), except in the case of white blood cell which are positively correlated. Note that the intraindividual variability of the considered exponents is low — that is, they are relatively stable with respect to their evaluation at different times.

Table 2.5 *Blood cells fluctuations of two sheep.* We report here the exponents obtained by de-trended fluctuation analysis, α, and for the power spectrum β. Let us recall that a scale free situation leads to $\beta = 3 - 2\alpha$. For comparison, a (standard) random walk leads to $\alpha = 0.5$ and white noise leads to $\alpha = 1.5$, the lowest value of correlation coefficient is 0.995 for 200 days (2 orders of magnitudes).

Cell categories	Sheep 1, α	Sheep 2, α	Sheep 1, β	Sheep 2, β
Red blood cell	0.98	1.00	0.97	0.91
Platelets	1.24	1.11	0.65	0.70
White blood cells	0.83	0.83	1.14	1.16

Cellular Respiration

In [Aon et al., 2008], time series associated to cellular respiration are analyzed. The two situations under study are a yeast culture and the mitochondrial network of cardiomyocytes.

For yeast cultures (*Saccharomyces cerevisiae* under controlled, approximately constant conditions), the time series under consideration are the concentration levels of dissolved O_2 and CO_2. The measurements were performed with a 12s resolution, and span 118h. The cultures seem dominated by a cycle of $\simeq$ 13h, whose periodicity, though, is unstable. The analysis of these time series provided a constant relative dispersion among scales in both cases, and a power spectrum exponent $\beta = 1.95$ for O_2. These results hold in spite of the aforementioned seemingly predominant cycle, which still generates some subharmonic components, and have been also tested by a cut off above its scale. i. e. by looking at the behaviors at time scales smaller than this particular scale. This means that the long-range behaviours are not caused by this cycle. For CO_2, however, the situation was somewhat more complicated since only small (time) scale behavior ($<$ 1 min) is associated with scale-free behavior $\beta = 1.45$, while larger scale are dominated by white noise. Notice also that considering a given, specific part of the $\simeq$ 13h cycle leads nevertheless to multi-scale behavior at large scales (beyond the 1 min limit).

Now, let us consider the cardiomyocytes studied in the same article, [Aon et al., 2008]. Cardiomyocytes, the cells of the cardiac muscle, rely heavily on oxygen supply for their intense biological activity. Respiration occurs via mitochondria, which are in a particularly high number in each cardiomyocyte (in comparison with other cells). The observed quantities are the mitochondrial membrane potential $\Delta\Psi_m$ (observed by two photon microscopy associated to a potentiometric fluorescent dye) and the amount of reactive oxygen species (ROS). The time resolution is 110ms while the duration is 7min. The corresponding time series have scale-free behaviours, associated to a power spectrum exponent of $\beta = 1.79$. This behavior is also confirmed by RDA. Interestingly, an isolated mitochondrium only displays such a behavior for a range of scales much more limited than in a collective situation: the fractional dynamic occur only at the shorter time scales. Accordingly, the power spectrum shifts towards a white noise behavior at larger scales and the RDA shows a loss of correlations across scales.

Lung Respiration

A variety of different quantities associated with lungs respiration can be measured. The two quantities which have been mainly studied are the inter-breath interval and the tidal volume. Results of DFA for these two quantities are reported in [Thamrin et al., 2010]. In all cases, the mean exponents α are below 1 and, except in one case, above 0.5, so between white and pink noises. The exception is the tidal volume in non-REM sleep. Indeed, in the decomposition of sleep in different phases, the first partition criterion is whether rapid-eye movements (REM) are observed or not. These two situations are physiologically very different in terms of activity. This is confirmed here by an exponent of 0.8 for REM sleep and 0.5 for non-REM sleep, in [Thamrin et al., 2010]. For inter-breath intervals [Peng et al., 2002], results are of 0.68 ± 0.07 for young men and of 0.60 ± 0.08 for elderly men. The measurement was performed for 2 h (roughly 2000 respiratory cycles), in an inactive awake state. Here, we see that the tendency is a decrease of the exponents with aging. This tendency is also reported for the women group, but it is more limited. This result qualitatively means that the inter-breath intervals tend to be more random with increasing age (closer to white noise) and in other words they tend to have less correlations.

Body Temperature

The structure of temperature variations follows also such a behavior. In [Stern et al., 2009], using DFA the authors find exponents of $\alpha \approx 1.5$ for rectal temperature fluctuations, with an increase of this exponent in correlation with age, from 1.42 at 4 weeks to 1.58 at 20 weeks. This tendency is valid both in the mean population and for individual trajectories of the same subject at different age. This growth, however, goes with variability, meaning that the exponent decreases for some infants. Another noteworthy aspect is the absence of correlation of α with room temperature, the mean body temperature or immunization. The exponent has been also evaluated for adults in [Varela et al., 2003], but with a different methodology (skin temperature). The authors found $\alpha \approx 1.3$. A positive correlation between the exponent and age was also noted (from 18 to 83 years).

Some More Cases

In [Labra et al., 2007], the fluctuations of the O_2 consumption log increment ($r = \log(VO_2(t + \Delta t)/VO_2(t))$) has been studied. The shape of the distribution found, depending on initial conditions, is a double exponential also known as the Laplace function. This leads to a characteristic "tent" shape when the log of the probabilities is plotted. This study considers different species, which allows to evaluate the allometric exponent for the standard deviation. The latter is -0.241 ± 0.103, which is consistent with a temporal allometry following the exponent $1/4$ for times. This structure corresponds to a power law distribution for the fluctuations which has a structure compatible with allometric relationships.

Other cases showing this kind of fluctuations are the blood pressure [Wagner et al., 1996], the colonic pressure [Yan et al., 2008], the exploratory behavior of rats [Yadav et al., 2010], the gait dynamic and the gut dynamics [West, 2006], ... Interestingly, in the case of ocular cascades and fixations on a text, using literate and illiterate subjects, [Shinde et al., 2011] show that the scale-free dynamic is broken in association with attentional behavior.

2.4.3 The Case of Cardiac Rhythm

The cardiac rhythm is probably the most studied case of biological time series, as cardiac rhythm clearly has important medical implications. In particular, a crucial motivation in the analysis of beat-to-beat time series is the analysis of properties that have diagnostic applications. The study of such time series is also highly relevant for understanding normal physiological dynamics, such as aging, ... Usually, the beat-to-beat interval is more precisely defined as the interval between two R wave, in an electrocardiogram. These intervals are the easiest to use considering their short duration and strong magnitude, resembling a Dirac function.

2.4.3.1 Structure of Cardiac Rhythm

The study of the beat-to-beat interval *increment* [Peng et al., 1993], which is the discrete derivative of the beat-to-beat interval[21], has shown anti-correlations, with an exponent of $\beta' \simeq -1.01 \pm 0.16$ in healthy cases and $\beta' \simeq -0.54 \pm 0.25$ in diseased cases. This results are obtained for over 10^5 heartbeats, on time series obtained for a routine behaviour.

An interesting point is raised in [Peng et al., 1995], using DFA on time series recorded with an ambulatory monitor. The feature observed, called crossover phenomena, is a shift of the exponent between α_1 for the short time scale (< 10 beats) and α_2 for the longer time scales. This shift is different in the healthy situations and in some pathological cases. In the healthy cases, we have $\alpha_1 > \alpha_2$ ($\alpha_1 = 1.20 \pm 0.18$ and $\alpha_2 = 1.00 \pm 0.13$), which means that the short time scale evolution is more regular than the dynamic at larger time scales. This aspect is interpreted by the authors as a regularization, at short time scales, by the interactions with the respiratory rhythm. However, for patients with congestive heart failure, the exponents found are such that $\alpha_1 < \alpha_2$ ($\alpha_1 = 0.80 \pm 0.26$ and $\alpha_2 = 1.13 \pm 0.22$). This differences are argued to be able to distinguish the healthy and pathological cases — which means that the overlap of the two behaviors is limited. The authors remark, however, that the shift in exponents is not observed in all subjects.

A study [Pilgram & Kaplan, 1999], using data from [Peng et al., 1995] over 24 h, focused on the power spectrum and its distribution when analyzed locally (in time). Their findings show that there is an instability of the exponent β for time windows of the same sizes, but at different temporal positions. This can be understood as a non-stationarity of the power spectrum, which goes beyond a possible bimodal distribution (for wake and non-REM sleep). This result is confirmed by multifractal analysis [Ivanov et al., 2001]. A larger singularity spectrum is found in the healthy case, leading to a greater complexity of healthy heart rhythm, beyond the sole value of the exponent described above. Compatible results are also found in [Makowiec et al., 2006], where healthy aging is also studied and lead to a decreased complexity in the above sense.

In [Kiyono et al., 2004], the authors focused on the fine structure of the (detrended) *distribution* of the time increment corresponding to n heartbeats. This structure and especially its dependence with the scale n allows to discriminate between

---

[21] This series was studied instead of the beat-to-beat interval because it is roughly stationary whilst the beat-to-beat interval is not (and DFA was not available yet). Notice that, under regularity hypothesis, the corresponding exponent for the beat-to-beat interval is $\beta = \beta' + 2$.

2.4 Elementary Yet Complex Biological Dynamics

models that have been proposed to understand the variability of the heart rate. These models are turbulence-like models, called multiplicative cascade, and models based on criticality. The authors have found non-Gaussian scale invariance for these distributions, with a shape that is also incompatible with the cascade turbulence like models, but is, however, a general feature of critical behaviors. These results have been found for both constant routine protocol and normal daily life, for measurements spanning 24 h.

2.4.3.2 Some Factors Associated with the Cardiac Rhythm Structure

Age is correlated with a decrease of heart rate complexity or, from another point of view, with a more regular heart rate [Pikkujamsa et al., 1999], see figure 2.6. However, as before, the variability, observed by the variations of exponents, remains high among different individuals. These measurements have been done among subjects with no observed heart disease. Note that the criterion for a healthy condition given above, $\alpha_1 > \alpha_2$, is also met for these results, and this at all ages. It is also noteworthy that the mean difference between these exponents is low for the elderly group. These results are not isolated and are confirmed by several other studies, such as [Iyengar et al., 1996].

Table 2.6 *Heart rate variability at various ages.* The results are from [Pikkujamsa et al., 1999], on 24 h measurements on healthy humans. Notice the increase of regularity (loss of complexity) with age. It is also relevant to emphasize the high level of variability of the measured exponents, which is given here by the standard deviation.

Quantity	Children < 15 year	Young Adults 15 year to 39 year	Middle-Aged 40 year to 60 year	Elderly > 60 year
β	1.15 ± 0.18	1.12 ± 0.19	1.32 ± 0.14	1.38 ± 0.17
α_1	1.06 ± 0.11	1.19 ± 0.14	1.19 ± 0.16	1.15 ± 0.16
α_2	0.98 ± 0.06	1.00 ± 0.08	1.07 ± 0.07	1.14 ± 0.07

In [Wilson et al., 2009], the DFA approach lead to significant differences between a group of healthy subjects having long term sedentary habits ($\alpha_1 = 1.20 \pm 0.16$) and another group doing regular aerobic exercise ($\alpha_1 = 1.02 \pm 0.20$). The evaluation of habits, in this study, is obtained by self-report. The criterion for the sedentary group is the reporting of less than 1 half an hour seance of exercises intense enough to break a sweat per month; the members of the active group have had 3 to 4 such seances per week. Both regimes have to be sustained for at least a year in order to be valid. This study is not based on a large statistical sample but it is still interesting. Another relevant point is that the classical linear statistical analyses do not allow to show a significant difference between the two groups. This tendency has been confirmed by different studies, see [Carter et al., 2003] for a review.

An approach of heart rate variability during sleep [Bunde et al., 2000] leads to DFA exponents of ~ 0.5 for both light and deep sleep, with less variability for the observed exponents when the sleep is artificial. Exponents of ~ 0.85 have been reported for REM sleep (Rapid eye movement sleep), with, again, a lower variability of the exponents when the sleep is artificial than when it is spontaneous.

In [Esen et al., 2001], the heart dynamics of young habitual smokers have been found to exhibit different characteristics than the control group. At rest, in the supine position, the mean exponents found are $\beta = 1.42 \pm 0.36$ for smokers and 0.96 ± 0.16 for non-smokers. By contrast, in a standing position, the exponents observed are 1.47 ± 0.25 and 1.27 ± 0.12 respectively. A noteworthy point is that the dynamic of the heart for smokers is characterized, in particular, by remarkably limited changes of exponents associated to the postural differences (in comparison with the healthy situation).

Note also that long-range variations in heart rate are found to be correlated to body movements, and both of them are also correlated to circadian rhythms [Aoyagi et al., 2000]. It has also been shown in [Song & Lehrer, 2003] that heart rate variability is changed when breathing rate is modified, which is obtained on humans by conscious control. More precisely, slower respiratory rates lead to a higher heart rate variability.

To sum this paragraph up, we have seen that all the following factors change the heart rate scaling properties: age, diseases, the activity habits, sleep stages, smoking, circadian rhythms, body movements, ...

2.4.4 Conclusion

We have seen that a wide class of biological time series has a particular kind of forms, that can be described as fractal-like and display long-range correlations. The associated characteristics lead to a change of perspective from the usual mathematical interpretation of homeostasis. The latter understands the biological regulation as a relaxation towards a mean and corresponding fluctuations as having a Gaussian shape. In [West, 2006], this point is especially emphasized and is argued to lead to paradigm change as for the understanding of biological regulation. In particular, this approach leads to a further characterization that may be added, typically, to autopoiesis, [Varela, 1979], which is based on an understanding biological systems as processes.

Note, however, that the relevance of the means, in physiology, has been criticized before these dynamics had been discovered, primarily by perspectives taking the point of view of the organism. For example, in [Canguilhem, 1972], health is understood as the ability of an organism to establish, change and adjust its norm. This perspective is coherent with the relatively recent accounts presented in this section. We thank also Denis Noble for pointing out to us that in the Chinese medical treaty, the Mai Jing ($\sim$ AD 300), changes of heart rhythm are considered as healthy while its stability is understood as pathological.

From the point of view of scale symmetries, we have seen that the exponents found experimentally are usually associated with an important amount of variability. In particular we have seen that aging and serious diseases lead in most cases to more regular time series, see also [Goldberger et al., 2002a,b]. In the case of the cardiac rhythm, we have seen that there is a variety of factors which influence the variability of the observed exponents. In particular, the long-term habits have a relevant effect on them (we have discussed smoking and exercising), but other factors are also relevant, such as circadian rhythms, body movements, ... Variability is also relevant at the intraindividual level and is found in the study of the variation of the exponent over time, in the same time series.

As a short conclusion to this section, we can safely say that a great variety of biological time series exhibit (time) scale symmetry and consequently long-range correlations — this is why these processes are also said to exhibit memory. However, in spite of the universality usually found in physics in these contexts, we found that the experimentally observed exponents, in biology are associated with a high degree of variability, which can, in part, be associated to various observable factors. In chapter 3 and especially in section 3.4.4, we will come back to these results, in combination with allometry, and provide an alternative framework for qualitative analysis, associated to a bidimensional time.

2.5 Anomalous Diffusion

The case of structures in *space-time* is more complex than the situations we discussed earlier. Indeed, a mathematical account of an object in space and time is very close to a physico-mathematical description in a proper phase space, by which we mean that it is very close to a physical full-fledged structure of determination. As a result, these accounts are usually given almost directly by physical models.

We will limit ourselves to the case of anomalous diffusion in this chapter, and discuss cases of criticality in chapter 7.

Diffusion corresponds to a macroscopic description of the dispersion, along time flow, of the spatial distribution of some entities. Microscopically, any diffusion is based on random paths (of particles or whatever composes the diffusing objects). Anomalous diffusion, as the name implies, corresponds to a situation where the macroscopic description has a singular form.

2.5.1 Principle

A classical diffusion is a diffusion based on a homogeneous space and time and with homogeneous probabilities of movement in space and time. From a microscopic point of view, the probabilities of movement are the same for each space-time point and for each direction, in a Euclidean space. In this situation, the macroscopic equation describing the evolution of a density ρ subject to diffusion (without sources, or in other words with a conservation principle) is the following:

$$\frac{\partial \rho}{\partial t} = D \sum \frac{\partial^2 \rho}{\partial x_i^2} \qquad (2.14)$$

where D is the diffusion coefficient and the x_i are the 3 space directions. Since the expectation of the distance traveled is 0, only the mean squared distance traveled is interesting. It has the form:

$$<(\Delta x)^2> = D\Delta t \qquad (2.15)$$

However, these homogeneity conditions are not always met. In particular, when the diffusion occurs in a "crowded" medium, space homogeneity can be lost, because of the space taken by other particles, especially if they have a wide size distribution. Another relevant situation is the case where space has a fractal structure. Similarly, in an interacting medium, time homogeneity can be lost, the interactions "gluing" temporarily the traveling particles. This is particularly relevant when the times of these "stops" have a power law distribution. Another aspect that can disrupt classical diffusion is active transports.

The simplest form allowing to describe such situations is the following:

$$<(\Delta x)^2> = \Gamma(\Delta t)^\alpha \qquad (2.16)$$

where Γ is called the generalized diffusion coefficient. The crucial quantity, here and as usual, is the exponent α, which determines the precise nature of the observed phenomenon.

$0 < \alpha < 1$ defines *subdiffusion*. This situation corresponds in particular to a diffusion in a crowded environment. Somewhat reciprocally, when subdiffusion is explained by crowding, that is for inactive molecules, the properties of subdiffusion can be used to assess the crowding of an environment [Weiss et al., 2004]. Alternatively, a recent review, [Berry & Chaté, 2011], argue that subdiffusion is better obtained by time trapping (with power law escape probabilities) than by homogeneous obstacles. Note also that diffusion on fractals [Renner et al., 2005] can lead to subdiffusion, with an exponent that is not given by the spatial fractal dimension alone.

This situation is associated with a greater confinement of particles for long-range displacement, see figure 2.9. At the limit $\alpha = 0$, the elementary objects are confined (their mean distance traveled is bounded). The other limit, $\alpha = 1$, correspond to a classical diffusion, see below.

$\alpha = 1$ corresponds to classical diffusion. classical diffusion is diffusion in an homogeneous media. Obstacles only change the speed of the diffusion but don't change the shape of the corresponding statistical distribution.

$1 < \alpha < 2$ defines *superdiffusion*. In this case the transport is faster for long distances than in the case of classical diffusion. Such a situation is analyzed as active transport, where energy is used to organize (order) the motion of particles. In the cell, this situation typically corresponds, for example, to the effect of molecular motors using energy to crawl on the fibers of the cytoskeleton. Another intracellular example is provided by shuttle streaming in Physarum polycephalum.

2.5 Anomalous Diffusion

$\alpha = 2$ corresponds to the case of classical transport. It is not really a case of diffusion, and there is a macroscopic velocity associated to this behavior, corresponding to a collective oriented movement.

When α is not an integer, diffusion is called *anomalous* since it corresponds to situations where the spatio-temporal structure is non-homogeneous. However, the deeper reason for this name is the singular nature of the situation, leading naturally to a description in terms of fractional derivation [West, 2010].

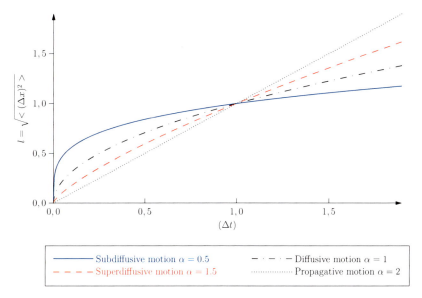

Fig. 2.9 *Different forms of diffusion.* Here, we represent the standard deviation of the distance to the initial position as a function of the time interval of diffusion. Notice that the long time behaviour corresponds to a greater confinement for small α. However, for short time scales a small α correspond to a faster dispersion.

Such behaviours can be empirically studied by single particle tracking (for example molecules with fluorescent labels by fluorescence microscopy). Somewhat reciprocally the spatial structure of the cytoplasm can be assessed by evaluating the form of the diffusion, for inactive molecules of various sizes.

From a theoretical point of view, anomalous diffusions change the probabilities of chemical reactions [Aon et al., 2004b] or from a different perspective, change their kinetics. This is easy to understand since subdiffusion leads to more compartmentalization than a classical diffusion. As a result molecules that are produced in the same region have higher probabilities of interaction. This is for example valid in the case of the structure of DNA [Bancaud et al., 2009].

2.5.2 Examples from Cellular Biology

The nature of diffusion in the bacterial cytoplasm has been recently studied, thanks to the recent progresses in single particle tracking with high spatio-temporal resolution. In [Golding & Cox, 2006], the motion of individual fluorescently labeled mRNA is studied in *Escherichia coli*. The observed motion of this molecule in the bacterial cytoplasm looks confined for periods and time ending by major jumps, which is a qualitative description of the behavior of anomalous diffusion. Indeed, the evaluation of the corresponding anomalous exponent lead to $\alpha \simeq 0.7$, which describes a subdiffusive behavior. This exponent is relatively stable, and robust with respect to the study of different situations, when, for example, considering mutated cells, with altered cytoskeletons. This result is confirmed by a power spectrum analysis of the space trajectories of molecules. This analysis leads to an indirect estimation of α, which is 0.77. The coefficient Γ, however, has a strong dependence on the specific molecule and cell involved, leading to a large variability. Overall, with this experimental methodology, based on the diffusion of mRNA, the source of anomalous diffusion is not assessed. In particular, this approach cannot identify the relative contributions of the space crowding on one side and of interactions on the other.

On the other side, [Weiss et al., 2004] uses the property of subdiffusion, measured for inactive molecules of various sizes, to evaluate the structure of the cytoplasm in HeLa cells[22]. In a similar way as for bacteria, anomalous subdiffusion has been observed, with α from 0.59 to 0.84, depending on the size of the used tracer molecule. Again, these results are robust with respect to the disruption of the cytoskeleton (actin filaments and microtubules).

For particles of bigger sizes, spontaneously appearing lipid granules have been studied in yeasts [Tolić-Nørrelykke et al., 2004], leading to a comparable exponent of ~ 0.75 over a wide range of time scales (10^{-4} s to 100 s).

Diffusion occurs also in plasma membranes. In [Smith et al., 1999], the properties of the diffusion of major histocompatibility complex molecules in the membrane have been evaluated. The result is a coefficient $\alpha = 0.49 \pm 0.16$, with, thus, a large variability.

In the case of eukaryotic nuclei, the anomalous diffusion correspond also to an exponent of ~ 0.79 for euchromatin and ~ 0.75 for heterochromatin [Bancaud et al., 2009]. In both cases, these exponents were found to be independent of the size of the molecules observed. However, when chromatin is not the main component of nuclei, this independence is no longer met. Here, we then have anomalous diffusion associated to a structure with a further symmetry of the anomaly, for different sizes of particles, associate to the fractality of the structure of DNA.

---

[22] HeLa cells are an established cell line from human cervical cancer cells, taken from Henrietta Lacks.

2.5.3 Conclusion

We have seen some examples of biological situations displaying subdiffusion. Note that research on anomalous diffusion has made significant progresses recently, thanks to the development of single molecule tracking methods. These behaviours have compelling consequences on the understanding of the chemical level (among other). In particular, subdiffusion leads to a fuzzy compartmentalization, beyond the stricter connexity loss associated with membranes. This phenomenon thus introduces a distributed heterogeneous spatial organization in the cell, as a crucial determinant of the interaction between its different (spatial) parts. Biologically, it means that the cell's cytoplasm and other structures organize the random movements of smaller components.

2.6 Networks

Approaches in the conceptual and mathematical frame of networks allow to consider large numbers of different entities, represented by nodes, with specific interactions among them, represented by edges. The edges may be considered as oriented or not, depending on the nature of the interactions. Corresponding to this description, both the structure and the dynamic of networks can be considered.

The subject of networks is an important and extremely active field of research, both theoretically and experimentally. We will restrict ourselves to very basic aspects of their description and empirical analysis. See, for example, [Lesne, 2006] for more theoretical frames.

2.6.1 Structures

The structure of a network can typically be approached by the statistical properties of its number of edge per node. A scale-free network is then a network where each node follow a statistic of the form $k^{-\gamma}$ for the number of edges k attached to it. Resulting from this form, the sub-graphs of a scale-free network have the same statistical properties than the whole network. There is a wide variety of processes (iterated network transformations, typically) that can generate such networks. One of the simplest processes of this kind is a growth with preferential attachment of new nodes on nodes having already many connections. Since various network structures can lead to the same statistics, the validity of these approaches should not be theoretically overemphasized, as it is not an evidence of common causal mechanisms [Fox Keller, 2005].

Another statistically interesting quantity is the clustering coefficient and its scaling properties. The local clustering coefficient C_i for a node i is the ratio of the number of edges among its neighbours N_i with the maximum possible number of such edges.

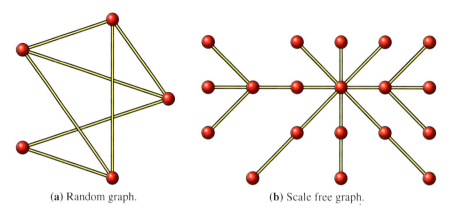

(a) Random graph. (b) Scale free graph.

Fig. 2.10 *Example of random and scale-free graphs.* In the random graph, in the sense of Erdös-Rényi, the probability of each edge to be present remains independent of the rest of the graph. In the scale-free case, however, during graph growth, the nodes that have more edges have a higher probability to get new edges (this is only one of many ways to illustrate the statistical properties associated to the situation).

Table 2.7 *Statistical properties of some biological networks.* These results are from [Almaas et al., 2007]. Note that the variance is high.

	Transcription		Metabolic		Protein Interaction
	E. coli	S. cerevisiae	E. coli	S. cerevisiae	S. cerevisiae
γ	2.1 ± 0.3	2.0 ± 0.2	2.0 ± 0.4	2.0 ± 0.1	2.4 ± 0.4
α	1.0 ± 0.2	1.0 ± 0.2	0.8 ± 0.3	0.7 ± 0.3	1.3 ± 0.5

$$C_i = \frac{2N_i}{k_i(k_i - 1)} \qquad (2.17)$$

This quantity, thus, evaluates the local connectivity, around nodes. Then, in a scale-free situation we have a distribution of the clustering which is expected to follow a power law: $C(k) \propto k^{-\alpha}$.

We present some empirical evaluations of α and γ in table 2.7, for different biological networks which are found approximately to be scale-free. The networks are evaluated by high-throughput techniques, and usually the results have a high degree of imprecision, which can have different forms depending on the technique used (prone to false positives or not and more or less prone to miss some links). The approaches used are generally technically based on nucleic acids screening techniques. For example, the metabolic networks, which represent the chemical reactions, are estimated by a combination of genetics and biochemical results [Jeong et al., 2000]. The protein interaction networks are in particular evaluated by screening for protein complexes, which means that only complexes stable enough to last throughout

2.6 Networks

the manipulation process can be seen. Transcription networks are estimates of the mutual interactions between gene transcriptions.

It is noteworthy that all these networks represent, in general, relatively arbitrary slices on the network of molecular interactions — inasmuch this object is well defined. In particular, all the observed networks interact with each other, but also miss wide classes of interactions. For example, mRNA or sugar are usually not taken into account. The regularity of the statistical properties evaluated is also challenged by some authors, on the basis of alternative statistical methods to assess them. In [Khanin & Wit, 2006], for example, scale-free behaviours are found to be valid only for a restricted range of scales, whilst this study is based on data that have been previously claimed to correspond to scale-free networks.

Notice also that the theoretical validity of such networks can be problematic, since these theoretical objects are not necessarily well defined. In particular, a theoretical reason to challenge their definition is the dependence of chemical reactions on spatial organization, as described in our section on anomalous diffusion. This means that the ability of molecules to interact in a cell does not depend only on the chemical structures of these molecules and their potentiality of interaction, even in a given cellular compartment, as given by membranes.

A recent study, [Rossberg et al., 2011], has found scaling in the structure of the trophic links of marine animals (fishes and squids). The approach of this study is, however, substantially different of what we have already discussed, since it is based on *weighted* links. Indeed, the considered function is, for a consumer species, the number of preys Z_c, which have, in its diet distribution, a weight that is higher that f ($0 < f < 1$). These quantities are evaluated by the study of stomach's contents, meaning that the exceedingly rare preys cannot be taken into account. Since f is bounded, a change of parameter is used in favor of the diet ratio, $r = f/(1-f)$, which tends to infinity when f tends to 1 (this change of variable does not change much the situation for small f). Then, the empirical results obtained follow approximately a power law, $Z_c \propto r^{-\alpha}$, with a mean exponent of 0.54. This exponent is consistent with the community average of Z_c for all data sets studied, except one. The variability for the diet partition function of individual consumer species is, however, high (this is a community regularity property, not a specie property).

2.6.2 Dynamics

The dynamic of the networks above has been extensively studied theoretically, especially along the line of the work by Kauffman. The question is typically that of the nature and the effect of a perturbation. The qualitative classification of the different situations depends on whether a perturbation leads to a return to the previous situation (stable situation) or whether it leads to a further departure from the initial situation (chaotic dynamic). Between these two cases, a critical point occurs, where complex dynamics arise [Kauffman, 1993, 2001, Nykter et al., 2008b].

Since it is technically difficult to assess directly the networks dynamics over time, an alternative way to show that there is an underlying critical dynamic, in the above sense, is to look at the time series of the states of the network's nodes.

The structure of such time series is, however, somewhat difficult to study. Its analysis requires to consider high dimensional states, where the metric (the notion of distance) is not obviously given, as one ignores the structure of the network. For example, in [Shmulevich et al., 2005, Nykter et al., 2008a], the system considered is the transcriptome, observed by microarrays time course experiments. In particular, this means that we are not considering individual trajectories (since the cell measured is destroyed in the process). The trajectories considered are thus collective behaviours, associated to external perturbations (antigens for macrophages). The state is then represented by a vector, which corresponds, at a given time point, to the presence (1) or absence (0) of each RNA. The distance the authors use is the normalized compression distance. It is a based on an effective estimation of algorithmic complexity (by the compressibility). More precisely, for Boolean vectors x and y, this distance is:

$$NCD(x,y) = \frac{C(xy) - \min\{C(x), C(y)\}}{\max\{C(x), C(y)\}} \qquad (2.18)$$

where $C(u)$ is the compressed size of u and xy is the concatenation of the vectors x and y. Recall now that Kolmogorov complexity of a sequence is the size of the shortest program generating that sequence — an incompressible sequence has thus a maximal complexity. Kolmogorov complexity, *stricto sensu*, is not computable. If Kolmogorov complexity were computable, then we could write a finite program which would generate sequences of any Kolmogorov complexity, by screening all possible strings by increasing sizes until finding one with the right complexity: a contradiction. Another proof may be given by showing that, when trying to implement a program which would compute Kolmogorov complexity, one stumble quite straightforwardly on the halting problem — the famous undecidable problem discussed in particular by Turing. Then an approximation is used, usually given by Lempel-Ziv algorithm, like in the papers we are discussing [Shmulevich et al., 2005, Nykter et al., 2008a].

Now, the idea used in these articles is to compare the distance between time points t and $t + \Delta t$ and the distance between time points $t + \Delta t$ and $t + 2\Delta t$. If this distance shrinks, it means that we have a situation of relaxation, or, in other words, of stability. On the opposite, if it grows, this means that we have a chaotic situation and that the perturbation gets amplified. If it stays roughly constant, then it is evidence that we are in a critical situation (the validity of this classification and the validity of the approximation is tested by the authors on data generated by simulation and where the network dynamics is therefore known). The authors finally find that macrophages exposed to different antigens have a critical dynamic at the level of their transcriptome.

A wider study [Balleza et al., 2008] shows similar results for specimen in four biological kingdoms: bacteria *E. coli* and *B. subtilis*, fungi *S. cerevisiae*, animal *D. melanogaster* and plant *A. thaliana*. The method to show critical behavior, here, differs from above, and is based on simulation of the result of perturbations, on the basis of the known (assumed) topology and dynamics of the graphs. Two cases are studied: either small well-known networks or larger network with heuristically estimated dynamics. *In fine*, the dynamics are found to be consistent with criticality.

2.6.3 Conclusion

In this brief description of scale-free networks and critical network dynamics, we have reported a few very interesting results. They are based on techniques that allow in particular to assess global properties in situations that can be described as collections of discrete entities with heterogeneous natures. As a result, these accounts tend to be a huge progress with respect to the methodologies that focus on the properties of elementary components only.

There are, however, limitations in the validity of these results, corresponding to both empirical (statistical) aspects and theoretical interpretations. Overall, this research field is recent; the study of "natural" scale-free networks is only approximately 10 years old, whilst the empirical analysis of networks dynamics is only in its infancy. The results obtained are nevertheless interesting, but have to be considered in a cautious perspective.

Let us finally remark that the quantitative relation between the different networks that are usually studied in cells is not straightforward. Indeed, a recent study [Taniguchi et al., 2010] has shown that the number of mRNA and the number of the corresponding proteins are not correlated at the level of a single cell. Therefore, one network cannot be used as a proxy for the other.

2.7 Conclusion

This survey of scale symmetries highlighted by experimental results or case studies has led us to discuss a variety of techniques and empirical situations. Generally speaking, a wide variety of biological processes is relatively well described by scale symmetries. By this, however, we only mean that the considered situations are better described as scale symmetric than having a characteristic scale. This distinction is mandatory because in almost all cases where the studies are sufficiently comprehensive, relevant (intraindividual, interindividual and/or interspecific) variability in the scaling exponents have been found. As a result, the scale symmetries under question should be taken cautiously.

Among these approaches, allometry is somewhat peculiar. It does not describe (directly) the scaling relations inside an object but compares objects of different sizes. We have seen that globally and most of the times, allometry follows approximately an exponent of $1/4$, which is a nontrivial situation. Accordingly, the metabolism scales globally with an exponent close to $3/4$. However, the variability of species with respect to these relations remains high and may correspond to ten fold variations in the case of the metabolism. Different approaches of the statistical aspect of the problem lead some authors to argue that other exponents are relevant.

Moreover, if one consider intraspecific allometry, associated with development, we see that a wide range of allometric relations can be described, including symmetry shifts (changes of exponent). The relevance of the 1/4 allometry is, however, still high in interspecific cases, since it is approximately observed for a wide class of rhythms and rates.

We have then considered morphological fractality in organisms. Again the situation is approximately described by a (statistical) scale symmetry, which typically corresponds to the definition of the fractal dimension. The reliability and stability of the results is, however, far from perfect, with respect to both interindividual variability and the intrastructural variations in scaling properties. However, as we have pointed out, the structures involved are highly discontinuous and intricate; thus, they loosely correspond to fractal situations. Moreover, their structures clearly lead to an instability of the measurement of Euclidean notions such as lengths, surfaces ..., which depend on the resolution. As a result, the fractal description is still a considerable progress, but it leads to results that should be taken cautiously.

The dynamics we have considered lead to similar lessons. Multi-scale dynamics are ubiquitous, and long-range correlations should be taken as the default assumption, which has profound consequences on the theoretical nature of biological regulation. From the point of view of the measurement, these results are crucial; they typically mean that the heart rate or the blood cell count change spontaneously and importantly over time. Even a measurement that is performed and averaged on relatively large time scales is not sufficient to obtain a stabilized value. There, again, variability of exponents is found both between individuals and inside a given dynamic.

It is noteworthy that, for both morphological fractal-like structures and biological fractional dynamics, the exponents found can be experimentally shown as correlated to specific factors. In particular, the age is relevant in both cases. For dynamics, aging is generally related to a more regular dynamic. Other relevant factors are the activity habits, diseases, ... Again we should recall that all these correlations, which are sometimes remarkably clear tendencies, are found on a general background characterized by an important variability. It is also noteworthy that, in some cases, other factors do not influence the observed exponents, whilst one would have expected they would have.

We omitted in this chapter cases which are analyzed as phase transitions, as they require a longer introduction. We introduce them in chapter 6, and discuss biological evidences in section 7.2.

As a final statement, we can say that the situations presented in this chapter are relatively subtle. Scale symmetries are relevant; they provide an unique insight on biological phenomena and allow to exhibit and solve instabilities of more classical measurement. However, they do not match exactly the empirical situation. Variability clearly appears almost ubiquitously, or at least as soon as the available data are sufficient. This variability can be put in relation with various factors, corresponding in particular to the history of the object considered.

In the following chapter, we will use two of the regularities discussed here. More precisely we will use temporal allometries to propose abstract geometrical schemes

2.7 Conclusion

for biological time. This will also allow us, among other aspects, to propose an original approach of the multi-scale structure of biological time series, as described in section 2.4 above. Finally, the pervasive evidence of symmetry breaking or changes, in the empirical results presented here, will further justify the theoretical role we give, in this book, to symmetries and their changes.

Chapter 3
A 2-Dimensional Geometry for Biological Time

Abstract. This chapter proposes a mathematical schema for describing some features of biological time. The key point is that the usual physical (linear) representation of time is insufficient, in our view, for the understanding key phenomena of life, such as rhythms, both physical (circadian, seasonal, ... rhythms) and properly biological ones (heart beating, respiration, metabolic, ...). In particular, the role of biological rhythms do not seem to have a counterpart in the mathematical formalization of physical clocks, which are based on frequencies along the usual (possibly thermodynamical, thus oriented) time. We then suggest a functional representation of biological time by a 2-dimensional manifold as a mathematical frame for accommodating autonomous biological rhythms. The "visual" representation of rhythms so obtained, in particular heart beatings, will provide, by a few examples, hints towards possible applications of our approach to the understanding of interspecific differences or intraspecific pathologies. The 3-dimensional embedding space, needed for purely mathematical reasons, allows to introduce a suitable extra-dimension for "representation time", with a cognitive significance.

Further aspects of the compactified time will be analyzed in section 7.5, where we will associate this approach to the forthcoming notion of extended critical transition[1].

Keywords: biological rhythms, allometry, circadian rhythms, heartbeats, rate variability.

3.1 Introduction

Living phenomena display rather characteristic and specific traits; among these, manifestations of temporality and of its role are particularly remarkable: development, variegated biological rhythms, metabolic evolution, aging, ... This is why we believe that any attempt at conceptualizing life — be it only partially — cannot avoid addressing such temporal aspects that are specific to it. We will examine this

---
[1] This chapter is a revised version of [Bailly et al., 2011].

question from different angles, including temporality, in view of providing a first attempt at a synthesis.

The intuitive "geometry" of physical time is, since Newton, the understanding of time as a "straight line". This was later enriched by the order structure of Cantor type real numbers, an ordered set of points, topologically complete (dense and without gaps). Thermodynamics and the theories of irreversible dynamics (phase transitions, bifurcations, transition to chaos, ...) have imposed an "arrow" upon classical time, by adding an orientation to the topological and metric structure. But it is with relativity and quantum physics that the theorization of time has led to rather audacious reflections. In the first case, by means of its famous causality cone, Minkovski space explains, within the framework of a unified geometry of space-time, the structure of any possible correlation between physical objects, in special relativity. This is only one example from a very rich debate which goes so far as to introduce a circular time (proposed by Gödel as a possible solution to Einstein's equations).

In quantum physics the situation is more complex or, in any case, less stable. We go from essentially classical frameworks to a sometimes two-dimensional time, in accordance with the structure of the field of complex numbers by which Hilbert spaces are defined, the theoretical loci of quantum description. Feynman's audacious temporal "zigzags" are also worth mentioning [Feynman & Gleick, 1967]. This approach is a very interesting example of intelligibility by means of a "geometric" restructuring of time: very informally, the creation of antimatter would cause within the *CPT* symmetry (charge, parity, time) a symmetry breaking in terms of charge, while leaving parity unchanged. The global symmetry is then achieved by locally inverting the arrow of time. Another approach, with similar motivations, is that of the fractal geometry of space-time, specific to the "scale relativity" proposed by [Nottale, 1993]: in this frame, time is reorganized upon a "broken" line (a fractal), which is continuous but non-derivable. Further interesting reflections, along similar lines, may be found in [Le Méhauté et al., 1998].

Physics however will remain but a methodological reference for our work, because the analysis of the physical singularity of living phenomena requires a significant enrichment of the conceptual and mathematical spaces by which we make inert matter intelligible, see chapters 1 and 7. One of the new features which we introduce is the use that we will make of the "compactification" of a temporal straight line. In short, we will try to *mathematically understand rhythms and biological cycles by means of the addition of "fibers" (a precise mathematical notion, introduced below) which are orthogonal to a physical time that remains a one-dimensional straight line.*

From our standpoint, a living being is a true "organizer" of time. By its autonomy and action, including internal activities, it confers to time a more complex structure than the algebraic order and the topology of the real numbers. In short, the time of a living organism, by its specific rhythms and its coupling to environment, intimately articulates itself with that of physics all the while preserving its autonomy. We would therefore like to contribute to making the *morphological complexity* of biological time intelligible, by presenting a possible geometry of its structure, as a two dimensional manifold.

3.1 Introduction

The first section introduces the theme of biological rhythms. One consequence of our approach is the possibility of giving, by some mathematically suitable structures and concepts, a more precise and relevant meaning to notions that are usually treated in a rather informal fashion and unrelated between one another, such as those of biological rhythms vs. representation time, physical time, ...

3.1.1 Methodological Remarks

Let us recall that physical theories, through history, were constructed largely on the grounds of major dimensional constants (gravitation, the speed of light, Plank's constant — with dimensions, respectively: acceleration, speed, action — that is energy × time). What is striking, in biology, is the presence of a few major invariants *with no dimension*, given by the rhythms which we will discuss below. We suggest here to start with these rare invariants, the constants and rhythms which are to be found in biology, because, beyond these examples and the physico-chemical activities, the *structural stability* of living phenomena is not so "invariant", physically speaking, as it is profoundly imbued with *variability*. We will discuss the latter in the section 3.4.4 and more generally in chapter 7.

Observe also that in physics, time is mostly described as a parameter of the state functions describing a system. The phenomena encountered in biology, however, seem to trigger the need of other theoretical strategies and this at many different temporal levels of organization (physiology, ontogenesis, phylogenesis, ...). We will provide a geometrical scheme of biological time that stresses the crucial role of time in biology and allows to understand some of the above features mainly through the use of two theoretical concepts.

The first one, which we will discuss in depth below, is the ubiquity of rhythms in biological temporal organization. There are indeed only a few features that are ubiquitous in biology and the iteration of similar processes seems to be one of them. We will however make a clear distinction between two type of cyclicity encountered in living systems.

The second concept is a way to understand the constitution and maintenance of biological organization, both in phylogenesis and embryogenesis, that has been formalized by the notion of anti-entropy in [Bailly & Longo, 2009], see also chapter 9 and 7. That approach allows the addition of a new theoretical aspect of time irreversibility in biological systems, that completes and adds up to the thermodynamical irreversibility driven by the notion of entropy. At the level both of evolution and ontogenesis, this irreversibility manifests itself by the increase of complexity of the organism (number of cells, number of cell types, cell networks — neural typically, geometrical complexity of the organs, constitution of interacting yet differentiated levels of organization, ...).

Methodologically, by a duality with physics, in [Bailly & Longo, 2009] time is understood as an operator (like energy in Quantum Physics), not as a parameter. We will go back to that approach in 9, as it turns time into a fundamental observable of biology (like energy in physics) and it gives meaning to time's key role in "biological organization", since rhythms organize life.

Besides the methodological relevance of our approach to biological time, the mathematical details contained in this chapter and in the next one are not necessary for an understanding of the other issues discussed in the rest of this book and may be skipped at first reading.

3.2 An Abstract Schema for Biological Temporality

3.2.1 Premise: Rhythms

We introduce now a second dimension of time associated with the endogenous internal rhythms of organisms as defined below. We represent this second dimension over a "circle", that is as a compactified dimension ($\mathscr{S}_1$ topology[2]). Thus, this second dimension of time is measured by θ, an angle over a circle with a radius R_i (where R_i is the proper biological time): this circle expresses the temporal circularity, the iterative component, that is specific to internal rhythms.

3.2.2 External and Internal Rhythms

We distinguish two types of rhythms associated with biological organization, each referring to a distinct temporal dimension (below, we will note them as t and θ, respectively):

(EXT). "external" rhythms, *constrained by* phenomena that are exterior to the organism, with a physical or physico-chemical origin and which may be imposed by the physical context upon the organism. So these rhythms are the same for many species, independently of their size. They express themselves in terms of physical periods or frequencies (seconds, s, frequencies, Hz). Thus the invariants are dimensional: they are described with respect to the dimension of physical time (in $\exp(\iota\omega t)$). Examples: seasonal rhythms, the 24 hours-cycle and all their harmonics and sub-harmonics, the rhythms of chemical reactions which oscillate at a given temperature, ... In short, the theoretical symmetries of these rhythms are strictly physical.

(INT). "internal" rhythms, of an endogenous origin, *specific to physiological functions* of the organism, depend on purely biological functional specifications. These rhythms are characterized by average periods which scale as the power $1/4$ of the organism's mass and, when given as a ratio with the life span of the organism, which scales in the same way, they are expressed as numbers. The invariants are therefore pure numbers, with no dimension, see section 2.2. We propose to describe them with regard to a new compactified "temporal" dimension θ, with a non-null radius. The numeric values then correspond to a "number of turns",

---

[2] The circle is the *compactification* of the real number straight line, by the addition of a point and its folding.

3.2 An Abstract Schema for Biological Temporality

independently of the effective physical temporal extension: this motivates their representation as an independent dimension, see also section 7.5. We will closely analyze some examples: heartbeats, breathings, cerebral frequencies, ...

Even if we may be somewhat repetitive, we will now describe how these rhythms take place in biological organization, which is precisely what we would like to provide account for:

- The external rhythms (Ext) are to be identified with physical time (typically measured by a clock) whose temporal features does not depend of the biological system we consider. Key examples include circadian, circannual or tidal cycles. The effects or the relevance of these cycles depend of course on the organism that we consider (with possible sexual dimorphism). For example, diurnal and nocturnal animals are in phase opposition, whereas tides are mainly relevant for marine organisms, and even more so in the foreshore. Whatever organism we consider, the period and the phase of these rhythms are the same as they are dependent on external physical events. In order to be a little more precise, these rhythms are generally associated with a double process: the physical process, outside the living system (and which can be very precisely predicted as it is associated to physical theoretical symmetries) and its "trace" inside the system which is kept synchronized by so-called "Zeitgeber" (light for circadian cycle for example). This distinction leads in particular to a specific inertia, encountered for example in the "jet lags" phenomenon.

    Simple chemical oscillations inside an organism will fall in this category as well, since their period is determined by physical principles, even though their phase may depend on a specific organism (a given trajectory) since the organism and the chemical system may be co-determined.

    As a result, this kind of rhythms, and their subharmonics, can be considered in the usual physical way and mathematically represented by terms like $e^{i\omega t}$.

- The second kind of rhythms, the endogenous biological cycles in (Int), does not depend directly on external physical rhythms. They could be called autonomous or eigen rhythms and scale with the size of the organism (frequencies brought to a power $-1/4$ of the mass, periods brought to a power $1/4$), which is not the case with constraining external rhythms as they impose themselves upon all (circadian or seasonal rhythms, for example). The internal rhythms are encountered when we consider the heart rate, the respiratory rate, the mean life span, ..., see chapter 2, section 2.2 or, for example, [Savage et al., 2004] or [Lindstedt & Calder III, 1981]. These rhythms are naturally associated with the number of their iterations (they can be seen as dual variables), and these numbers provide a natural way of speaking of the age of a biological system, yet different from the time measured by a clock. As an elementary example, a two year old dwarf hamster is old, if not senile, while a dog is a young adult and a human is a toddler.

    We stress that this kind of rhythms leads to the use and the study of pure numbers (the total number of heartbeats during the life or an organism, say), instead of quantities with a physical dimensionality (such as intervals of physical time). The point is that these numbers seem, at least in most cases, to have invariant properties[3]. A clear and impressive example of this is the mean number of heartbeat (or respiration) during life, which is basically invariant among mammals. We will further justify this and more examples below.

In summary, endogenous biological rhythms:

- are determined by pure numbers (number of breathing or heart beats over a lifetime, for example) and not, in general, by dimensional magnitudes as is the case in physics (seconds, Hertz, ...);
- depend on the adult mass of the organism that we consider, by following the allometric law $\tau_i \propto W_f^{1/4}$ (for heterotherms, the temperature is involved too);
- in our approach, they are analyzed and put into relation to each other by adding an additional compactified "temporal" dimension (an angle, actually, like in a clock), in contrast to the usual physical dimension of time, a line, non-compactified and endowed with dimensionality.

Since these endogenous rhythms co-exist with physical time, we consider a temporality with a topological dimension equal to 2 formed by the *direct product* of the non-compactified part, the real straight line of the variable t (the physical time parameter) and, as a fiber upon the latter, the compactified part, a circle $\mathscr{S}_1$, where the variable θ ranges between 0 and 2π, with $0 \cong 2\pi$. Since we consider a two-dimensional time, whose second dimension is associated with specific biological invariants, our approach is very different of the usual approach of biological time in terms of dynamical systems, which allows to tackle different kind of questions, like synchronization or stability (see for example the noteworthy book [Winfree, 2001]), but do not deal with these approximate invariants.

The idea of using supplementary compactified dimensions in theoretical physics has been introduced by Kaluza and Klein [Overduin & Wesson, 1997], and is still widely used in unification theories (string, superstring, M-theory, ...). There are of course major differences between these uses of compactified dimensions and ours. First, they concern mostly the addition of *space*-dimensions; second, these dimensions are not observable in physics, whereas they are very much so in biology. In our approach, the "projection" of this second dimension on physical time leads to quantities that have the dimension of a time; their mean follows the allometric law, as such they are parameterized by a mass (or, equivalently, by an energy[4]).

We insist that the endogenous rythmicities and cyclicities are not physical temporal rhythms or cycles as such, as they are *iterations* of which the total number is set independently from the empirical (temporally physical) life span. As we said, they are pure numbers, a few rare constants (invariants) in biology. Our aim is that of a geometric organization of biological time which, by the generativity specific

---

[3] There is still some variability, but this variability appears "naked" when considering these numbers, whereas the mass and temperature's effects come first when considering dimensional quantities.

[4] One may see again the dual role of energy vs. time as parameter vs. operator, the duality w. r. to Quantum Physics we mentioned and that has been extensively used in [Bailly & Longo, 2009] and in chapter 9 below.

3.3 Mathematical Description

to mathematical structures, would also enable us to *derive* meaning and to *mathematically correlate* diverse notions. The text itself constitutes the commentary and the specification of the following schemata, which are meant to "visualize" the two-dimensionality which we propose for the time specific to living phenomena.

3.3 Mathematical Description

We first consider both external and internal rhythms; later, we will mainly focus on internal rhythms of organisms. The heart rate of mammals can be considered as a paradigmatic, running example. We begin by providing a qualitative draft of our scheme to show its geometrical structure in figure 3.1, then we will quantify its parameters and explain more extensively their meaning.

3.3.1 Qualitative Drawings of Our Schemata

Following the aforementioned ideas, biological time is a (curved) surface: thus, it will be described in 3-dimensions (the embedding space). Note that, if we were considering only biological rhythms, our 2-dimensional manifold would be a cylinder: the (oriented) line of physical time *times* the extra compactified dimension. The situation is more complicated, in view of the further, physical rhythms we want to take into account. They do not require an extra dimension, but they "bend" the cylinder, by imposing global (external) rhythms. Thus, a proper biological rhythm is represented by what we may call a "second order helix", that is, a helix that is obtained (is winding) over a cylinder, $\mathscr{C}_i$. This cylinder, in turn, is winding around a bigger cylinder, $\mathscr{C}_e$, whose axis is the line (τ). As basic reference, we choose orthogonal Cartesian coordinates. Physical time, which is oriented by thermodynamic principles of irreversibility and is measured by a clock as in classical or relativistic physic, will be the first axis (t) of our reference system and will enable the characterization of instants and the measurement of durations. The second axis, (t'), will be associated with the proper irreversibility of biological time (for example the irreversibility of embryogenesis or, just, of "living", see 3.5). As such, it will represent the *biological age*, or the internal irreversible clock of the organism we consider. It is thus physical time, but measured by different internal clocks according to the organic: having lived 10,000 heartbeats for a mouse or an elephant spans very different physical time length, about $1/50$ for one over the other, say, they cover roughly the same proportion of their respective life span though. The (t) and (t') axis are oriented in the usual way $((t)$ towards the right and (t') pointing upwards). The third axis, (z), (see 3.1) is generated by the mathematical need of a 3-dimensional embedding space; yet, we claim that it has a biological meaning that will become clear later, in section 3.3.2.

The surface of the cylinder $\mathscr{C}_i$ is parameterized by t (the physical time) and $\theta \in [0, 2\pi]$ (the compactified time).

Let's then take a further step by gradually making explicit the functional dependencies.

82   3 A 2-Dimensional Geometry for Biological Time

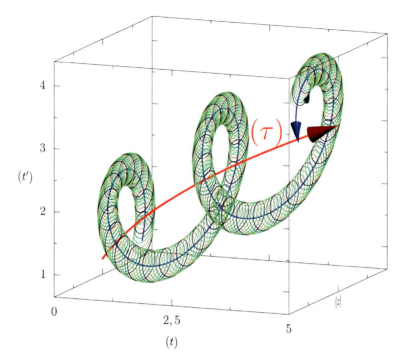

Fig. 3.1 *Qualitative illustration of our geometric scheme, as a 2-dimensional manifold.* In red, the global age of the organism τ, in blue its modulation by the physical rhythm ϕ. Here the surface is suggested by the accumulation of helices.

- The average progression with respect to (t') will be represented by a function $\tau\left(\frac{t-t_b}{\tau_i}\right)$. t_b is the physical time of a biological event of reference (time of fecundation for example). τ_i is a characteristic time of the biological activity of the adult: for example, the mean "beat to beat" interval under standardized conditions (other reference systems can be chosen such as the mean time taken to attain 98% of adult mass, life expectancy, respiratory interval, ...). This value represents, as a function of physical time, the age of the system inasmuch this age is biologically relevant (see figure 3.2a: the graph of τ lies on the $(t \times t')$ plane). τ is a growing function due to the irreversibility of biological time, and has a decreasing derivative due to the decrease of activity during development and aging.

3.3 Mathematical Description

We set then:

$$\vec{\mathfrak{F}}_{\tau_i}(t,\theta) = \begin{pmatrix} t \\ \tau\left(\frac{t-t_b}{\tau_i}\right) \\ 0 \end{pmatrix} \quad (3.1)$$

- We next consider a physical (external) rhythm of period τ_e (its pulsation is then $\omega_e = \frac{2\pi}{\tau_e}$) that affects the activity rate of the organism — the circadian rhythm, for example, leads to $\tau_e = 24$ hours. This produces a winding spiral or helix, $\mathscr{C}_e$ (see figure 3.2b: here we need the third dimension (z) for the embedding space of our manifold). In the definition of $\vec{\mathfrak{G}}_{\tau_i}(t,\theta)$, R_e represents the impact of this physical rhythm on biological activity:

$$\vec{\mathfrak{G}}_{\tau_i}(t,\theta) = \vec{\mathfrak{F}}_{\tau_i}(t,\theta) + \begin{pmatrix} 0 \\ \frac{R_e}{\omega_e \tau_i}\cos(\omega_e t) \\ \frac{R_e}{\omega_e \tau_i}\sin(\omega_e t) \end{pmatrix} \quad (3.2)$$

The term $\frac{1}{\omega_e \tau_i}$ is proportional to the number of iterations of the compactified time during one period of the physical rhythm, as such it can be considered as the temporal weight of this rhythm for an organism (mean number of heartbeat during a day, for example), it allows to understand that a year is more important for a mouse than for an elephant. As a consequence the radius of $\mathscr{C}_e$ is proportional to both the impact R_e of this rhythm on biological activity, and on the weight of this rhythm in terms of number of iterations of the endogenous rhythm considered during one period of the physical rhythm[5].

- We can finally add a biological (internal) rhythm, which depends on an increasing function $s_{\tau_i}(t)$ (see figure 3.3). $s_{\tau_i}(t)$ has a proper biological meaning: for example, if we impose $s_{\tau_i}(t_b) = 0$, with $t = t_b$ when the heart begins to beat[6], $s_{\tau_i}(t)$ is the number of heartbeats of the organism at time t, and thus the mean maximum of s, obtained when death occurs, does not depend on the organism we consider (among mammals, typically). Set then, for $\vec{\mathfrak{G}}_{\tau_i}(t,\theta)$ as in equation 3.2:

$$\vec{\mathfrak{T}}_{\tau_i}(t,\theta) = \vec{\mathfrak{G}}_{\tau_i}(t,\theta) + \begin{pmatrix} 0 \\ R_i\cos(2\pi s_{\tau_i}(t) + \theta) \\ R_i\sin(2\pi s_{\tau_i}(t) + \theta) \end{pmatrix} \quad (3.3)$$

---

[5] We did not discuss here the phase of this rhythm, as it is not required in the rest of our development. Le us just mention here that, in full generality the argument of the periodic term should be $\omega_e t + \varphi$, where $\varphi = \varphi_e + \varphi_s$. φ_e is the physical phase of the rhythm and φ_s is the specific phase shift of a biological object considered, allowing to accommodate diurnal and nocturnal organisms for example.

[6] Let's remark that, unlike in physics — classical, relativistic or quantum— biological time has an origin, whatever level of organization we consider. As a result there is no time-symmetry for translations, a fundamental property, in (relativistic) physics for the constitution of invariants, e.g. energy conservation.

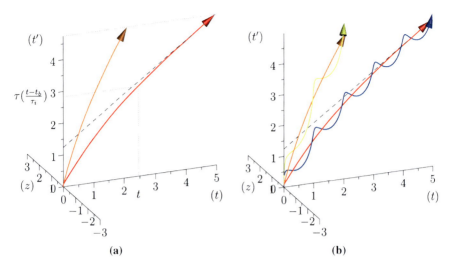

Fig. 3.2 *Qualitative Illustration of the first components of our model.* LEFT, figure (a), the function $\tau\left(\frac{t-t_b}{\tau_i}\right)$, which represents the global age of an organism: this age increases at a greater pace during development and slows down progressively, see section 3.5. In orange a small mammal (a mouse for example) and in red a bigger one (an elephant). The life span of the first is shorter than the one of the second. RIGHT, figure (b), in blue (and yellow), a physical rhythm has been added (this rhythm is very slow for illustrative purposes). Notice that this physical rhythm has the same period for both animals, but one of its iteration has a greater (physical time) weight for the smaller animal.

3.3.2 Quantitative Scheme of Biological Time

Now, a straightforward way to define more precisely s is to use $\vec{\mathcal{O}}_{\tau_i}$ and more precisely the length of the curve defined by $\vec{\mathcal{O}}_{\tau_i}$. We obtain then for the instantaneous pulsation, where τ' is the derivative of τ (thus $\frac{d}{dt}\tau\left(\frac{t-t_b}{\tau_i}\right) = \frac{1}{\tau_i}\tau'\left(\frac{t-t_b}{\tau_i}\right)$) and the other components are the derivative of the remaining coordinates in equation 3.3:

$$\frac{ds_{\tau_i}(t)}{dt} = \sqrt{\alpha^2 \times 1^2 + \left(\frac{\tau'\left(\frac{t-t_b}{\tau_i}\right)}{\tau_i} - \frac{R_e}{\tau_i}\sin(\omega_e t)\right)^2 + \left(\frac{R_e}{\tau_i}\right)^2 \cos(\omega_e t)^2} \quad (3.4)$$

The term α is here for (physical) dimensionality reasons: since our metric has the dimension of a frequency, and $\frac{dt}{dt} = 1$, then the derivative of the first component of the vector in equation 3.2 has no dimension and we need to introduce this coefficient whose dimension is a frequency[7].

---

[7] This kind of reasoning is commonplace in physics.

3.4 Analysis of the Model

When $\alpha = 0$ we can simplify 3.4 to:

$$\frac{ds_{\tau_i}(t)}{dt} = \sqrt{\left(\frac{\tau'\left(\frac{t-t_b}{\tau_i}\right)}{\tau_i}\right)^2 + \left(\frac{R_e}{\tau_i}\right)^2 - 2\frac{R_e\tau'\left(\frac{t-t_b}{\tau_i}\right)}{\tau_i^2}\sin(\omega_e t)} \quad (3.5)$$

Now, if we consider hibernating animals, or frozen organisms, we have situations where the physical time flows normally but where the biological time almost stops or even totally stops. For $\alpha \neq 0$, even in the frozen case, biological time would flow with $\frac{d}{dt}s_{\tau_i}(t) \geq \alpha$. It seems then natural to propose that $\alpha = 0$. Moreover, for $\alpha = 0$, we go back to allometric relations, since, in this case, $\frac{d}{dt}s_{\tau_i}(t)$ is proportional to $\frac{1}{\tau_i}$. Now, τ_i is proportional to $W_f^{1/4}$, by allometry, and, thus, $\frac{d}{dt}s_{\tau_i}(t)$, which is a frequency, to $W_f^{-1/4}$, as it should be.

Another way to express this is to say that physical time *per se* does not make biological organization get older: it is only when there is a biological activity (which in return is of course always associated with physical time) that aging appears.

We can now even give a meaning to the third axis, (z): since $\tau\left(\frac{t-t_b}{\tau_i}\right)$ is on the $(t \times t')$ plane, a positive (z) corresponds to a positive $\sin(\omega_e t)$, by equation 3.3, and it is associated with a slowdown of biological activity (sleep, for example), whereas the negative values are associated to a faster pace (wake for example).

As a fundamental feature of the model that we will analyze next, we assume that the speed of rotation with respect to the compactified time is constant, which leads to a radius $R_i = $ Cst.

$$\left\|\frac{\partial \vec{\mathfrak{T}}_{\tau_i}(t,\theta)}{\partial \theta}\right\| = R_i(t,\theta) = \text{Cst} \quad (3.6)$$

This assumption "geometrizes" time even further: acceleration and slow-down will be *seen* as contraction and enlargement of a cylinder in §3.4.4.2. In that section, as an application, we will develop a geometrical analysis of biological rate variability, and, as an empirical example, we will consider the heart rate of humans. Note that this radius R_i is the dimension accommodating the biological rhythms, thus it is not a physical dimension (it is a pure number). Our assumption is consistent with the idea that each iteration along the compactified time contributes equally to aging.

3.4 Analysis of the Model

In this section we will explore the various biological aspects that our approach allows to put together, mainly on the questions of interspecific and intraspecific allometry and on (heart) rate variability.

3.4.1 Physical Periodicity of Compactified Time

Since $\frac{d}{dt}s_{\tau_i}(t)$ provides the frequency of the biological rhythm, it is interesting to look for a simple analytic expression of the period associated. To do so, we perform a Taylor development (under the hypothesis $\tau'\left(\frac{t-t_b}{\tau_i}\right) \gg R_e$) of the inverse of equation 3.5, and as a result we obtain an approximation of the physical time associated with an iteration of the compactified time (the time between two heartbeats for example):

$$\frac{1}{\frac{ds_{\tau_i}(t)}{dt}} \simeq \tau_i \left(\frac{1}{\tau'\left(\frac{t-t_b}{\tau_i}\right)} + \frac{R_e}{\tau'\left(\frac{t-t_b}{\tau_i}\right)^2} \sin(\omega_e t) \right) \quad (3.7)$$

We can observe several things here. First, for adults we have $\tau'\left(\frac{t-t_b}{\tau_i}\right) \simeq Cst$ and this constant does not depend on the size of the organism we consider. In this case, the result has the form $\tau_i(a + b\sin(\omega_e t))$. As a consequence, when we consider different species, there is no variation of the ratio ($\frac{\tau_i b}{\tau_i a}$) between the continuous and the periodic (with respect to physical time) components of the biological rates. Alternatively, the ratio between the rates of the biological rhythm during the slow period of the physical rhythm (sleep for example) and during the fast period (wake) does not depend on the species either. It is noteworthy that this prediction holds experimentally (see for example [Savage et al., 2004] and [Mortola & Lanthier, 2004]).

On the other side, the relationship between this two rates is not linear in intraspecific variations (i.e.: when τ' is not constant, mainly during development), and the variation of the coefficient of the rhythmic component $R_e \tau'\left(\frac{t-t_b}{\tau_i}\right)^{-2}$ is far greater than that of the steady (continuous) component $\tau'\left(\frac{t-t_b}{\tau_i}\right)^{-1}$. This mathematical deduction agrees with experimental results, as for the tendency, since in [Massin et al., 2000], for example, it is shown that the continuous component varies like $t^{0.16}$ while the sinusoidal part (associated with the circadian rhythm) varies like $t^{0.75}$ for humans (between 2 months and 15 years).

3.4.2 Biological Irreversibility

We can now look more precisely at the second axis, (t'), of our reference system. Since this aspect of biological time is irreversible and flows in the same direction than physical time ($\tau(t)$ is an increasing function of t), $\vec{\mathcal{G}}_{\tau_i}$ in equation 3.2 should increase with respect to this direction. When this condition is met, we will say that these times are "*cofluent*". This can be easily mathematized by looking at the partial derivative of the (t') component of $\vec{\mathcal{G}}_{\tau_i}$ (obtained with the dot product by the unitary vector $\vec{e}_{t'}$) with respect to t:

3.4 Analysis of the Model

$$\frac{\partial \vec{\mathfrak{G}}_{\tau_i}(t,\theta)}{\partial t} \cdot \vec{e}_{t'} = \frac{1}{\tau_i} f'\left(\frac{t-t_b}{\tau_i}\right) - \frac{R_e}{\tau_i}\sin(\omega_e t) \tag{3.8}$$

We obtain then three possible different scenari, assuming that $\tau'\left(\frac{t-t_b}{\tau_i}\right)$ tends to be a constant for adults (and seniors), written $\tau'\left(\frac{t_\infty}{\tau_i}\right)$. We then use equation 3.5 to derive their *observable* consequences:

$\tau'\left(\frac{t_\infty}{\tau_i}\right) > R_e$. In this case, biological age and the physical clock are cofluent, and the minimum biological rate is achieved during adult sleep (figure 3.3a and 3.3d).

$\tau'\left(\frac{t_\infty}{\tau_i}\right) \simeq R_e$. In this case, the two times are minimally cofluent, the derivative tends to zero (during night or winter) when the organism grows older, that is the rate of the biological rhythm tends to 0 during the (physical) time of little biological activity. It seems to be particularly relevant for hibernation (figure 3.3b and 3.3e)....

$\tau'\left(\frac{t_\infty}{\tau_i}\right) < R_e$. in this case they are no longer cofluent, the nullification of the biological rate would appear during development, and, as a result, the slowest biological rhythm would appear during sleep of young individuals (figure 3.3c and 3.3f).

This abstract cases can be tested with empirical data and the first two cases have actual biological occurrences (see for example [Hellbrugge et al., 1964, Cranford, 1983]). We believe that theoretically biological time should be always cofluent so that the third case should not be realized. Indeed, the existing data, which are mostly given for humans, confirm that case 3 does not hold (young individuals have slow rhythms, during sleep typically, which are faster than adults slow rhythms).

It would be nice if our theoretical deduction, which excludes the third mathematical possibility as irrelevant, like in physical reasoning, were empirically confirmed in large phyla. Conversely it would be also interesting if this theoretical derivation leads to the discovery of species where also the third case is realized. In either case, the generative role of this simple mathematics would follow a paradigm very often implemented in physics, through history.

3.4.3 Allometry and Physical Rhythms

When we consider organisms with different adult masses (W_f), we obtain a variation of τ_i according to the scaling relationships ($\tau_i \propto W_f^{1/4}$), whereas ω_e does not change. As a result, this change corresponds to a dilatation of the (t) axis (as far as f is concerned) whereas the physical rhythm modifies the geometry of biological time because the variations it triggers are anchored to the physical value ω_e (see figures 3.2a, without physical rhythm, and 3.2b, with physical rhythms.).

Then, it is the interplay between physical rhythms and biological ones that breaks the symmetry (by dilatation) between organisms of different (adult) masses that have the same temporal invariants (most mammals for example). As a result, in this

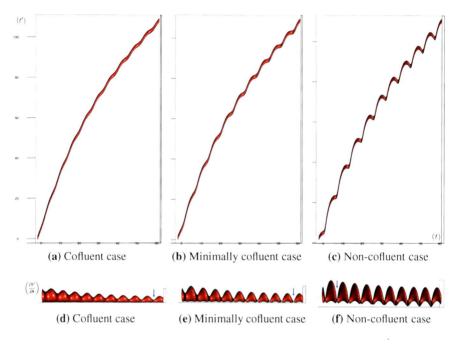

Fig. 3.3 *Illustration of the three scenari.* TOP, figures (a, b, c): the scheme $\vec{\mathfrak{T}}_{\tau_i}(t,\theta)$ and BOTTOM, figures (d, e, f), its time derivative $\frac{\partial \vec{\mathfrak{T}}_{\tau_i}(t,\theta)}{\partial t}$. FROM LEFT TO RIGHT: Confluent case, minimally confluent case and non-confluent case. Since the radius of the compactified time is proportional to its physical rate when looking at $\frac{\partial \vec{\mathfrak{T}}_{\tau_i}(t,\theta)}{\partial t}$ (see §3.4.4.1), the bottom pictures allows to see when the slowest rate occurs (i.e.: when the radius is the smallest, blue arrow. Here respectively: for adults in figure 3.3d and 3.3e and for infants in 3.3f).

situation, the physical conditions can be seen as constraints or frictions on biological temporal organization. Our point of view can be compared to the (physically) dimensionless time in [West & Brown, 2005], but this latter paper only considers the autonomous aspect of biological time, thus it does not take into account this important interplay.

Of course, another way to illustrate these aspects is to count the lifelong number of iterations of cycles: as for biological cycles, this number does not vary much when considering different species, the invariance at the core of our approach, whereas it is strictly proportional to life span, as for physical cycles.

3.4.4 Rate Variability

Let us first introduce informally the applications and viewpoint we will hint to in this section, where the data are obtained from a medical database. Our approach to biological time allows naturally, as we will further specify, to a representation by a

3.4 Analysis of the Model

cylinder whose radius is proportional to the cardiac *rate*. If we assume that n heartbeats yield a complete rotation around the cylinder, then a faster heart rate would appear as a circular outgrowth (a sudden increase in the radius). In this representation, a healthy individual has a complex cardiac dynamics during the day, with frequent rhythms' accelerations of varying length (from seconds to many hours), see section 2.4. This shows up in the figures by the many circular outgrowths of different radii. On the contrary, an individual with an artificially regulated cardiac rate (with a pacemaker, say) gives a relatively smooth cylinder. We will also show a case which corresponds to a sudden cardiac death, without other particular symptom than the altered variability of the rate.

Of course we do not provide a *theoretical determination* of spontaneous biological rate variability, but just a *geometrical representation*. As a matter of fact in our framework, it is quite straightforward to explore the *structure* of biological rhythms and of their variations. More precisely, we can easily and effectively represent raw datas (for example the series of "beat to beat" interval over time). As a result, we obtain more than a qualitative schema: it is a theoretical grounded representation of the "anatomy" (including pathological anatomies) of biological time and this anatomy is infused with variability.

3.4.4.1 Renormalization

First we need to describe how we can use scales in our framework.

If we want to consider n iterations of the compactified time θ as an iteration of an other view of compactified time $\tilde{\theta}$ we obtain $\tilde{\theta} = \frac{\theta}{n}$ and $\tilde{s}_{\tau_i} = \frac{s_{\tau_i}}{n}$, then by a "renormalization" using the principle of constant speed for the compactified time, one has:

$$\left\| \frac{\partial \vec{\mathfrak{T}}_{\tau_i}(t,\theta)}{\partial \theta} \right\| = \frac{\tilde{R}_i}{n} = \text{Cst} \tag{3.9}$$

So $\tilde{R}_i = R_i n$. This result is exactly (modulo a global dilatation of the (t') and (z) axis by a factor $\frac{1}{n}$) what we obtain if we construct directly our system at the level of n iterations.

3.4.4.2 More on Rate Variability

If we look at the function obtained by taking the derivative of $\vec{\mathfrak{T}}_{\tau_i}(t,\theta)$ with respect to t, we obtain:

$$\frac{\partial \vec{\mathfrak{T}}_{\tau_i}(t,\theta)}{\partial t} = \begin{pmatrix} \frac{1}{\tau_i}\tau'\left(\frac{t-t_b}{\tau_i}\right) \\ -\frac{R_e}{\tau_i}\sin(\omega_e t) - 2\pi R_i s'_{\tau_i}(t)\sin(2\pi s_{\tau_i}(t) + \theta) \\ \frac{R_e}{\tau_i}\cos(\omega_e t) + 2\pi R_i s'_{\tau_i}(t)\cos(2\pi s_{\tau_i}(t) + \theta) \end{pmatrix} \tag{3.10}$$

Here, instantaneous heart rate, $2\pi s'_{\tau_i}(t)$, appears directly as the radius of compactified time, which now has the physical dimension of a frequency.

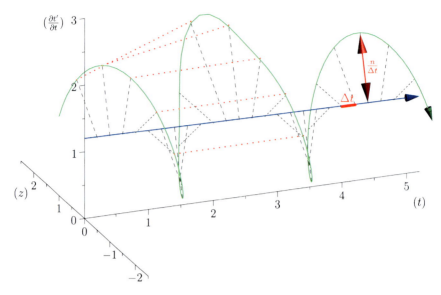

Fig. 3.4 *Renormalization and principles of variability representation.* Here, we consider $\frac{\partial \vec{\mathfrak{T}}_{\tau_i}(t,\theta)}{\partial t}$ and we renormalize the compactified time by $n = 10$. A change of speed for the iteration m of the original compactified time appears as a sharp contrast between this iteration and its neighbors: iteration $m-1$, $m+1$, $m-10$, $m+10$. As a result, if there is a coherence for 10 successive iterations, we obtain a fully circular outgrowth or contraction (for an acceleration or a slowdown respectively).

If the experimental time of each heartbeat is given in a list $(t(m))_{1 \leq m \leq M}$, we obtain a discrete empirical version of $\frac{\partial \vec{\mathfrak{T}}_{\tau_i}(t,\theta)}{\partial t}$, renormalized by n:

$$\frac{\partial \vec{\mathfrak{T}}_{\tau_i}(m)}{\partial t} = \begin{pmatrix} t(m) \\ \hat{A} - \hat{R}\sin(\omega_e t(m)) - 2\pi \frac{n}{t(m+1)-t(m)} \sin\left(\frac{2\pi m}{n}\right) \\ \hat{R}\cos(\omega_e t(m)) + 2\pi \frac{n}{t(m+1)-t(m)} \cos\left(\frac{2\pi m}{n}\right) \end{pmatrix} \quad (3.11)$$

where $\hat{A}$ is an estimation of $\frac{n}{\tau_i}\tau'\left(\frac{t-t_b}{\tau_i}\right)$ which may be soundly considered constant during the few days of the measure. $\hat{R}$ is an estimation of $n\frac{R_e}{\tau_i}$. Both of these values are estimated by using equation 3.5 and $(t(m))_{1 \leq m \leq M}$. We obtain a 2-dimensional structure by using triangles between adjacent points, that is to say for $m \leq M - n - 1$, the triangles $(m, m+1, m+n)$ and $(m, m+n, m+n+1)$. It is worth mentioning that this approach allows to obtain an empirical version of $\vec{\mathfrak{T}}_{\tau_i}(t,\theta)$ too.

The renormalization by n allows to observe directly the correlations between n consecutive heartbeats (a full circle) and the contrasts between a group and its neighbors (see figure 3.4), thus discriminating easily between the sleep situation (no correlations wider than $\simeq 100$ heart beat) and the healthy wake state (correlations at

3.4 Analysis of the Model

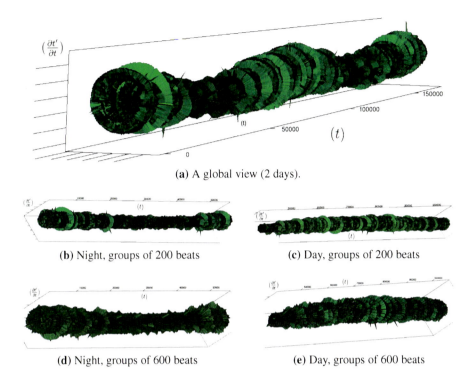

Fig. 3.5 *Comparison of the situations during sleep and wake.* The point to notice here, is that the structure tends to become a regular cylinder during night at high scales, whereas the wake is always complex. (Sample s20011 from The Long-Term ST Database, [Goldberger et al., 2000]). The series of beat to beat intervals provided by this database is used directly, in our framework, to estimate the few parameters we need and more importantly to provide the radii involved (each heartbeat is represented).

each scale). The latter is indeed characterized by a succession of randomly spaced outer circle (see figure 3.5).

Moreover this representation may be useful to study cases of heart diseases and even aging, since these situations are characterized by an alteration of heart rate variability. We illustrate this alteration in cases of sudden cardiac death in figure 3.6 computed with data from the The Sudden Cardiac Death Holter Database, see [Goldberger et al., 2000]. This figure evidentiate the anatomy vs. the pathological anatomy of heart rhythms and suggests the extension of this approach to other biological rhythms which are less explored.

- Figure 3.6a is an example of a healthy case, which is characterized by a complex temporality during wake.
- In figure 3.6b, (intermittent) pacing leads to an excessively regular cylinder, with very limited heart rate variability.

- Atrial fibrillation (a kind of arrhythmia, see comments in figure 3.6), in the figure 3.6c, leads to an "hairy" structure, which represents a strong short term randomness (few correlations between successive heartbeats).
- Last but not least the figure 3.6d is not associated with a specific diagnosis (put aside sudden cardiac death at time 9000) but it clearly shows a very simpler structure than the healthy case.

Our approach allows to discriminate all these various cases by rather striking geometrical differences. Wavelet analysis is often used for the same purpose, but the wavelet approach is based on a massive reorganization of the data, through a decomposition in various components, whereas we only perform a geometrical and synthetic composition of them.

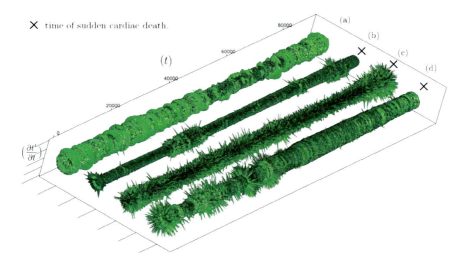

Fig. 3.6 *Comparison between a healthy situation and cases of sudden cardiac arrest.* (a) Healthy case, cf figure 3.5. (b) Female aged 67 with sinus rhythm and intermittent pacing. (c) Female, 72, with atrial fibrillation. (d) Male, 43, with sinus rhythm. (The data are from samples 51, 35 and 30 from The Sudden Cardiac Death Holter Database, see [Goldberger et al., 2000]).

3.5 More Discussion on the General Schema 3.1

3.5.1 The Evolutionary Axis (τ), Its Angles with the Horizontal $\varphi(t)$ and Its Gradients $\tan(\varphi(t))$

The central line (τ), see figure 3.1, is the "result" of the various components (physical time, external and internal rhythms) and we propose that it refers to a "physiological" time associated to the temporal progression of the organism over the course

3.5 More Discussion on the General Schema 3.1

of its life. In order to better understand the different chronological parts of life, this "axis" may be decomposed in distinct segments, each being characterized by their angle, φ, with regard to the abscissas (the φ angle under consideration then becomes that of the tangent), connected by zones with a change of curvature around specific times $(t_0, t_1, t_2, \ldots)$. We will qualitatively distinguish five parts (with unequal lengths). Note that this angle depends on the choice of unit for the physical time. We can overcome this limitation by choosing the adult angle as 45°, or similarly but not identically the period of the adult rhythm.

I Around t_{00} (which would correspond to the fertilization of the egg that will form the organism), a new segment begins with a very large angle (80° for example) and consequently with a very high gradient. This segment will correspond to *embryogenesis*.

II Around t_0, there occurs a first curvature of the axis in order to initiate a segment of which the angle (and the gradient) still remains high (at 60°, for example). Time t_0 would correspond to birth[8] and the following segment to *growth* (development).

III Around t_1, we would have a new curvature generating a medium sized angle (45° for example) with a gradient approaching 1. t_1 would correspond to the appearance of the reproductive faculty (age of puberty[9]) and to the entering into the phase of *adult maturity*.

IV Around t_2, we would have another curvature generating a small angle segment (30° for example) with a weak gradient. t_2 would correspond to the period of loss of fecundity (menopause, andropause)[10] and to the beginning of *aging* as such.

V Around t_3 the axis becomes horizontal ($\varphi = 0$, $\tan(\varphi) = 0$) and is definitely broken. t_3 represents the time of *death*.

Concerning the various durations (namely that of the life span $t_3 - t_0$), we know by the above mentioned laws of scaling generally encountered in biology, that these durations scale according to the organism approximately by $W_f^{1/4}$, where W_f is the mass of the adult organism.

If we now consider $v_t = \tan(\varphi(t))$ as being the "speed" of evolution of the physiological time (τ) with regard to the physical time t, we would make the following remarks which motivate the various gradients of (τ):

- between t_{00} and t_0 this speed is very high: initial cell divisions, morphogenesis, setting in of the first functionalities;
- between t_0 and t_1, the speed remains high; it corresponds to growth, to development, to the completion of the setting in of functionalities, to a high metabolism;
- between t_1 and t_2 the speed is moderate; it corresponds to the regularity of the metabolic reactions, of cellular renewal, etc., that are characteristic of adult age;

---

[8] At germination, for plants.
[9] At the moment of flowering or of fruit-bearing, for plants.
[10] At the end of production, for plants.

- between t_2 and t_3, the speed is low: lowering of the metabolic rate, of cellular regeneration, of activity; this corresponds to aging;
- after t_3 the speed is null: it is the death of the organism.

Note that this is consistent with the analysis in terms of rates (like heart rate), that we have performed in the previous section. The point of this discussion is to relate our approach to the key step in the evolution of the organization of an organism. It is important to keep in mind that in biology the representation by a mathematical function, τ in our quantitative scheme, usually masks this kind of qualitative changes, that we discuss here as sudden changes of the angle.

3.5.2 The "Helicoidal" Cylinder of Revolution $\mathscr{C}_e$: Its Thread p_e, Its Radius R_i

In our qualitative analysis (see 3.1) we have a cylinder of revolution $\mathscr{C}_e$, with a radius R_i, which is winded as a helix having a thread of p_e around the (τ) axis, without touching this axis but faithfully following its changes of direction.

The thread p_e of this helicoidal cylinder can be assimilated to a period; it corresponds to the *external* cyclical rhythms imposed upon the organism by its environment (annual, lunar, circadian cycles, for instance, see §3.2.2(EXT)), which are independent physico-chemical rhythms that we have taken into account in the first paragraph; they are essentially of a physical origin and are imposed upon all organisms exposed to them. The $R_i = 0$ case will be discussed below.

3.5.3 The Circular Helix $\mathscr{C}_i$ on the Cylinder and Its Thread p_i

This circular helix $\mathscr{C}_i$, with a thread p_i, is winded around the surface of the cylinder $\mathscr{C}_e$ (it is a "second order" helix because the winding cylinder is also helicoidal). We consider the thread of this helix (which is also a period) to refer to the compactified time θ (the circle which generates this cylinder) introduced here and associated to the *internal* biological cycles of the organism. As we stressed several times, these are relatively independent from the environment. This is the case, for example, of cardiac and respiratory rhythms, of the rhythms of biochemical cascades, etc. (see §3.2.2(Int)). Let us also recall that the period associated to these cycles also scale by $W_f^{1/4}$, at least from t_1 (and also, in practice, from t_0).

3.5.4 On the Interpretation of the Ordinate t'

Let us return now to the issue of the interpretation we can give to the ordinate t'. In a mathematical sense, it is *generated* by the compactified fiber of the temporal rhythms specific to living phenomena. More specifically, it is mathematically necessary as a component of the three-dimensional embedding space of helices produced by the direct product of the physical time t and of the compactified time θ, which are, according to our hypothesis, two independent dimensions. We already hinted

3.5 More Discussion on the General Schema 3.1

to a possible biological meaning of the (z) coordinate. But first and more precisely, what could the ordinate t' correspond to, from a biological standpoint?

If we define a speed for the passing of time τ comparatively to t' in a way that is similar to the definition of $v_t = \tan(\varphi(t))$, we will have $v_{t'} = \cotan(\varphi(t))$; at the inverse of v_t (we have $v_t v_{t'} = 1$), this speed is small at first but continues to grow when t (or τ) grows.

In the case where the organism under consideration is the human being, an interpretation promptly comes to mind. The velocity $v_{t'}$ would correspond to the *subjective perception* of the speed of the passing of the "specific" or physiological time τ: at first very slow, and then increasingly rapid with aging. In such case, t' would be the equivalent of a *subjective time*. One will notice that, from the quantitative standpoint, if between t_1 and t_2 (the interval of the adult phase) we confer φ with the value of 45° approximately, as we have already indicated above, the speed of the passing of time τ with regard to objective physical time (v_t) coincides more or less with the subjective perception of the passing of this time $(v_{t'})$ (in fact, $\tan(\varphi) \simeq \tan \varphi \simeq 1$).

As it is matter, here, of human cognitive judgment of the time flow, we are aware of its historical contingency. The remarks below, thus, are just informal preliminaries to forthcoming reflections, where the historicity of young vs. old age perception of time, for example, should be relativized to specific historical cultures and social frames. We then leave the reader to have any reflection regarding the subjective perception of time during youth and old age. We can imagine that such thoughts will coincide with ours, if we belong to the same "culture" (time which passes slowly while young and, later, very quickly...).

In what concerns organisms other than human beings, of which we do not know if they have a subjective perception of the speed of the passing of physiological time τ, it is more difficult to assign a clear status to this dimension of t' (although certain relatively evolved species seem likely to express impatience, for example, or to construct an abstract temporal representation by exerting faculties of retention and especially of protention). So would this dimension not begin to acquire a concrete reality only with the apparition and development of an evolved nervous system (central nervous system, brain)? But then what about bacteria, amoebas, paramecia...?

Actually, it may be possible to somewhat objectivize the approach by advancing a plausible hypothesis regarding the general character of t': we could consider that it is a question of a "temporality" that is associated to the "representational" dimension. Let us explain.

Since living organisms are endowed with more or less capacity for retention and *protention* (possibly pre-conscious "expectation"), we tentatively propose the following qualitative interpretation: the element of physiological time $d\tau$ is associated to the element of physical time dt and to dt' by the evident relation $d\tau^2 = dt^2 + dt'^2$. It stems from this that dt'^2 can be written as $dt'^2 = d\tau^2 - dt^2$ or as

$$dt'^2 = (d\tau - dt)(d\tau + dt) \qquad (3.12)$$

It is then tempting to see in the first factor the minimal expression of an element of "retention" (for physiological time, relatively to physical time) and in the second the corresponding expression of an element of "protention". The product of the two would generate the temporality component of a "representation" which borrows from the "past" and from the "future", as constitutive of the flow of biological time. As all living organisms appear to be endowed with both a capacity for retention — as rudimentary as it may be — and with a protentional faculty (even more rudimentary maybe), the generality of the dimension t' would be preserved and the "representational" capacity (at least in this elementary sense) appears as being a property of living phenomena. We will develop this idea, of protentional and retentional capacity of living organisms in the next chapter. This property, for conscious thought, could even be extended to subjectivity, in accordance, in the specific case of the human being, with the phenomenological analysis by which we began: dt'^2 would be a form, as elementary as infinitesimal, of the "extended present", in the Husserlian tradition, described by other analyzes, such as the coupling of oscillators in [Varela, 1999].

Finally, it would be the two-dimensionality $t \times t'$ — (physical time) × (representation time) — which would enable to *mark out* the temporality of living phenomena, which may be represented in the geometrical way as we have described in this chapter.

3.5.4.1 Conclusion

In summary, in addition to the objective physical time t (evidently still present and relevant), we presented a general geometric schema of time from a biological standpoint, which includes:

- a general temporality of ontogenesis (τ) (the axis);
- a temporality associated to the external rhythms (the helicoidal cylinder winded around this axis from a distance) that are characterized by the thread p_e;
- a temporality associated to the internal rhythms involving a compactification of time: the helix with a p_i thread at the surface of the cylinder.

Our purely theoretical representation allowed though an apparently effective drawing of rhythms, such as the cardiac ones. Images may matter a lot in organizing knowledge and it may be worth noting that they came to our mind only after realizing the purely mathematical need for a second compactified dimension as for biological time.

As a final deductive remark, we should note that if the radius R_i of the helicoidal cylinder becomes null, it will be reduced to a helix winded around (τ) and the internal cyclicity will tend to disappear as such (there remains only the external rhythms that are physical). The general schema we have presented concerns mainly the properties of the animal world, while, we claim, this last case, where $R_i = 0$, mainly concerns plants. With rare exceptions, which should be more closely analyzed, the rhythms of plants (metabolic, chlorophyllian, of action — activation of organs...) seem completely subordinated to the physical external rhythms. Thus, the non nullity of R_i, that is, the two-dimensionality of the cylindrical surface, should be associated to the greater autonomy — the rhythms of the central systems, typically

3.5 More Discussion on the General Schema 3.1

— and to the autonomous motor capacity which the animal enjoys comparatively to plants, the two being obviously correlated.

Of course, there is no clear-cut transition, no well-defined boundary between animal and plant life forms, in particular in the marine flora/fauna. For this reason, we find the representation of the passing from the one to another in the form of a continuum to be adequate: the continuous contraction of the helicoidal cylinder which tends towards being a helix, which is a line (the time of plants). The non observability of the difference between animal and plant, in some "transitional" cases, would correspond to an interval of biologically possible measurement, with no phase transition (of the type of life form) that is clear or discontinuous. Once the limit, the helicoidal line, is reached, even the three-dimensional embedding space can be collapsed onto the two dimensions: the rhythm becomes the oscillation of one measurement (of chlorophyllian activity, for example) with regard to the axis of oriented physical time (the spiral is flattened into a sine wave, for example) as is the case in many periodic physical processes.

Chapter 4
Protention and Retention in Biological Systems

> Husserl uses the terms protentions and retentions for the intentionalities which anchor me to an environment. They do not run from a central I, but from my perceptual field itself, so to speak, which draws along in its wake its own horizon of retentions, and bites into the future with its protentions. I do not pass through a series of instances of now, the images of which I preserve and which, placed end to end, make a line. With the arrival of every moment, its predecessor undergoes a change: I still have it in hand and it is still there, but already it is sinking away below the level of presents; in order to retain it, I need to reach through a thin layer of time.
>
> M. Merleau-Ponty

Abstract. This chapter proposes an abstract mathematical frame for describing a peculiar feature of cognitive and biological time. We focus here on the so called "extended present" as a result of protentional and retentional activities (memory and anticipation). Memory, as retention, is treated in some physical theories, such as relaxation phenomena, which will inspire our approach, while protention (or anticipation) seems outside the scope of physics. We then suggest a simple functional representation of biological protention. This allows us to introduce the abstract notion of "biological inertia"[1].

Keywords: Memory, Cognition, protention, retention and biological time.

4.1 Introduction

The notions of "memory" and "anticipation" are analyzed and formalized here from a temporal perspective. By this, we propose a simple mathematical approach to *retention* and *protention* that are apparently shared by all organisms, albeit rudimentarily. Moreover, in life phenomena, memory is essential to learning and it is oriented towards action, the grounding of protention. Our approach will allow to propose the notion of "biological inertia", a form of "continuation" of ongoing action. The frame is purely mathematical and abstract: only practitioners will be able to give values to our coefficients and develop, possibly, concrete applications of the approach, from cell biology to human cognition. Our aim is to give a precise and relevant meaning

---

[1] This chapter is a revised version of [Longo & Montévil, 2011b].

to notions that are usually treated in a rather informal fashion and unrelated between one another, such as those of time of representation, time of retention and time of protention.

A long phenomenological tradition introduces an important distinction between memory and retention, on the one hand, and anticipation and protention on the other. In short, the common meaning of "memory" seems to essentially refer to a "conscious reconstruction" of something that was experienced (very well put by [Edelman & Tononi, 2001] as a "brain which sets itself back into a previously experienced state"). Anticipation would be its temporal opposite — the awareness of an expectation, of a possible future situation. Memory and anticipation do not, a priori, have a biological *characteristic time*, a notion which is essential to our analysis. In our approach, instead, pre-conscious retention, as a biological phenomenon, is to be seen as an extension of the present; it is the present which is "retained", during a brief interval of time, which we will call a characteristic time, for the aim of the action and of perception, it is a form of extension of the immediate past into the present. For example, when listening to a word or a phrase, we retain the part which has already occurred for a certain (characteristic) duration of time. The mental duration of a phrase, particularly of a musical "phrase", is needed for grasping meaning or a melody (see for example [Perfetti & Goldman, 1976, Nicolas, 2006]): it is the present which leaves a trace the time necessary for action or, possibly, for subsequent awareness. But protention (as preconscious anticipation) is also essential to appreciate a melody or understand a phrase. When reading, the analysis of saccadic eye movements demonstrates that we first look at least at half of the word following the one we are reading, see [Wildman & Kling, 1978]. This protentional behavior participates in the reconstruction of meaning: we appear to make sure of the meaning of the word we are reading by making a partial guess upon the following word.

Technically, protention will be given by a temporal mirror image, as it extends retention forwards into time. Protention is, above all, the *tropism* inherent to action performed by any life form. This point is at the center of our approach: we call retention and protention these particular aspects of memory and of anticipation that are specific to all life forms — a sort of present which is extended in both directions. Thus we do not limit our analysis to the phenomenological use of these words, inasmuch it limits their meaning to situations that can be examined through conscious activities. We believe that this extension to pre-conscious activities remains compatible with (and helps to understand) its classical usage, particularly such as described by [Van Gelder, 1999] and [Varela, 1999], who develop the concepts of intentionality, retention and of protention, introduced and discussed in length by Husserl in his analysis of human consciousness.

In this chapter, we propose an elementary mathematical model of these inevitably fuzzy notions, one which is as rudimentary as possible, but one that can nevertheless support discussions regarding their precise conceptualization and their increasingly thorough mathematization. As always in our approach, the theory matters more than the mathematics. Yet, the introduction of the notions of "biological inertia" and "global protention" are, typically, a consequence of the generative power of mathematics.

4.1 Introduction

For the purposes of our theoretical understanding, we will define some basic principles and more specific notions, after some methodological preliminaries.

4.1.1 Methodological Remarks

Let us recall that in our general approach, also presented and followed in [Bailly & Longo, 2011, 2008, 2009, Bailly et al., 2011], our attempted aim is not to reconstruct the physico-mathematical complexity of some aspects of biology, but to propose firstly and above all a proper biological perspective. We believe that the *theoretical differentiation* between theories of inert and of living phenomena requires, among other things, a change in the relevant *parameters* and *observables*. This chapter is a further example of our approach, and a short general discussion may help to further clarify it.

As long as the actions of living organisms, including their cognitive performances, which occurs the moment that life appears (in this sense, we speak of protention and of retention in the amoeba or the paramecium), are analyzed within physical space-time and physical observables, the physico-mathematical takes precedent over the specificity of the biological.

In order to further specify our methodological frame, we go back to an example. The remarkable mathematics of morphogenesis, a broad issue and references we already discussed in the introduction to this book, from phyllotaxis to the analysis of the dynamic structuring of organs, organize the growth of living organisms according to physical geodesics. This growth is shaped within physical space-time, by optimizing, for example, the occupation of physical space, the exchange of energy by a surface within a volume, ... In these cases, the spatio-temporal and energetic parameters and observables enable a very interesting and often technically very difficult analysis. This is a relevant approach of the *physical complexity* of living phenomena and of its material structures. We could also say the same of analyses of networks of cells, of which the most complex are neural networks. Informational interaction, often a gradient of energy, enables to develop a theory, now very rich from the mathematical standpoint, of these formal networks of which increasingly important applications are being considered for the construction of machines that are somewhat intelligent (at last). However, this mathematical approaches are far from providing a "theory of organisms", as they focus on local properties, largely disregarding either the role of the organism which regulates and integrates both organs' formation and networks' dynamics, or both. Moreover, cells' networks form tissues which form organs which, in turn, integrate both of them, while even the dialogue between the two scientific communities is far from developed. In short, the relevant information obtained still misses the integration in an active organism.

Our modest attempt, instead, always tries to propose a perspective for the organism as a whole, even though, each time, from a specific point of view: so far, we discussed scaling laws, rhythms, now protention and retention, further on it will be a matter of global (extended) criticality, of pertinent phase spaces and enablement, of organismal complexity .... A conceptual and mathematical integration between the relevant morphogenetic and networks' analyses or alike and our perspectives would be an interesting project for further work.

In this chapter, the mathematics used will not go beyond a few equations which could be presented to high school students. What is interesting, in our view, is approaching biological time according to its own specificity, by starting with some invariants which appear to be specific to living phenomena, as we did in the previous chapters, or with properties that are not treated by current physical theories, such as protention. In chapter 3, we proposed a two dimensional representation of biological time as a mathematical frame to accommodate the autonomous (internal) biological rhythms (cardiac, respiratory, metabolic rhythms, ...). In the perspective of this chapter, a conjectural link may be made: one may understand the expectation or anticipation of a rhythm to iterate, as a minimal form of protention. Once rhythms are installed, the organism is "tuned" to (and "expects") their iteration. That is, biological rhythms may be seen as a least form or a possible origin of protensive capacities.

Before developing a further geometrization of biological-time, we will face yet another taboo of physicalism in biology: the inverted causality specific to protention. We will not present a physical theory of teleonomy, but will use as data and principle the evidence of protentional behaviors that may be observed in any life form. When the paramecium, encircled by a ring of salt, tries after many attempts to break through the obstacle, risking its own life and possibly even succeeding [Misslin, 2003], we can take note of the retention-learning phenomenon and of the ensuing teleonomic gesture (a protention) and develop an adequate theory (see also [Saigusa et al., 2008], as for the amoeba). Likewise, when we understand that the brain, prior to a saccadic eye movement prepares, by a clear anticipation, the corresponding primary cortex which is apt to receive the new signal (see [Berthoz, 2002]), we propose to frame this fact by a suitable conceptualization and possibly some mathematics. In either case, there is certainly an underlying physico-chemical mechanism which will one day enable to grasp the phenomenon by means of physical causality, yet, perhaps, an adequate form of causality which may need to be invented, as often in the history physics — needless to insist that causes in the relativistic light cone or as quantum probability's correlations, were major novelties.

For the moment, let us consider these phenomena as a form of protention to be analyzed (correlated, formalized, ...) by a theory specific to living phenomena, even if it has no correspondence nor meaning within current physical theories. The mathematics to be found in the following pages will give us the advantages of formalization: it forces to specify concepts and to stabilize them as much as possible — this is what mathematics is first about. Maybe that which follows is false, but it should then be possible to say so in relation to a precise formulation.

4.2 Characteristic Time and Correlation Lengths

The notion of "characteristic time", which we inherit here from physics, appears to be very important in biology as well: it concerns the unity of the living individual

4.2 Characteristic Time and Correlation Lengths

because, for example, fluxes and their transport entail lengths and, therefore, relevant transport times. We will also speak of characteristic times for retention and protention.

For example, according to the size of the organism, there appears to be two sorts of transport processes. For large organisms, it is of a "propagative" type (v_p velocity, along networks and "channels") with a typical correlation length of $L_p = v_p \tau$, where τ represents the characteristic time. For smaller organisms (cells, for example), it is rather of a "diffusive" type (diffusion coefficient D, due to molecular diffusion processes) wiht typical correlation length $L_d = (D\tau)^{1/2}$.

We stress the difference regarding dependency in function of time: linear in one case, as a power of $1/2$ in the other. Note that anomalous diffusions can also occur, see section 2.5, which yield different exponents, usually smaller (subdiffusion) and sometimes bigger, when energy is used for active transport.

Two complementary remarks:

- The size of the organism also affects structures determining the mode of transport, for example the respiratory function (oxygen transport): in the case of small organisms (insects, for example) the transport is performed by *trachea* (or even pores), multitudes of little cylinders where the air diffuses in order to reach the cells. In the case of large organisms (fishes, mammals), transportation and exchanges are performed by means of *gills* or of *lungs*, centralized anatomic structures which present the fractal geometries we evoked above and which enable to conciliate difficultly compatible constraints (efficiency, steric limitation, homogeneity), and then by various sorts of vascular systems. Transportation, in this last case, is also much more of a "propagative" type (even if diffusion does play a role, namely in bronchioles).
- These considerations essentially apply to various *structural* aspects responding to identical functions. The *functional* aspect responds for its part very generally to common scaling laws (the metabolism which corresponds particularly to oxygen intake, the various rhythms, the relaxation times, ...). It therefore appears that the modes of transport associated to identical functions can be different and can correspond to different anatomic structures (trachea, gills, bronchial trees/lungs). This is the well-known phenomenon of analogy of structures in evolutionary biology.

Finally, account taken of these remarks, since the characteristic times τ mostly scale as $W_f^{1/4}$, where W_f is the mass of the intended adult organism (see chapter 2, section 2.2 or [Lindstedt & Calder III, 1981, Savage et al., 2004]), it is necessary to expect the correlation lengths to scale differently according to the mode of transport: respectively L_p in $W_f^{1/4}$ and L_d in $W_f^{1/8}$, following the definitions of L_p and L_d.

In the sequel, our characteristic times will more precisely refer to "relaxation times", still in analogy to physics (see next footnote), yet in properly biological frame, that is in relation to retention and protention.

4.2.1 Critical States and Correlation Length

The physics of criticality and self-organized systems has massively entered the domain of biology since early ideas by [Nicolis & Prigogine, 1977, Bak et al., 1988, Kauffman, 1993]; .... In chapter 7, we will extend this approach, in direct reference to far from equilibrium systems, by considering living entities as being in an "extended critical situation", beyond the pointwise analysis of critical transitions proper to physical theories.

Before getting into the details of that matter, it is interesting to consider that physical criticality is associated with a so-called critical slowdown (see for example section 6.2.3 for a brief introduction and references): the relaxation time of a system tends to infinity when it goes near the critical point. The qualitative meaning of these situations in biology is that the effect of a stimuli would take a relatively long time to stabilize (or, more generally, the organism would take, in principle, a long time to "react" or "adjust" with respect to physical systems), if one views life as close to or in an (extended) critical state – this will be our perspective in chapter 7. In particular, criticality would lead to very slow cognitive reactions at least under the assumption that reaction needs a stabilization.

More generally the elaboration/reaction time should necessarily be slow in an organism with long correlations in space and slow characteristic time of the individual components of the system. However, organisms and especially metazoans must often react quickly and are able to do so. Consequently, biological organization provides a solution to this paradox. This solution is to compensate this slowness by preparing the organism to a forthcoming stimulus *in advance*.

Our analysis of protection and biological inertia will try to provide a simple framework to approach these properties. Note that, in this perspective, perception itself is co-determined by this protentional activity — we mentioned the case of listening, to a tune or also language, at least when "meaning" is involved. More generally, we understand perception not as an issue of "input elaboration", but as the interference of an ongoing protensive activity with the ecosystem.

4.3 Retention and Protention

4.3.1 Principles

We describe *retention R* by specifying it under the form:

$R_k(t_0, t)$ at an instant t of an anterior "event" e of nature k at time t_0,

For short and if needed, we will pose that $e_0^k = e^k(t_0)$ (where $t_0 \leq t$).

Virtual protention, of an event of the same nature $e_1^k = e_{t_1}^k$ at moment t of an ulterior instant t_1 ($t \leq t_1$) will be noted $V_{Pk}(t, t_1)$. However, (actual) protention will be considered as a function also of retention R_k because, and this is an essential principle of our approach, in *the absence of the retention of an event of nature k there will be no possible protention for an event of such nature*. We will therefore have $P_k(R_k, t, t_1) = 0$, for $R_k = 0$. For the sake of simplicity, we described this dependence of protention on retention as a linear dependence and our (actual) protention,

4.3 Retention and Protention

$P_k = R_k V_{Pk}(t, t_1)$, will express this[2]. Moreover, in conformity with our previous analyses, we will pose that this protention is a monotonous increasing function of the retention in question, that is $\frac{\partial P_k}{\partial R_k} \geq 0$.

4.3.2 Specifications

On the basis of the distinction made above, we have thus introduced the notions of retention and of virtual protention, as "immediate" and "passive" memory and anticipation. This is meant to express the fact these phenomena do not stem from the intentionality related to a conscious activity of a subject (generally endowed with a more or less elaborate nervous system), but are proper to simple processes of biological reaction/stimuli/response, of which many primitive organisms in relationship to their environment are the locus. To the aim of developing this point of view, we now introduce distinct concepts by means of simple mathematical functions, mainly *relaxation functions* and their combinations[3].

More specifically, we will first define the retention function:

$$R(t_0, t) = a_R \exp\left(\frac{t_0 - t}{\tau_R}\right) \qquad (4.1)$$

t_0 is the time of occurrence of an event which is the object of the retention, t is the present moment ($t > t_0$); τ_R is the characteristic time associated to the decrease of the retention as we move away form the occurrence of the event. Notice that when τ_R tends to 0, $R(t_0, t)$ tends to 0. a_R is a coefficient which can be associated to an individual or to a species, for example, in comparison to others of which such faculties are more or less developed.

We propose to use relaxation functions, because the loss of retention, by moving away from the moment of the intended event, for example the beginning of a phrase or, more generally, from the beginning of any action (including listening), can be considered as a sort of gradual "return to equilibrium" — no more affected by the (sudden) event. This, obviously, does not preclude us from maintaining a memory of a more long-term past (the initial part of a discourse, for instance): we limit ourselves to an analysis of the local, pre-conscious effect which contributes to the extended present of an ongoing activity.

---

[2] After reading a draft of this text, L. Manning gave us references to IRM data confirming the neurophysiological and neuroimaging evidence for protention and the dependence of protention on retention: [Szpunar et al., 2007, Botzung et al., 2008]. Further, more specific experiments would be required in order to quantify the coefficients we introduce here and check/adjust the linearity of this dependence.

[3] Relaxation functions are among the simplest decreasing functions enabling to define a characteristic time τ in physics, they often represent the basic model for the return to equilibrium of a system that was initially brought out of equilibrium. The speed at which the system returns to the equilibrium f_e of the system's f function, $\left(\frac{df}{dt}\right)$ remaining proportional to this interval, that is $\frac{df}{dt} = -\frac{|f - f_e|}{\tau}$, where τ is the characteristic time.

How may we now formally define virtual protection? We propose to make it mathematically intelligible by means of a *temporal symmetry* with regard to R, that is, by changing sign to time t. So we define, by a symmetry adjusted by two new parameters, a_P and τ_P, a *virtual* protection. Now, time t_1 is the time of the event to be anticipated and which is in the future of the present instant t ($t_1 > t$) in the function:

$$V_P(t,t_1) = a_P \exp\left(\frac{t-t_1}{\tau_P}\right) \qquad (4.2)$$

The different parameters, a_P and τ_P, play the same *mutatis mutandis* role as those which intervene in R (cf fig. A). In particular, $\tau_P = 0$ leads to $V_P(t,t_1) = 0$.

Finally, as we claim that protection depends also on retention, we define (actual) *protention* $P(t,t_0,t_1)$ by the product RV_P:

$$P(t,t_0,t_1) = R(t_0,t)V_P(t,t_1) = a_P a_R \exp\left(\frac{t_0-t}{\tau_R}\right)\exp\left(\frac{t-t_1}{\tau_P}\right) \qquad (4.3)$$

The (linear) dependence of P on R, according to the principles stated above, emphasizes that such a capacity can only exist, phenomenologically speaking, if there exists, in one form or another, a sort of "memory" R (retention) relative to the event of which the reiteration or something resembling it is to be anticipated. We are aware that we are making a strong but empirically plausible hypothesis here, see footnote 2. Thus, the specific traits of this "expectation" of an unknown future, protention, is not exactly symmetrical with regard to the retention of a known past. And this by the fact that protention depends on retention — and not conversely — and that, by its nature, it remains "potential" (it is the expectation of a "possible" event).

In the case where $R = 0$ (complete absence of retention), the protention is canceled out by the fact that there no longer exists any referent enabling to anticipate the expected event.

Still from the phenomenological standpoint, we will expect that in general $\tau_P \ll \tau_R$, that is, that the characteristic time of retention be greater than that associated to protention P (in order to "anticipate", it is first necessary to "remember", as stressed above). So the contribution of V_P in the definition of P (the second exponential in τ_P^{-1}), evolves more rapidly than that of retention for a same concerned duration. And we will always have $P \leq a_P R$, as a function of time t, and this for any value of τ_P and τ_R. $P = a_P R$ is achieved only in the very moment that the time to be anticipated is the actual present, that is for $t = t_1$ and hence $\exp\left(\frac{t-t_1}{\tau_P}\right) = 1$.

To make the role of the parameter t more explicit, with regard to the interval (t_0,t_1) and to the characteristic times τ_P, τ_R, some simple algebraic manipulations enables to put the expression P in the form of the product of a function of t and of two coefficients solely dependent on t_0 and t_1, that is:

$$P(t) = a_R a_P \exp\left(\frac{\tau_R - \tau_P}{\tau_R \tau_P}(t-t_0)\right)\exp\left(\frac{t_0-t_1}{\tau_R}\right)\exp\left(\frac{\tau_R-\tau_P}{\tau_R \tau_P}(t_0-t_1)\right) \qquad (4.4)$$

4.3 Retention and Protention

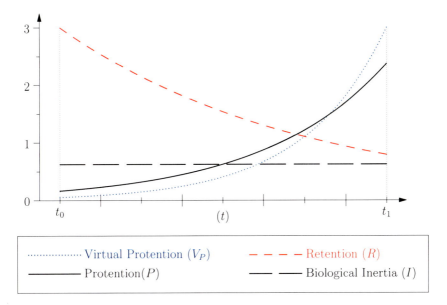

Fig. 4.1 *Illustration of the basic quantities we define.* Notice that protention is a growing function of time.

4.3.3 Comments

First, we should notice that $\frac{\tau_R \tau_P}{\tau_R - \tau_P}$ is an interesting quantity: it has the dimension of a time and is the characteristic time of $P(t)$.

When τ_P tends toward τ_R, this characteristic time tends to infinity, and respectively $\frac{\tau_R - \tau_P}{\tau_R \tau_P}$ tends to 0. This means that when τ_P is close to τ_R, $P(t)$ is almost stationary as a function of t.

On the contrary, when $\tau_R \gg \tau_P$, minor changes in time strongly affect $P(t)$. More precisely, $P(t)$ is small when far from t_1 (and close to t_0), while it is very sensitive to (small) changes of t, when t is close to t_1. This means that, in this condition, the vicinity of the virtual event is where the effect of protention is important, see figure 4.2.

It is crucial, however, to understand that protention, for example in the case of a cognitive situation, is not empirically associated with a change of behavior, but with the *speed* of this change of behavior. This suggests a way to approach these quantities empirically by a comparison of the reaction time between situations where the event associated with retention (at time t_0) occurs and when it *does not*: in the first case, a more sudden change is then to be expected close to the expectation time t_1.

Alternatively, situations where the event at time t_0 occurs but where the event (at time t_1) does not occur allow to evidentiate the presence of protention and to see a part of its effects. This is, for example, the case of amoeba in [Saigusa et al., 2008]. However, in many situations, the effect of protentional action will consist in a "sensitization" to the virtual stimuli with the preparation of a response. This may

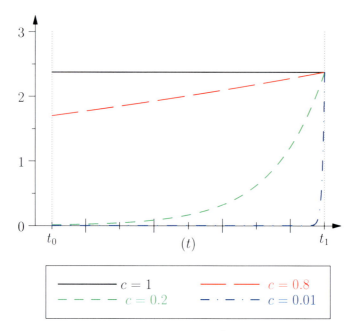

Fig. 4.2 *Protention* for various values of the ratio $c = \frac{\tau_P}{\tau_R}$. We observe that small value of c leads to a sharp curve near t_1 whereas value close to 1 are flat in the interval. We will discuss the biological meaning of this case in section 4.4.

lead to no behavioral change when the virtual stimuli does not happen, but leads to a change of organization associated with the preparation of the response (including at the sensory level) and possibly to a greater sensitivity to noise.

4.3.4 Global Protention

One may wonder when protention is maximal for a given individual. In our approach, the first possible answer is given by looking at the diagram in figure 4.2: this quantity is maximal close to t_1. However, we can refine the question (and the answer) by looking at the global amount of protention along the intended interval $[t_0, t_1]$. As protention is both variant and contravariant in the interval of $[t_0, t_1]$ (see definition 4.3), this question has a non-obvious answer.

For this purpose, we define the notion of *global protention*, which is the sum (the integral) of protention over time, between t_0 and t_1.

$$G_P(t_1 - t_0) = \int_{-\infty}^{\infty} P(t)dt \tag{4.5}$$

$$= \int_{t_0}^{t_1} P(t)dt \tag{4.6}$$

4.3 Retention and Protention

Following our specification of protention we can compute the global protention, which has the following form:

$$G_P(t_1 - t_0) = \int_{t_0}^{t_1} P(t) dt \qquad (4.7)$$

$$= \frac{a_{Rap} \tau_R \tau_P}{\tau_R - \tau_P} \exp\left(\frac{t_0 \tau_P - t_1 \tau_R}{\tau_R \tau_P}\right)$$

$$\times \left[\exp\left(\frac{\tau_R - \tau_P}{\tau_R \tau_P} t_1\right) - \exp\left(\frac{\tau_R - \tau_P}{\tau_R \tau_P} t_0\right)\right] \qquad (4.8)$$

$$= \frac{a_{Rap} \tau_R \tau_P}{\tau_R - \tau_P} \left[\exp\left(\frac{t_0 - t_1}{\tau_R}\right) - \exp\left(\frac{t_0 - t_1}{\tau_P}\right)\right] \qquad (4.9)$$

This quantity has a maximum for:

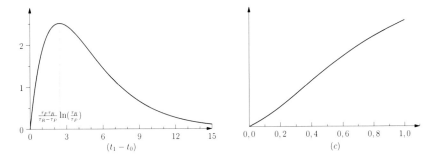

Fig. 4.3 *Global protention.* When considered as a function of the length of the time interval (LEFT), there is a maximum which corresponds to the greater effect of the couple Protention/Retention. RIGHT, we see the global protention as a function of $c = \frac{\tau_P}{\tau_R}$.

$$t_1 - t_0 = \frac{\tau_P \tau_R}{\tau_R - \tau_P} \ln\left(\frac{\tau_R}{\tau_P}\right) \qquad (4.10)$$

This maximum is a compromise between the need to give the protention time to have effect (covariant dependence on the size of $[t_0, t_1]$) and the need to have instants in $[t_0, t_1]$ that are close *both to* t_0 *and* t_1 (contravariance). This result means that there is a specific duration between the past event and the future event which optimize the protentional effects. This seems to be consistent with the results in [Saigusa et al., 2008], since these authors found that a specific value of the delay $t_1 - t_0$ (in our notation) leads to a greater protentional effect, that is the functional dependency on this interval of time has a maximum (a non-obvious fact). In section 4.4 we will go back to the relevant ratio $c = \frac{\tau_P}{\tau_R}$.

4.4 Biological Inertia

Consider now a relaxation phenomenon in physics. It will typically be given by $\Phi(t) = d\exp\left(\frac{t_0-t}{\tau_R}\right)$. If time $t_1 > t_0$ is given, one may decompose $\Phi(t)$ as

$$\Phi(t) = d\exp\left(\frac{t_0-t_1}{\tau_R}\right)\exp\left(\frac{t_1-t}{\tau_R}\right) \qquad (4.11)$$

The coefficient, not depending on t, that is $d\exp\left(\frac{t_0-t_1}{\tau_R}\right)$, is the "residual" at time t_1 and it may be understood as a form of "inertia" of the intended relaxed quantity: typically, it corresponds to "what remains" at time t_1 of a compound which decay with characteristic time τ_R. This coefficient is constant in the interval and decreases for increasing t_1.

In eq. (4.4), the following factors do not depend on t:

$$a_R a_P \exp\left(\frac{t_0-t_1}{\tau_R}\right)\exp\left(\frac{\tau_R - \tau_P}{\tau_R \tau_P}(t_0 - t_1)\right) \qquad (4.12)$$

The first exponential term corresponds to a physical inertia, as it only depends on τ_R, the characteristic time of retention, analyzed as a relaxation phenomenon; let us call it $I_\varphi(t_0, t_1)$. Then, we can consider that the other coefficient of protention represents a *biological inertia*, in the interval $[t_0, t_1]$, depending on the biological constants a_R, a_P, τ_R and τ_P:

$$I(t_0, t_1) = a_R a_P \exp\left(\frac{\tau_R - \tau_P}{\tau_R \tau_P}(t_0 - t_1)\right) \qquad (4.13)$$

In other words, protention in eq. (4.4) may be considered as a product of a function of time t, $\exp\left(\frac{\tau_R - \tau_P}{\tau_R \tau_P}(t - t_0)\right)$, modulated by constants and characteristic times, of a physical inertia $I_\varphi(t_0, t_1)$ and of a "biological inertia" $I(t_0, t_1)$. This last coefficient is also independent of t, but depends on the specific organism by the various indexed constants.

The physical inertia represents the "passive" decay of a physical relaxation phenomena, which makes a perturbation disappear during the return to equilibrium. On the contrary, the biological inertia coefficient is to be understood as a capacity to "carry over" the protensive effect. Their names are freely inspired by the inertial mass as a coefficient of acceleration. Thus and very informally, biological inertia would be the biologically pertinent coefficient of protention. In section 4.5, by references and a discussion, we will say more about this new concept. First a few technicalities.

We have to check whether our definitions depend on the specific reference we choose. That is to say if a time origin change:

$$t_0 \leftarrow \widetilde{t_0} = t_0 + \Delta t \qquad t_1 \leftarrow \widetilde{t_1} = t_1 + \Delta t \qquad t \leftarrow \widetilde{t} = t + \Delta t \qquad (4.14)$$

4.4 Biological Inertia

changes the way we split P in three parts, in equation 4.4. It it then straightforward to see that:

$$\exp\left(\frac{\tau_R - \tau_P}{\tau_R \tau_P}(t - t_0)\right) = \exp\left(\frac{\tau_R - \tau_P}{\tau_R \tau_P}(\tilde{t} - \tilde{t}_0)\right) \qquad (4.15)$$

$$\exp\left(\frac{t_0 - t_1}{\tau_R}\right) = \exp\left(\frac{\tilde{t}_0 - \tilde{t}_1}{\tau_R}\right) \qquad (4.16)$$

$$a_{RaP}\exp\left(\frac{(\tau_R - \tau_P)}{\tau_R \tau_P}(t_0 - t_1)\right) = a_{RaP}\exp\left(\frac{(\tau_R - \tau_P)}{\tau_R \tau_P}(\tilde{t}_0 - \tilde{t}_1)\right) \qquad (4.17)$$

This means that each of this quantities have a sound biological meaning. In summary, inertia introduces a coefficient which is independent of t. It is, in general, much smaller than a_{RaP} (and always smaller than a_{RaP}). This coefficient contributes to the dependence of P in function of t. In particular, it contributes in an essential manner to the decrease of the protention according to the temporal distance.

4.4.1 Analysis

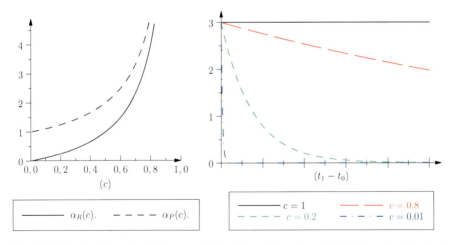

Fig. 4.4 *Biological inertia*. LEFT: we plot the factor of the characteristic time of biological inertia seen as protention (or retention). RIGHT: biological inertia as a function of the length of the time interval for various values of c.

In order to better understand the sense we attribute to this "inertia" of living objects, given our preceding remark regarding orders of magnitude of characteristic times, we may focus on the ratio c of figure 4.3, that is on c such that $\tau_P = c\tau_R$. We consider $0 \leq c \leq 1$ and rewrite I in the equivalent form:

$$I(t_1 - t_0) = a_{RAP} \exp\left(\frac{1-c}{c\tau_R}(t_0 - t_1)\right) \tag{4.18}$$

$$= a_{RAP} \exp\left(\frac{1}{\alpha_R(c)\tau_R}(t_0 - t_1)\right) \quad \text{with } \alpha_R(c) = \frac{c}{(1-c)} \tag{4.19}$$

Then I has the *form* of a "long term retention" if $c > 0.5$ or a "short term retention" if $c < 0.5$. Conversely, and maybe even more intuitively, inertia can be also interpreted (by writing $\tau_R = \frac{\tau_P}{c}$ and eliminating this time τ_R) as a "long term virtual protention":

$$I(t_1 - t_0) = a_{RAP} \exp\left(\frac{1-c}{\tau_P}(t_0 - t_1)\right) \tag{4.20}$$

$$= a_{RAP} \exp\left(\frac{1}{\alpha_P(c)\tau_P}(t_0 - t_1)\right) \tag{4.21}$$

$$\text{with } \alpha_P(c) = \frac{1}{(1-c)} \tag{4.22}$$

Biological inertia would then be both an extended retention, eq. (4.18), and a virtual protention, eq. (4.20), which are both *independent* of the time t of the action: in fact, it depends only on the instants that are relevant to the event retained and occurring in t_0 or which is the object of an expectation (protention towards t_1). It is therefore an inertia which "carries over" the life form from t_0 towards t_1, by the preservation of its own structure and its relationship with the environment (see section 4.5).

The $\tau_R = \tau_P$ Case

In the case where the characteristic retention and virtual protention times are equal ($\tau_R = \tau_P = \tau$ and the c from the equation above is equal to 1), the protention P becomes $a_{RAP} \exp\left(\frac{t_0 - t_1}{\tau}\right)$. It is therefore independent of the present observation time t. This holds, of course, within the interval between the moment of the occurrence of the event in question and the moment t_1 where it is mobilized again (since we still have $t_0 < t < t_1$). But then, according still to hypothesis $c = 1$, one has $P = a_{RAP} \exp\left(\frac{t_0 - t_1}{\tau}\right)$, with $I(t_0, t_1) = a_{RAP}$. Thus, when ($\tau_R = \tau_P$), only inertia is present in protention.

We can also clarify this situation by considering that, if the observation time t is close to the instant t_0 of the occurrence of the event (recent retention), then the temporal interval for a virtual protention, $(t - t_1) \simeq (t_0 - t_1)$, increases; conversely, if time t is far from t_0 (remote retention), the temporal interval involved in this virtual protention and within which the latter plays its role (the future of the observation moment t) reduces in magnitude, given that, in this case, the protention P as such remains independent of t.

These remarks are meant to highlight the fact that, in the latter case, the *intensity* P of the protention remains invariant, whereas the duration upon which virtual protention takes place — the future of t — can change in size: $t_1 - t$.

4.5 References and More Justifications for Biological Inertia

We have come to propose a mathematical notion of biological inertia through an apparently arbitrary play of symmetries and calculations, of which we would now like to better explain the meaning and the objectives. To emphasize the importance of the concept, but without wanting to make excessive and uncontrolled analogies with immensely illustrious precedents, let us note that modern physics started off with a good analysis of inertia, as a "pursuing a state" without aim nor teleology: Galilean inertia[4]. Later on, Newton viewed inertial mass as a coefficient of acceleration, in presence of a force.

In biology, this notion can already be found, although rarely, under various forms. For [Vaz & Varela, 1978] "the lymphoid system has an *inertia*, which resists attempts to induce sudden and profound deviations in the course of events". So this is a weak notion of inertia, close to the "persistence" of structural stability. Likewise, we could talk about inertia in the case of the notion of "dynamic core" presented in [Edelman & Tononi, 2000], because it also refers to the continuity/persistence of individuation (see also [Le Van Quyen, 2003]). This theme is also used by [Varela, 1997], where the term of inertia appears also in the attempt to grasp the "force", specific to any organism, enabling its "bringing forth of an identity".

In our approach, which is inspired by the methods of physics without identifying with it, we firstly define retention by a relaxation function, which is a physical notion — and can even be considered as adequate to describe the "memory" to which some often refer in relation to certain physico-chemical activities. Virtual protention is given then by a temporal symmetry, *modulo* some adjustment coefficients; this notion, which has no analogy in physics, is by this, and at least, the "projective" reflection of retention. Protention follows, as a linear combination of these two values, in function of time. Then, by a very simple algebraic calculation, we separated the part containing the temporal variable from the functional definition: what remains is a constant, a function of all other parameters (characteristic times, specific constants, interval range), which we called biological inertia. As we pointed out, when retention and protention have the same characteristic times ($\tau_R = \tau_P$), inertia coincides with protention. We would then say that this is the simplest situations from a cognitive viewpoint: the organism can only anticipate by means of inertia. In any case, the proposed notion of inertia appears to clearly specify the informal idea of "bringing forth of an identity", with the reference to retention and to protention, at the minimal cognitive level.

But why would this inertia not simply correspond to the fact of following a geodesic trajectory, like in physics? Some will say that the amoeba, the paramecium, etc., follow a gradient in the same way that a physical object follows the

---

[4] Without forgetting Giordano Bruno who had an informal yet quite relevant notion of inertia, a few years prior to Galileo. It then became possible to understand planetary movements without God being required to push the planets around at all times. We similarly aim at a concept of inertia for living phenomena with no reference to "vital impetus" or divine thrust.

trajectory dictated by the Hamiltonian, through the principle of least action. It may appear that such is the case in *in vitro* experiments where, within a highly purified environment, the unicellular organism is exposed to one or two very specific gradients (chemical, thermal, ...). On the other hand, in an *in vivo* situation, in the ecosystems preferred by such animalcules (and which are very polluted, from our standpoint) they must "arbitrate" between qualitatively different stimuli: several physico-chemical gradients, an edible and close bacterium that is not too large, another smaller one, etc. Now the paramecium, say, appears to "learn" (see [Misslin, 2003]), that is, it enjoys at least retention, which contributes to protection (and, after reading Misslin quoted above and references, one could posit for it $\tau_R > \tau_P$, or even $\tau_R \gg \tau_P$)[5]. And it is difficult to conceive of learning without error, or without several attempts and without the memory of these attempts (retention), even if such memory is extremely rudimentary. The subsequent action is therefore one among many possible ones, from the standpoint of the ecosystem, because it also depends on the specificity of individual retention (experience). Among these many possible trajectories, the one it follows has only to be *compatible* with the ecosystem. No gradient or physical geodesic is adequate to describe this plurality of possibilities of evolution, phylogenesis, ontogenesis and of action, which also depends on the specificity, hence on the history, of the species or of the individual (retention and biological inertia). Our modest inertial attempt tries to do this, in a way that is as preliminary as mathematically simple.

In this perspective, we can interpret an increase of $(\tau_R - \tau_P) \geq 0$ as a greater cognitive "complexity". It appears that the protection, when $\tau_R \gg \tau_P$, must account for more "experience" in order to achieve the objective of the action; it depends upon a greater amount of lived and retained history, and hence on a greater specificity (individuality) of the living object. So it better participates to the incessant process of individuation, which is a play between the richness of retention and the diversity of possible future trajectories.

Another way to associate a growth of complexity to the growth of $(\tau_R - \tau_P) \geq 0$, is to consider cases where the *global* protection is constant. Then the increase of $(\tau_R - \tau_P) \geq 0$ means that protection is more localized near t_1, with the same global effect. Then this situation is more "complex", since the preparation to the virtual event occurs when it is closer — and the organism must be "quickly ready". In this case, it is easier for it to protend another event t'_1, with t'_1 between t_0 and t_1, since the organism is not yet fully focused on t_1 (the P grows very slowly "for long" and fastly increases only close to t_1). This situation allows the organism to have longer times of correlation: during the early part of these extended protentional activities, it may prepare also for other events.

---

[5] A paramecium manages the movements of about 2,000 cilia during highly complex swimming activities; some of its cilia also serve to direct food towards a "mouth" (opening upon the membrane), by means of very articulate movements.

4.6 Some Complementary Remarks

In this section, we will discuss further aspects associated to protection and retention. These aspects consitute mostly open issues, which could lead to further investigations.

4.6.1 Power Laws and Exponentials

We have chosen here to approach protection and retention by relaxation functions, which are crucially associated to a particular time scale. However, some aspects of biological "memory" are often associated to power laws, which define typically scale-free situations, see for example [Werner, 2010] for cognitive aspects or chapter 2 above for other aspects.

We will next describe four relevant issues, which further justify and allow to develop our approach to biological time.

- The decay of memory, measured in psychophysics, can take power laws forms, see [Werner, 2010]. We should notice that this behavior is associated to relatively long term tendencies; it describes the tail of distributions and not their short term behavior. In general, such distribution have a small scale cutoff, which also allows integration, and in this chapter we are mostly interested in short term effects. Interestingly, [Shinde et al., 2011] experimentally finds that the attentional behavior are typically associated with a specific time scale, whilst situation not associated to attention are scale free (these results are obtained by observing saccadic eye movements). In all cases, our approach is compatible with these empirical results.
- However, the various results on the scale free fluctuations of biological time intervals can lead us to propose that the time constants (τ_R and τ_P) can have non-stationary behavior, and fluctuate in a scale free manner (as the heart's rhythm). In this case we would have local (in time) scale dependent behavior, with temporal properties that vary with time.
- The allometric behavior of this temporal quantities is a crucial issue. When associated with usual "physiological" rhythm, protection and retention logically should follow allometric scaling with exponent $1/4$. However, when associated to neuronal activities the situation can be more difficult to analyze, see section 2.2.
- *In fine*, our default assumption is that at least a part of protentional activity is naturally associated to the temporal structure described in chapter 3. This is also consistent with the idea that the minimal protensive behavior is associated with the iteration of internal rhythms, as a "bet" of the organism on its own physiological stability.

4.6.2 Causality and Analyticity

This subsection is a preliminary analysis, which shows a potential line for future research.

Under certain conditions, strict causality can be formalized in the following way. Consider an input $I(t)$ and an output $G(t)$, and, under the assumptions, in particular, of time translation invariance of the response kernel K, which describes the system, and of superposition of the answer, one obtains that the relation between them has the form:

$$G(t) = \int_{-\infty}^{\infty} I(t')K(t-t')dt' \qquad (4.23)$$

Then, strict causality is the following condition $t' > t \Rightarrow K(t-t') = 0$, which simply means that the future does not influence the output at a given time. This condition can be simplified by simply saying that $\tau < 0 \Rightarrow K(\tau) = 0$

Under this assumption, one can show that we can derive properties of analyticity of the transfer function in Fourier space [Toll, 1956, Bros & Iagolnitzer, 1973]. More precisely we have:

$$I(t) = \frac{1}{\sqrt{2\pi}} \int_{-\infty}^{\infty} \hat{I}(\omega)e^{\iota \omega t} d\omega \qquad (4.24)$$

$$G(t) = \frac{1}{\sqrt{2\pi}} \int_{-\infty}^{\infty} \hat{K}(\omega)\hat{I}(\omega)e^{\iota \omega t} d\omega \qquad (4.25)$$

Then, $\hat{K}(\omega)$ is the real boundary of an analytic function of the strict upper half plane (with $\Im(\tilde{\omega}) > 0$, $\hat{K}(\tilde{\omega})$ is analytic). More precisely, for $\tilde{\omega} = \omega_r + \iota \omega_i$,

$$\hat{K}(\tilde{\omega}) = \hat{K}(\omega_r + \iota\omega_i) = \frac{1}{\sqrt{2\pi}} \int_{-\infty}^{\infty} K(\tau)e^{\iota \omega_r \tau} e^{-\omega_i \tau} d\tau \qquad (4.26)$$

The analycity is then associated to the exponential decrease allowed by the cancellation of $K(\tau)$, for negative τ, and the corresponding decrease of $e^{-\omega_i \tau}$ in the upper half-plane, where $\omega_i > 0$.

Physically, this corresponds to the frequency viewpoint, and Fourier transform, for ω_r, and to Laplace transform and decay viewpoint, for ω_i (the combination is usually called a generalized Laplace transform). Analyticity allows to derive the Kramers-Krönig relation between the real and imaginary part of a response function, see for example section 10.9 of[Sethna, 2006].

The most straightforward analogy with protection and retention is in the form of a quadratic response (where protection is considered as exact, which is conceptually incorrect):

$$\int_{-\infty}^{\infty}\int_{-\infty}^{\infty} P(t,t_0,t_1)I(t_0)I(t_1)dt_0 dt_1 \qquad (4.27)$$

$$= \int_{-\infty}^{t}\int_{t}^{\infty} a_R I(t_0)\exp\left(-\frac{t-t_0}{\tau_R}\right) a_P \exp\left(-\frac{t_1-t}{\tau_P}\right) I(t_1) dt_1 dt_0 \qquad (4.28)$$

This interpretation corresponds to a situation where the values taken at different time points yield distinct protentional activities. Another interpretation is, however, that the whole quantity I is relevant, which gives:

$$\int_{-\infty}^{t} R(t,t_0)I(t_0)dt_0 \times \int_{t}^{\infty} P_v(t,t_1)I(t_1)dt_1 \qquad (4.29)$$

In this case, we have a causal and an anti-causal transfer function, which correspond to analyticity and non-analyticity in different complex half-plane for the transfer functions. If we consider that the product should be a sum (which would be the cumulative effect of retention and protention), then we should have singularities of an unique transfer function in both half planes.

This discussion in terms of transfer function allows us to illustrate our concepts from another viewpoint, which is also used in the study of critical phenomena. Moreover, in this context, we see that this viewpoint allows us to associate protention to singularities of the transfer function.

4.7 Towards Human Cognition. From Trajectory to Space: The Continuity of the Cognitive Phenomena

The continuity of space-time, which the mathematics of continua proposes and structures in a remarkable way, from Euclid to Cantor, follows — and does not precede — the "perceived" continuity of a figure, of a contour or of a trajectory. Euclidean geometry is not a geometry of space, it is only a geometry of figures, with continuous edges, that is of figures made out of continuous lines, constructed by means of ruler and compass and submitted to translations and to rotations. It is much later, with Descartes, that geometry finds its constitutive environment in an abstract space, underlying and independent from the figures which evolve within. The analytic reconstruction of Euclidean geometry will follow, by means of this ideal framework, an algebraico-geometrical continuum, organized in Cartesian coordinates. Then, since Cantor, we have a fantastic reconstruction *by points* of the underlying continuum, a possible reconstruction, though, not an absolute (see [Bell, 1998] for an alternative topos-theoretic approach, with no points).

Let us now try to grasp a possible constitutive path or even a cognitive foundation of this *phenomenal* continuum which is the privileged conceptual and mathematical tool for the intelligibility of space, on the basis of our analysis of retention and of protention.

The recent analysis of the primary cortex (see [Petitot, 2008] for a survey) highlight the role of intracortical synaptic linkages in the perceptual construction of edges and of trajectories. Neurons correlate themselves locally, along "association fields" [Field, 1987, Field et al., 1993] composed of smooth (differentiable) curves that "are grouped together only when alignment fails along particular axes" [Field et al., 1993]. These neurons are sensitive to "directions": that is, they activate when detecting a direction, along a tangent. Then they (pre-)activate other neurons in the association field (they prepare in advance the spike which is not yet fired). This preactivation of associated neurons is, in our view, a component of the protensive activity. Then, neuronal activation follows a specific direction which (re-)constructs the pertinent line [Petitot, 2008].

Thus, the continuity of an edge or of a trajectory is constructed by "gluing" together fragments of the world, in the precise geometrical (differential) sense of gluing. In other words, we *force* a "continuation", that is the unity by continuity of an edge by relating neurons which are pre-associated and are, locally, along particular axes.

Thus, in our view, the phenomenological reconstruction of this continuum is based on the retention and the protection of a *non-existing* line, a trajectory of a physical body. In other words when following, possibly by an ocular saccade, a moving body, a contour, we "integrate", in the mathematical sense, the tangents that are locally associated in the field. The related inertial phenomena of the activation/deactivation of neurons may be one of its constitutive elements, with inertia as a coefficient of protention.

Thus, ocular movements or saccades which follow a moving body, an edge, play a crucial role: a retentive/protensive phenomenon originates also in the muscles enabling the saccades or in the neurons managing them. As we already mentioned, a protentional displacement in the receptor field of the cortical neurons *precede* the saccades [Berthoz, 2002] and the retention/protentional activity seems based on the saccadic systems. The brain prepares itself and anticipates a moving object, of which the movement is perceived following an ocular saccade, or of which the trajectory or edge is perceived by running the eye along or over it. This is, in our view, the keystone of a fundamental protentional activity.

Let's summarize the cognitive and philosophical consequences of our approach. First, it should be clear that, for us, the World is not continuous, nor discrete: it is what it is. Since Newton and Cantor, by specific tools, or, now, in Quantum theories and Topos Theoretic approaches, we mathematically organized it in various ways, possibly over different "backgrounds" from Euclid's or Cantor's continua, [Bell, 1998]. In our view, the phenomenal continuity of trajectories, of an edge, is due to the *retention* of that trajectory, edge, scanned by the eye, which is *"glued"* with the *protention* by the very unit of the cerebral and global physiological activity (the vestibular system, for example, has its own retention and inertia).

In the case of contours, the specific saccades along the direction of movement or towards the extreme of a reconstructed segment (for example in Kanizsa's triangles, see [Petitot, 2008]) stimulate a specific activation in the association field (a specific connection between neurons in the field).

It would then be this "gluing" — a mathematically solid concept (at the center of differential geometry, of which Riemannian geometry is a special case) — that entails the cognitive effect which *imposes* continuity upon the world: the image of the object and of its past position is reassembled (glued by the conjunction of protention and retention) with that of the object and of its expected position or a contour is made continuous even when non existing (as in Kanizsa illusions).

We could indeed imagine that an animal with no fovea (the part of the eye which enables a follow up of a target by a continuous focus), a frog for example, and which takes spaced out snapshots of an object in movement, would not have the impression of a continuous movement in the way in which we, the primates, "see" it.

By measuring relaxation and (pre-)activation times of associated neurons it should be possible to quantify our coefficients in these specific phenomena. Inertial coefficients in particular would yield different values according to the different protentional capacities in different species (frogs for example may have little inertia w.r. to these phenomena, if our understanding above is correct).

So the continuity of a trajectory or of an edge is, in our opinion, the result of a spatio-temporal reassembling of the retentions and protentions that are managed by global neural activity in the presence of a plurality of activities of such type: muscles, vestibular system, ... but also the differentiable continuity of the movement or gesture participates by means of its own play of retention/protention. In short, by a cognitive process of gluing, we attribute continuity to phenomena which are what they are (and which a frog surely sees quite differently). Then, by a remarkable conceptual and mathematical effort having required centuries, we have even come to theorize, continuous abstract lines, surfaces and their edges, first, and then even the continuity of environing space, as the background of these structures. And this in different ways: Euclid, Cantor, Topos Theory today. The phenomenal continuity of lines and background spaces is the consequence, we believe, not the cause of the cognitive/perceptive continuity of the movement and of the gesture, which is grounded on the unity of protention and retention. Of course, in this perspective, the continuity of an edge or of a surface would also be the continuity of a movement: the movement of the saccade or of the hand caressing it, both retained and protended.

Let us note that, in our attempt towards spatialization of time for living phenomena, in this chapter and in chapter 3, — a spatialization/geometrization which, although schematic, should contribute to its intelligibility — we have proceeded, in this section, along the opposite approach: a sort of temporalization of space. Its apparent continuity would be the result of a cognitive activity *on* time, the extended present obtained by retention and protention.

In the previous chapter and in this chapter, we have seen two different, original aspects of biological time. The first is associated with rhythms, physical and internals; the second concerns the abnormal local structure of biological time.

In the following chapter, we will provide some more background on symmetries and symmetry breaking in physics. In chapter 6, we will review physical phase transitions. These two chapters are mostly meant to be a technical introduction to the subsequent chapters. In chapter 7, we will return to biology and, from the point of view of biological time, we will undertake the question of biological historicity and give a specific meaning to it. This should provide a deeper insight on the biological structure of determination we are proposing in this book.

Chapter 5
Symmetry and Symmetry Breakings in Physics

> Since the beginning of physics, symmetry considerations have provided us with an extremely powerful and useful tool in our effort to understand nature. Gradually they have become the backbone of our theoretical formulation of physical laws.
>
> Tsung-Dao Lee

Abstract. Symmetries play a major theoretical role in physics, in particular since the work by E. Noether and H. Weyl in the first half of last century.

We first present a few examples of how symmetries allow to objectivize physical phenomena and then a short survey of conceptual and technical aspects of symmetries and symmetry breakings, beginning by the role played by Galileo's group in the construction of space in modern physics. Then, we provide an account of Noether's theorem, which relates invariant quantities and symmetries. A short, but general classification of symmetry breaking in physics follows. Our purpose is to describe some aspects of spontaneous symmetry breakings, in order to introduce critical transitions in the next chapter, where these breakings apply. Last, we will propose a link between randomness and symmetry breaking.

This brief overview will allow us to draw some preliminary conclusions as well as hints towards our project, in particular concerning the distinctive consequences associated to continuous and discrete symmetries and their breakings. We will also understand, at least partially, why symmetry breaking is associated to specific features, such as singularities, and the loss of "standard" behavior in classical physical dynamics. This will help us to open towards our analysis of the critical singularity of life.

Keywords: symmetry, invariant, symmetry breaking, criticality.

5.1 Introduction

In Physics, objectivity is obtained by the co-constitutive use of experiments and mathematized theories. In order to make further progress towards highly mathematized theories in biology, in particular towards theories of the "living object" or of the organism as a system, it would help first to understand how such a feat was achieved in physics.

Physical theories have very general characteristics in their constitution of objectivity, and in particular in their relationship with mathematics. In order to define space and time, as well as to describe physical objects, physicists ultimately use the notion of symmetry. Physical symmetries are the transformations that do not change the intended physical aspects of a system, in a theory. As we shall see, symmetries allow to define these aspects in a non-arbitrary way.

We do not aim here at an exhaustive review, but we will nevertheless provide a sufficiently complete account of fundamental physico-mathematical results which have global consequences in physics, for the purposes of our work in biology. Thus, this chapter is an introduction to the work in chapter 7, see also [Longo & Montévil, 2011a], and is a preparation to the notion of "extended critical transition", at the core of our approach in biology. We will also provide some background on models used in other chapters, which are related to symmetry breaking.

In other words, we will not develop an exhaustive analysis of the role of symmetries in the foundation of physical theories, which can be found in many books, such as [Van Fraassen, 1989, Bailly & Longo, 2011], but we will provide an overview of this role, with diverse levels of details. we will also focus on some interesting, partly technical aspects relevant for the following chapters.

5.2 Symmetry and Objectivization in Physics

We will first discuss some examples to show how symmetries allow to define physical objects by the use of mathematics. Then and on this basis, we will discuss more conceptually the constitution of physical objectivity, in order to contrast it with biological situations in chapter 7.

5.2.1 Examples

The following very simple examples show how symmetries may be used to define and objectivize physical objects. This will hep to specify for the non physicist what we mean by "theoretical symmetry" and their role in a theory and in an equational determination. Recall that symmetries form a set of transformations that have a group structure; that is 1) the set contains a transformation, called identity, which doesn't change anything, 2) two symmetries applied successively yield a symmetry and 3) a symmetry can be inverted.

5.2 Symmetry and Objectivization in Physics

5.2.1.1 Chemical Concentrations

We will first consider the example of an elementary reaction in chemistry (for the sake of simplicity, we will put temperature and other thermodynamic aspects aside):

$$A + B \longrightarrow D \qquad (5.1)$$

Which has the following dynamic:

$$\frac{d[D]}{dt} = -\frac{d[A]}{dt} = -\frac{d[B]}{dt} = k[A][C] \qquad (5.2)$$

What is it that allows to write such equations?

- Firstly we need to define chemical species: A, B and C. From a molecular viewpoint, their definition requires that all molecules called A behave the same and differently from B and C. This allows to make the theoretical step from a molecule A_1, and a second one, A_2, to two molecules of A; that is to say we can use $A_1 + A_2 = 2A$ and so forth. This assumption is a hypothesis of symmetry by permutation within the set of molecules of species A. Another necessary symmetry is associated to the conservation of matter, which means that only (chemical) transformations occur and there is no creation or annihilation of atoms. This is a time translation symmetry, which leaves the quantity of each kind of atoms invariant. This justifies that the chemical reaction has to be balanced. Both symmetries are needed for equations 5.1 and 5.2 to be meaningful and valid.
- Now, the number of molecules of each species are not sufficient to provide the theoretical determination of the system. Let us assume that the number the reaction occurring is determined in a given volume and small time interval by the molecules which are there. It means in particular that any transformation which does not transform these quantities, does not change the dynamic. For example, the date or the location of an experiment are irrelevant, inasmuch the local properties of the molecules do not change. This defines what is relevant and irrelevant, all irrelevant aspects can be transformed without consequences, which means that they also invariantly undergo symmetry transformations (they define symmetries, for short).
- We assumed above that localization in space is crucial. If the compounds are in two different vials nothing will happen. Therefore, a symmetry of the distribution of the compounds over space (homogeneity) is used both theoretically and experimentally (by mixing) so that the global description by concentrations depicts what happens locally (this can be further justified by the notion of entropy).
- We mentioned in introduction that the symmetries allow to determine the specific trajectory of a given system. In order to obtain this, a further assumption is needed: the number of molecules produced during a very short time is proportional to the number of reagent of each kind. From another point of view, the probability for a molecule of A to react is independent of $[A]$, the number of elements of A, but proportional to $[B]$. As a result, when $[B]$ doubles, everything else

being kept constant, the same dynamic occurs but twice faster. This symmetry relates the states and their changes.

We see on this specific case that symmetry assumptions allow to define the space of description (concentrations) and the specific trajectory inside this space. Therefore, it is at the core of the constitution of the object as such. In particular, they allow to define a system as determined by its state and show how a change of (initial or not) state leads to a transformation of the trajectory.

5.2.1.2 Classical Space

As a second example, we briefly discuss how the intelligibility of (classical) physical space is obtained.

The first point that we have to address is how to define (and quantify) the position of an object in the physical space, conceptually (almost) from scratch. We don't aim here to be exhaustive, but to pinpoints some elementary ingredients that are used to define space.

- First of all, we need to assign quantities to space. In order to do that, we can start from a ruler. A ruler is basically a segment (an edge) that has a symmetry by rotation (which enforces that it is straight). With it, we can define a unit of length. Now, the distance between two points that are far away can be assessed as the minimal number of time this ruler has to be put in order to draw a *continuous* curve from the first point to the second (or equivalently a straight curve, that is to say with a symmetry by rotation). In order to measure smaller distances, fractions of this ruler can be obtained with a compass, which provide fractional lengths.

    The fundamental operation that is performed here is to compare space and a given object, then to assimilate this part of space to this object: they have the same length and the same direction. By transitivity the same is then said between two different parts of space.

    Note that this kind of manipulations are not sufficient in order to generate a mathematical continuum. In order to understand the physical need for it, it is however sufficient to assume that the measurement of a length with increasing precision always ultimately converge to a number.
- Now, in order to get the position of an object, more than just distances is required. We need a reference system. To do so, we need a first point, corresponding to an object and which will be the origin, and three other points, defined by other objects, which define the directions of space. These points can be taken at a distance of one "unit" from the origin. In order to define the position on the basis of this points, the last useful concept is that of orthogonality. A definition of orthogonal crossing is that both lines are axis of symmetry for the figure of the crossing.

    Now the position of any point can be defined as the algebraic (oriented) distance between the origin and the orthogonal projection of this point along each axis.

- It is crucial to note that the quantities obtained this way are largely arbitrary: they depend on an arbitrary reference system. Another observer will choose another reference system and therefore obtain completely different results. There is two ways to overcome this kind of difficulty. Either by enforcing a conventional unique reference system (which is, by the way, the solution used for physical units like meters), but this choice would be arbitrary.

  The second solution is to make both results compatible. In order to do that, one uses the first reference system to describe the second one, that is the length of the second unit of measure and the position of the origin and of the three objects defining the axes. Assuming that space is linear (euclidean), there is a unique linear function which transform the objects of reference of the first system to the ones of the second. Coordinates in the second reference system have this function implicitly applied from the viewpoint of the first reference system. Therefore, the correct coordinates in the latter can be obtained by applying the inverse of this function on the coordinate in the second reference system.

  Here, the arbitrariness of one viewpoint is not overcome by an absolute viewpoint, but by the unique way to go from one viewpoint to another, and backwards. In other terms these viewpoints are symmetric.

- Now, fully meaningful physical aspects should not depend on a specific viewpoint that is chosen arbitrarily. Thus, fundamental equations should have the same form in any reference system. In other terms they have to be symmetric by the transformation of one reference system to another. That does not only means that the trajectory of an object determined by the equations in one reference frame should be the same than the one determined in another reference frame, modulo the change of reference frame. it also means that two scientists who don't share any specific reference point should be able to work on the same equations. Reciprocally, if a fundamental equation depends on specific point in space, then it means that that space has an origin of some sort that has a real physical meaning (or that the physicist is making a mistake).

With this elementary examples, we understand already that physics proceeds by the identification of the non-identical, whether this is different parts of space or different molecules. This operation is however performed only when the identification is compatible with the intended understanding. For example, the assimilation of positions with coordinates is valid inasmuch we keep in mind the above relativity of these coordinates. Similarly, the definition of concentrations is valid only inasmuch it is a relevant quantity to describe the chemical reaction. In a heterogeneous system, like a cell, concentration may not be the relevant quantity.

5.2.2 General Discussion

Now, we will go back to full fledged physical theories and discuss things with more generality. Galileo's theory provides a simple and historical example of the role of symmetries that we want to evidentiate. For scholastic physics, the speed at which a body falls is proportional to the space traveled. Galileo instead proposed that it is proportional to the time of the fall and that it is independent of the nature (including the mass) of the empirical object considered (Galileo's law of gravitation). This idea together with the "principle of inertia" has been a starting point for the constitution of *space* and *time* in classical physics. More precisely, as a consequence of the analysis of inertia and gravitation, the geometry of space and time was later described by the Galilean group[1].

A change of this symmetry group, for example by adopting the Poincaré group, can lead to a very different physical situation, in the case of this example we get the physic of special relativity involving massive conceptual and physically meaningful changes, starting with a maximum speed: the speed of light[2]. We will explain now what motivated this change.

The general "principle of relativity" states that the fundamental laws of physics, as equational forms and constants, do not depend on the reference system, as we hinted in the example of space; they are actually obtained as invariants with respect to the change of reference system. Now, a specific speed (the speed of electromagnetic waves in the void) appears in Maxwell's equations of electromagnetism. To overcome this contradiction, Einstein modified Galileo's group and by this transformed this speed into an invariant of mechanics. This change is not benign; among other things it turns lengths and durations into relative quantities (they depend on the reference system)[3].

Since the 1920s, due to Noether's theorems (see below), symmetries allow the mathematical intelligibility of key physical invariant quantities. For example, symmetries by time translations are associated with energy-conservation, and symmetries by space rotations are associated with the conservation of angular momentum. Thus, conservation laws and symmetries are in a deep mathematical relation (see section 5.3 below for technical details). Consequently, the various *properties* that define an object (mass, charge, etc.) or its *states* (energy, momentum, angular momentum, etc.) are associated to specific symmetries which allow these quantities to be defined. Depending on the theory adopted, this conceptualization allowed to understand why certain quantities are conserved or not: for example, there is no

---

[1] Galileo's symmetry group is the group of transformations that allows to transform a Galilean space-time reference system into another. It is interesting to notice that Galileo measured time by heartbeat, a biological rhythm; the subsequent theoretical and more "physical" measurement of time were precisely provided by classical mechanics, his invention.

[2] The symmetry group of a Euclidean space is the Euclidean group of automorphisms, while Poincaré's group corresponds to the automorphisms that define Minkowski's spaces [Catoni et al., 2008].

[3] The equation for a change of reference frame in special relativity is the following (for an object moving along the x axis).

$$t' = \gamma(t - vx/c^2) \quad (5.3)$$

$$x' = \gamma(x - vt) \qquad \text{with } \gamma = 1/\sqrt{1 - v^2/c^2} \quad (5.4)$$

5.2 Symmetry and Objectivization in Physics

local energy conservation in general relativity. This explicit reference to the theory adopted is required in order to produce "scientific objectivity", *independently* of the arbitrary choices made by the observer, such as the choice of time origin, the unit of measure, etc, but *relatively* to the intended theory, see the examples above. Thus, we say that symmetries provide the "objective determination" in physics [Bailly & Longo, 2011].

Because of the role and implications of symmetries, most contemporary challenges in theoretical physics lead to the search for the right symmetries and symmetry changes. The work aiming at the unification of relativistic and quantum theories is largely focusing on this. In moving from physics to biology we suggest here to apply a similar approach. That is, we plan to discuss the theoretical symmetries and their changes that are relevant for biology.

The symmetries that define physical properties allow to understand the physical object as *generic*, which means first that any two objects that have the same properties can be considered as physically *identical*; in a sense, they are symmetric or invariant (interchangeable) in experiments and in pertinent mathematical frameworks (typically, the equations describing movement). For example, for Galileo, all objects behave the same way in the case of free fall, regardless of their nature. Moreover, symmetries allow the use of the *geodesic principle*, whereby the local determination of trajectories leads to the determination of the full trajectory of physical objects, through conservation laws. For example, the local conservation of the "tangent" (the momentum) of movement, typically yields the global "optimal" behavior of a moving object; that is, it shows that it must go along a geodesic.

By this, in classical or relativistic mechanics, a trajectory is uniquely given and fully deterministic (formally determined). In quantum mechanics the evolution of the state or wave function (roughly, a *probability distribution*) is fully deterministic as well — and determined by Schrödinger's equation — while measurement follows this probability distribution (and here appears the indeterministic nature of quantum mechanics). In conclusion, by symmetries, the trajectory of a generic classical or quantum physical "object" corresponds to a critical path, a geodesic. In this precise sense, physical trajectories are *specific*, while, as we said, physical objects are *generic*.

It is crucial to understand that the specificity of the trajectory is needed to objectivize a given generic object: objects can be defined arbitrarily, it is the ability of a definition to frame the behaviour of an object which makes this definition an objectivation of the intended phenomena. Reciprocally, a specific trajectory only makes sense with respect to a generic object as abstractly defined and practically measured in a space of description (phase space).

To better understand the problem of *general* mathematical theorizing in biology, let's further analyze how, in physics, a concrete problem is turned into robust models and mathematics. To begin with, physicists try to define the right theoretical framework and the relevant physical quantities (properties and states) which are constituted by proper symmetries. As a result, typically, a mathematical framework is obtained, where one can consider a generic object. In classical mechanics, this is given by a pointwise object of mass m, speed v and position x, where these quantities are generic. Now, a generic object will follow a specific trajectory determined by its invariants and obtained by calculus. A measurement is then made on the experimental object to determine the quantities necessary to specify where this object is in this mathematical framework, namely, what is its mass, initial position and speed. And finally, what specific trajectory will the object follow ... at least approximately. In classical or relativistic physics, to a given measurement will correspond generic objects localized near the measurement due to the limited accuracy of this measurement. The measurement can have, in principle, an arbitrary high precision but never perfect. In quantum mechanics, as we recalled above, the equational determination (Schrödinger's equation) yields the dynamics of a probability law[4].

In classical dynamics, we face a well-known problem: the specific trajectories, which mathematically start within the same measurement interval, can either remain close to each other or disperse very rapidly. The linear situation corresponds to the first case, whereas the second situation is called "sensitive to initial conditions" (or chaotic, according to various definitions). Note that even the latter situation may lead to the definition of new invariants associated to the dynamics: the attractors that have a precise geometrical structure[5]. In both cases, these trajectories have robust properties with respect to the measurement.

In quantum physics, the situation is more complex because measurement leads to non-deterministic behavior. Yet, when approximations on the state function are performed, it leads to usually stable, robust statistics.

In all cases, a "robust measurement" means invariant or approximately invariant in a definite mathematical sense, as this concerns the measurement of states and properties of generic objects along specific trajectories (in quantum mechanics, as given by Schrödinger equation). Thus, we can finally say that all generic objects, which lead to a specific measurement, *behave* in the same way or approximately so, sometimes statistically. Note that this property of robustness, allowed by the genericity of the object, is mandatory for the whole framework to be relevant, as proof in natural sciences comes *in fine* from the empirical. We insist that both genericity for objects and specificity for trajectories (geodesics) are mathematically understood on the ground of symmetries.

In conclusion, in the broadest sense, symmetries are at the foundation of physics, allowing objective definitions of space and time and the constitution of objects and trajectories. In their genericity (an interchange symmetry), these objects follow specific trajectories (a consequence of symmetries in the equations and in their operatorial treatment). Genericity of objects and specificity of trajectories are mutually dependant. Finally, these trajectories are associated with empirical results that are robust with respect to measurement.

---

[4] In quantum physics, "objects" do not follow trajectories in ordinary space-time, but they do it in a suitable, very abstract space, a Hilbert space (a space of mathematical functions typically). Yet, as we said, what "evolves" is a probability distribution.

[5] The notion of attractor proposed, typically, a new object of knowledge, in physics, by the departure from linearity.

5.3 Noether's Theorem

Noether's theorem [Noether, 1918] has been first formulated to understand why energy, which is conserved in classical mechanics, is not locally conserved in general relativity (see [Byers, 1999] for a historical account[6]). In order to face this problem, the theorem proves a relationship between *continuous* symmetries, with a finite number of infinitesimal generators, and invariants. The relevance of this relation is far more general in theoretical physics than the particular yet fundamental issue of general relativity and energy: it concerns the relation between conservation properties and symmetries in the equations. Indeed, this approach has been adapted to various frameworks, and, as a result, is one of the standard tools used in theoretical physics, both in order to understand the consequences of original theoretical propositions and to investigate specific phenomena.

We will now formulate this result in the relatively simple context of classical mechanics. Let us consider a classical system, governed by its Lagrangian $\mathscr{L}(t,q_1,\dot{q}_1,\ldots,q_n,\dot{q}_n)$, where $\dot{q}_i = \frac{dq_i}{dt}$. The state is then described as a $2n$ dimensional vector, we will write such a state as $\bar{q}(t)$. In order to simplify the notations, we will write in the following:

$$\mathscr{L}(t,q_1,\dot{q}_1,\ldots,q_n,\dot{q}_n) = \mathscr{L}(t,q_i,\dot{q}_i)$$

Then we define the *action*. The variational principle applies on this quantity (that is to say, this quantity is stationary for the actual trajectory, see appendix A.2):

$$\mathscr{S} = \int_{t_1}^{t_2} \mathscr{L}(t,q_i(t),\dot{q}_i(t))dt \tag{5.5}$$

Theorem 5.1 (Noether, classical Lagrangian mechanics). *For the above Lagrangian, let us suppose that $\mathscr{S}$ is preserved under the action of a one parameter continuous group $\mathfrak{G}$ with infinitesimal generator $v = \tau\frac{\partial}{\partial t} + \phi_\alpha\frac{\partial}{\partial q_\alpha} + \psi_\alpha\frac{\partial}{\partial \dot{q}_\alpha}$. Then, the quantity:*

$$C = \tau\mathscr{L} + \frac{\partial \mathscr{L}}{\partial \dot{q}_\alpha}(\phi_\alpha - \dot{q}_\alpha\tau) \tag{5.6}$$

is an invariant of the dynamic (that is to say a quantity with a null derivative with respect to time).

Example 5.1 (Space translations). We consider space translations along vector u: $x \mapsto x + \varepsilon u$. For now, we have left the spatial structure of the system implicit (it was handled by the possible structure of the Lagrangian). To show the effect of space translations, we will assume that space has 3 dimensions and that the coordinates are ordered in the following way: q_1, q_2, q_3 are the three coordinates of a material point,

---

[6] Our presentation contains some (mild use of) mathematics in the appendix A.2, yet the reader, who would prefer to skip it, may just try to grasp the introduction and the statement of the theorem that we present here. This may help to understand the role of symmetries in physics and the conceptual transition we aim at in biology.

then q_4, q_5, q_6 is the coordinate of a second point, etc. We will then write ji ($j = 1, 2$ or 3) the various coordinates that correspond to the j direction. Reciprocally, for any coordinate α, the corresponding direction is $j = \alpha[i]$ (α modulo i)

The generator of the group is then $v = 0\frac{\partial}{\partial t} + u_{\alpha[3]} \frac{\partial}{\partial q_\alpha} + 0\frac{\partial}{\partial \dot{q}_\alpha}$. Applying Noether's theorem, we get the following conserved quantity:

$$C = 0\mathscr{L} + \frac{\partial \mathscr{L}}{\partial \dot{q}_\alpha}(u_{\alpha[3]} - \dot{q}_\alpha 0) = u_{\alpha[3]} p_\alpha \qquad (5.7)$$

Thus, the symmetry of the action with respect to space translations along vector u leads to the conservation of the momentum in the u direction.

Then, it is straightforward, by taking a basis of $\mathbb{R}^3$, that the invariance of the Lagrangian by space translations leads to the conservation of momenta.

Example 5.2 (Time translation). We consider time translations $t \mapsto t + \varepsilon$. The generator is then $v = 1\frac{\partial}{\partial t} + 0\frac{\partial}{\partial q_\alpha} + 0\frac{\partial}{\partial \dot{q}_\alpha}$. Applying Noether's theorem, we get the following conserved quantity:

$$C = 1\mathscr{L} + \frac{\partial \mathscr{L}}{\partial \dot{q}_\alpha}(0 - \dot{q}_\alpha 1) = \mathscr{L} - \dot{q}_\alpha \frac{\partial \mathscr{L}}{\partial \dot{q}_\alpha} = -\mathscr{H} \qquad (5.8)$$

Thus, the symmetry by time translations lead to the conservation of energy[7].

A couple of remarks are worth making now. First the continuity of the group under consideration is crucial. Indeed, it is this continuity that allows to preserve, infinitesimal step by infinitesimal step, the conserved quantity that we obtain as a result of the theorem. Note also that the boundary conditions are also crucial: it is with respect to these boundary conditions that the quantities are conserved. In the Lagrangian classical mechanics, this corresponds typically to the initial conditions.

From a theoretical perspective, the classical Lagrangian (or other formalisms) is not sufficient *per se*. The further description of the Lagrangian, besides the symmetries of space and time, is in general given by other symmetries, which are the symmetries of the fields handling the interactions. The field theoretic version, given in appendix A.2.2 allows to understand such fields (here, without quantification). The quantum field theory's analog of Noether's theorem, the Ward–Takahashi identity, shows also that quantities such as the electric charge (in quantum electrodynamics) are associated to symmetries. It is noteworthy then to understand that this approach captures an intuitive aspect of the otherwise fuzzy notion of matter (consistent with the notion of charge used in the field theoretic version for the conserved quantity): the conservation property, which leads to conservative flows of the quantities considered and is made possible by the continuity of the transformations.

Note also that, in spite of the wide generality of Lagrangian formalism, the situation described above is associated to specific symmetries, which correspond here to the symplectic nature of the geometry of the phase space. It appears, for example, in equation A.17 and leads to nontrivial relationships.

---

[7] Let us recall that, if the energy of a system is $\mathscr{H} = \frac{1}{2} p_\alpha \dot{q}_\alpha + V(q_i)$ then the Lagrangian is $\mathscr{L} = \frac{1}{2} p_\alpha \dot{q}_\alpha - V(q_i)$

5.4 Typology of Symmetry Breakings

After this digging into symmetries, we present in this section some details on the main different types of symmetry breakings encountered in physics. As already hinted, our approach in biology gives a major role to symmetry breakings. We thus provide a classification on the basis of the way in which a symmetry is broken. This classification is based on [Holstein, 2000] and on further considerations in [Strocchi, 2005].

EFFECTIVE SYMMETRY BREAKING. It corresponds to a situation where the symmetry considered is, in fact, an approximation and where there are relevant perturbations. As a result, in particular, the corresponding Noether charges are not exactly conserved (they have a divergence corresponding to the perturbation). A slightly different version is also called explicit symmetry breaking. It corresponds to a situation where terms are added (external fields for example), which explicitly breaks the symmetries of the Lagrangian. In both point of view, the ground state is changed because of the change of Lagrangian (or Hamiltonian) symmetries.

SPONTANEOUS SYMMETRY BREAKING (WEAK VERSION). This situation, also called degenerate ground state, occurs when a problem (typically the potential or the Hamiltonian) has a symmetry that the state cannot have, because the symmetric state does not minimize the potential, for example. The resulting state then breaks this symmetry.

A typical example of such situations is the Mexican hat potential $V(\phi) = 2a|\phi|^2 + |\phi|^4$, illustrated in figure 5.1. This potential indeed has a rotational symmetry, since it only depends of the absolute value of ϕ. Its minimums obey the necessary condition $a|\phi| + |\phi|^3 = 0$, which yields $|\phi| = 0$ or $|\phi| = \sqrt{-a}$ when $a < 0$. As a result, when $a < 0$, the minimums correspond to the second equation, $|\phi| = \sqrt{-a}$. In this case, all minimums are generated by the rotations of an arbitrarily chosen minimum. For example, in dimension 1 we have two states $\phi_0 = \sqrt{-a}$ and $\phi_0 = -\sqrt{-a}$; in dimension 2, we obtain a circle of radius $\sqrt{-a}$, as illustrated in figure 5.1; in dimension 3, we obtain a sphere; ...

SPONTANEOUS SYMMETRY BREAKING (STRONG VERSION). In [Strocchi, 2005], a sharp distinction is made between two kinds of spontaneous symmetry breaking, in the above sense. Indeed, depending on the situation, the different states corresponding to a broken symmetry can be *physically* changed into each other or not. The latter case will be qualified as a (strong) spontaneous symmetry breaking. This case is obtained typically when there is an infinite number of degrees of freedom. In this context, the breaking of the symmetry can lead to different behaviours, which live in different domains of the phase space. These domains are formalized as Hilbert space sectors in [Strocchi, 2005], which are in particular stable with respect to the time evolution and finite fluctuations.

The line of reasoning behind this notion is that of physically possible transformations are essentially localized (i.e. of finite size) so that different configurations at infinity cannot be physically interchanged. They can, however, be unstable with respect to symmetry transformations (which can be a symmetry of

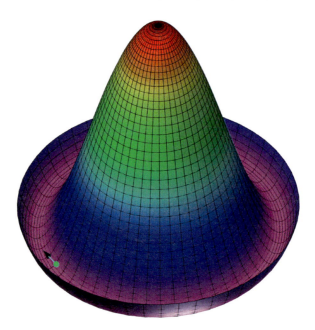

Fig. 5.1 *Mexican hat and (weak) spontaneous symmetry breaking.* The Mexican hat potential is a classic example of a spontaneous breaking of a continuous symmetry. The potential has a rotational symmetry; however, its minimum is not at the center but at a distance, thus on a circle because of the rotational symmetry. As result, minimizing the potential for a pointwise object leads to breaking the symmetry (green point), but energetically free fluctuations can occur (cyan arrow). In particular, if the system is subject to additive Gaussian noise and has also some dissipation, then the system will follow an additive Gaussian Brownian motion along the symmetric circle.

the Hamiltonian), which is then spontaneously broken. In particular this situation leads to an alternative version of Noether theorem, where the continuous group involved has to be a symmetry of a given sector (if it is not the case, the symmetry cannot physically generate the conserved quantity). We will briefly discuss it in the next chapter in terms of ergodicity breaking associated to a symmetry breaking phase transition.

ANOMALOUS SYMMETRY BREAKING. This case corresponds, in the framework of quantum field theory, to a situation where there is a symmetry of the classical action that the quantum field does not manage to maintain after any regularization. Depending on the nature of the symmetry considered, it can be theoretically acceptable (chiral anomaly for example) or not (gauge anomaly). More generally, the terminology is used when breaking a symmetry but making this breaking (for example, the order parameter) tend toward zero does not lead again to the symmetric situation. In other word, situations when tending to the limit is not the same as at the limit, the situation is then not continuous.

5.4.1 Goldstone Theorem

Here, we will present a result that shows that a spontaneously broken continuous symmetry leads to long-range fluctuations (in the language of condensed matter physics), called Goldstone modes. In the language of particle physics, this result shows that we obtain massless particles called Goldstone bosons.

The basic idea behind Goldstone theorem can be explained quite simply. Let us consider a system with a continuous symmetry group G. For example, a lattice where the state of each spin is described by an angle θ_i, in which case G is the set of rotations changing all the θ_i simultaneously by the same angle. Since the transformations of G are symmetries, they do not change the energy of any part of the system. The symmetry breaking means that the state of the system does not have this symmetry (there is a privileged direction θ_b). Then, a small variation following the (former) symmetry group (an infinitesimal rotation of all the spins in our example) does not need energy at the first order, because of the corresponding symmetry of the potential. In particle physics, one say that this kind of field structure constitutes a massless particle, because there is no energetic "resistance" to it (the Φ^2 term vanishes).

These fluctuations are peculiar: they basically involve the whole system (in our example, all spins are *simultaneously* rotated) which also correspond to the lack of an Hamiltonian term which would impose a particular length (see the dimensional analysis in 6.2.1.5). Conceptually, this kind of situation describes a form of stiffness: in a fluctuation, almost all the system is transformed simultaneously and in the same way. The most practical example is the case of a crystal; the Goldstone theorem explains its mechanical rigidity, see for instance [Sethna, 2006] for more illustrations.

This result is somewhat more difficult than Noether theorem because it applies to situation with an infinite number of degree of freedom. Moreover the theorem has specific hypotheses, which are not met in certain physically very relevant situations.

We use the classical version given in [Strocchi, 2005]. As discussed briefly in section 5.4, we are in a situation of strong spontaneous symmetry breaking, where a distinction should be made between different physically valid sectors of the phase space. These sectors, called Hilbert Space Sectors, are stable with respect to additive finite perturbations and the time evolution.

Theorem 5.2 (Goldstone). *Let us consider G a finite continuous group of symmetries of a model. We suppose that this symmetry group is spontaneously broken to G_{Φ_0}, by a solution Φ_0 which is an absolute minimum of the potential. Let us write H_{Φ_0} the sector of this solution. Then for any infinitesimal generator v^α of G_{Φ_0} with $v^\alpha \neq = 0$ we have:*

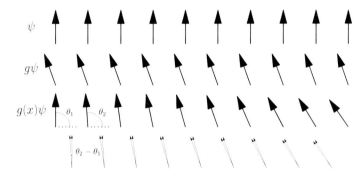

Fig. 5.2 *Goldstone modes.* The states ψ and $g\psi$ are symmetric and thus have the same energy. A long-range fluctuation $g(x)\psi$ which leads to very small local angle discrepancies needs a small amount of energy. As a result, at the limit, an infinite size fluctuation cost no energy. Typical examples are the vibration of a crystal, spin waves in a magnet, ...

- scattering configurations (behaviour for infinite time) associated to solutions in H_{Φ_0}, which are solutions of the free wave equation (they are the Goldstone modes).
- solutions in H_{Φ_0} that in arbitrary large finite volume (in space and time) behave like free waves (Goldstone-like solutions).

Beyond its predictive aspect, this very general result allows to hypothesize an underlying spontaneous symmetry breaking when long wavelength modes are observed. Because of its generality, Goldtstone theorem should be considered when discussing pointwise (continous) symmetry breaking and their properties, including afar from the critical point.

As a general comment, we see that (finite) continuous symmetries are peculiar. They allow, by continuity, to propagate invariant quantities (Noether's theorem) and generate massless fluctuations (Goldstone theorem).

5.5 Symmetries Breakings and Randomness

In this section, we will propose a preliminary and informal remark, which may turn out to be important when stressing the role of randomness in biology. Namely, we propose that in all existing physical theories each random event is associated to a symmetry change, as symmetry breaking and reconstruction (new symmetries may be formed).

A random event is an event where the knowledge about a system at a given time does not entail its future description; thus, the event is unpredictable, relatively to the intended theory. In physics though, the description before the event determines the complete list of possible outcomes: thus, what is unpredictable is a *numerical value* in a pre-given space of observables — modulo some finer considerations as the ones we will discuss in chapter 8 as for QM and statistical physics, on the dimensions of the phase space, typically. Moreover, in most physical cases, the theory provides

5.5 Symmetries Breakings and Randomness

a metric or, more generally a measure (of probabilities or other measures) which determines the observed statistics. That one may say: random or unpredictable, but not so much, as we know a probability distribution. Kolmogorov's axiomatic system for probabilities works this way and provides probabilities for the outcomes. The various physical cases can be understood and compared in terms of symmetry breaking.

QUANTUM MECHANICS: the unitarity of the quantum evolution is broken at measurement, which amounts to say that the quantum state space assumes privileged directions (a symmetry breaking).

CLASSICAL DYNAMICS: the intended phase space contains the set of all possibilities. Elements of this set are symmetric inasmuch they are possible, moreover the associated probabilities are usually given by an assumption of symmetry; for example, the sides of a dice or the regions of the phase space with the same energy. These symmetries are broken at the occurrence of the intended event, which singles out a result.

ALGORITHMIC CONCURRENCY THEORY: the theory gives the possibilities (a finite list) but does not provide, a priori, probabilities for them. These may be added if the physical event forcing a choice is known (but computer scientists, in programming theory and practice, usually "do not care" — this is the terminology they use, see [Longo et al., 2010]). The point here is to have a programm that works as intended in all cases.

We thus associated a random event to a symmetry breaking, in the main physical frames (plus one of linguistic nature: networks' programming). In each case, we have several possible outcomes that have therefore a symmetrical role, possibly measured by different probabilities. After the random event, however, one of the "formerly possible" situations is singled out as the actual result. Therefore, each random event that fits this description is based on a symmetry breaking, which can take different yet precise mathematical forms, depending in particular on the probability theory involved (or lack thereof). In this line of reasoning, randomness leads to a distinction between the possible and the actual result ("possible" and "result" have different specific meaning depending on the theory). The symmetry is then between the different possibilities and this symmetry is broken when one result is obtained out of them. This scheme of randomness seems quite general to us.

In the case where probabilities are defined, let us better specify the symmetries we are talking about. Let's consider an event X which can be either A, with probability p or B with probability $1 - p$. Then, we can consider $f_A(X) = 1/p$ if $X = A$ else $f_A(X) = 0$ and $f_B(X) = 1/(1-p)$ if $X = B$ else $f_B(X) = 0$. we see then that f_A and f_B have exactly the same expectancy. It is precisely this symmetry that experimenters try to show empirically, and that legitimates the probability values.

Note that random situations following this pattern define a before and an after, with the symmetry breaking separating them. This before and after may be epistemic (for chaotic dynamics) or intrinsic (quantum mechanics).

Let's now review more closely, in a schematic way, how the random events are associated to symmetry breakings:

QUANTUM MECHANICS: the projection of the state vector (measurement); non-commutativity of measurement; tunneling effects; creation of a particle ....

CLASSICAL DYNAMICS: bifurcations, for example, correspond typically to symmetric solutions for periodic orbits. Note that in classical mechanics, "the knowledge of the system at a given time" involve the measurement (inasmuch it limits the access to the state) and not only the state itself.

CRITICAL TRANSITIONS: the point-wise symmetry change lead to a "choice" of specific directions (the orientation of a magnet, the spatial orientation of a crystal, etc.). The specific directions taken are associated to fluctuations. Also, the multi-scale configuration at the critical point is random, and fluctuating.

THERMODYNAMICS: the arrow of time (entropy production). This case is peculiar as randomness and symmetry breaking are not associated to an event but to the microscopic description. The time reversal symmetry is broken at the thermodynamic limit.

ALGORITHMIC CONCURRENCY: The choice of one of the possible computational paths (backtracking is impossible).

If this list is exhaustive, as it seems, it is fair to say that random events, in physics, are correlated to symmetry breakings (and programming follows this pattern). Note that among all these cases, one doesn't fit completely in our qualitative discussion and has a more complex structure: the case of thermodynamics. Indeed, from a purely macroscopic viewpoint, there is no particular form of randomness associated to the theory, and provided that a trajectory is defined, it will be deterministic (except for critical transitions or similar situations). Randomness appears at the microscopic level, either understood as chaotic classical dynamics or classical probabilities (statistical mechanics). Both then correspond to the analysis of their respective categories. However, this doesn't explain the arrow of time, which is the interesting symmetry breaking in this situation. The evolution of a thermodynamic system is towards a symmetrization of the system, since it tends towards the macroscopic state to which correspond the greatest number of microscopic states (they are symmetric from a macroscopic viewpoint). That is, it tends towards the greatest entropy compatible with other constraints. In this case, randomness is used to explain a dispersion in the microscopic phase space, therefore it is a process of symmetrization which breaks the time symmetry but doesn't create macroscopic randomness. Macroscopic randomness may still appear if there are different minimums for the relevant thermodynamic potential, like in phase transitions.

All these symmetry changes and the associated random events happen within the intended phase space, or, in other words, within the set of possibilities given by the intended physical theory. The challenge we will be facing in biology (see 8), is that randomness manifests itself at the very level of the observables. Critical transitions are the closest physical phenomenon to the needs of the theoretical investigation in biology and we will discuss them in next chapter.

Chapter 6
Critical Phase Transitions

Abstract. In this chapter, we first present the basic principles of a relatively new area of physics, the analysis of critical phase transitions and more generally the theory of criticality. Then, we will introduce some mathematical methods that set the physics of criticality on robust grounds. We will also discuss briefly some variation on the theme of criticality such as self-organized criticality, often used in theoretical approaches to biology. Following the current analyses in physics, we present them here as point-wise transitions, with respect to (usually) one control parameter. This will constitute an opening towards the approach to criticality in the following chapters seen as an "extended" phenomenon in biology, that, we propose, is ranging on a non-trivial interval of definition.

Keywords: criticality, phase transition, fluctuations, symmetry breaking.

6.1 Symmetry Breakings and Criticality in Physics

In previous chapter, we observed that symmetries are at the core of the definition of physical objects and of their properties, states and *in fine* theoretical determination. Thus, a *symmetry change* (that is, the breaking of some symmetries and/or the formation of new ones) means a qualitative change of the object considered, or even a change of physical object, the object being understood as co-constituted by the theory (and its symmetries) and empirical observations. For example, a well-known research project in cosmology considers a single force to have existed in the universe right after the big bang. Then, the four fundamental forces are assumed to appeared by successive symmetry breakings, whereby some transformations, which were symmetries, did not preserve the object invariance anymore[1]. In other words, with the cooling of the universe, the system moved to a smaller symmetry group and qualitatively different forces appeared.

---

[1] The Higgs mechanism is an example of this situation; in this case, the symmetry breaking in the abstract electroweak space (for example) leads to different masses for bosons and as a consequence to a very short range for weak interaction and a long range for electromagnetism.

Closer to the scale of biology, materials like water or iron are able to show different properties in different situations. Depending on the temperature and pressure, water may be a solid, a liquid, or a gas. When liquid, there is no privileged direction (the system is isotropic, that is to say symmetric by rotations), whereas ice has a crystalline structure with spatially periodic patterns. This implies that the system is no longer symmetric by continuous rotations: it has a few privileged directions determined by its crystalline structure which means that it has a smaller symmetry group (less spatial symmetries). Similarly, iron can have paramagnetic behaviour (the system is not magnetized) or ferromagnetic behaviour (it is magnetized). In most cases, one can distinguish a more disordered phase at high temperature, where entropy dominates, and a more ordered phase, where energy dominates. These situations can be discriminated by an *order parameter* which is 0 in the disordered phase and different from 0 in the ordered phase[2]. The physics of criticality focuses on this kind of phase transitions, i.e., state changes (see [Toulouse et al., 1977, Binney et al., 1992], two classics among many texts on this matter, also quoted below).

To sum the situation up, one may understand certain changes of states in terms of spontaneous symmetry breaking, see section 5.5. That is, one may give a group of transformations G, which yields the symmetries of the initial situation, the disordered phase, and no longer of the subsequent one, the ordered phase, where the group of transformations that are symmetries changes. Here, order precisely means that a specific "direction" has been "chosen" and do not respect the initial symmetry. The direction(s) of a crystal or the poles of a magnet. The symmetry of the second phase is restricted to a subgroup G_1 of G. In general, then, the definition of the macroscopic state requires the introduction of a supplementary variable, called the order parameter, which gives the strength and direction of the symmetry breaking, for example the global field of a magnet. Typically, this parameter takes uniformly the value 0 in the disordered phase, which is symmetric by G and is different of 0 in the ordered phase, which is no longer symmetric by G.

Now, in physics, the change of state, or *phase transition*, occurs always mathematically at a point of the parameters' space. This point is called the *critical point*. This point-wise nature stems, among other reasons, from the Boolean nature of the validity of symmetries: the symmetry of the ordered phase is macroscopically valid, before the phase transition, then suddenly it gets broken. The critical point is associated with a sudden change of behavior due to the change of symmetry between the disordered and ordered phase. At the critical point, between these two states, a peculiar behavior appears which is due to the singularities of the state functions. For example, the order parameter is non-analytic because it goes from a *constant* 0 to a finite quantity, *by a finite change*. More technically, the *critical point* represents a singularity in the partition function describing the system[3]. In the case of iron's

---

[2] Here, order means low entropy, or less symmetries, and disorder means high entropy, and more symmetries, where symmetries are counted in terms of macroscopically equivalent "microstates".

[3] This function is non-analytic at the critical point, which means that the usual Taylor expansions, linearizations or higher order approximations do not actually provide an increasing approximation, we will go back to this point in section 6.2.1.2.

6.1 Symmetry Breakings and Criticality in Physics

paramagnetic-ferromagnetic transition, this allows to deduce the divergence of some physical observables, such as magnetic susceptibility. The reader should note that this form of *singularity*, associated to the specific shape of the divergence towards the infinite quantities at the critical *point*, is a core aspect of physical criticality.

This peculiar situation leads to a very characteristic behavior at the critical point [Jensen, 1998]:

1. Correlation lengths tends to infinity, and follow a power law, as for continuous phase transitions. That is, for a vector x and an observable N, if we note by $< . >_r$ the average over point r in space, then $< N(r+x)N(r) >_r - < N(r) >_r^2 \sim \|x\|^\alpha$. This is associated with fluctuations at all scales leading in particular to the failure of mean field approaches (in these approaches, the value of an observable at a point is given by the mean value in its neighborhood or, more precisely, its mathematical distribution is uniform). We will come back to this below, in section 6.2.1.5.
2. Critical slowing down: the time of return to equilibrium of the system after a perturbation tends to infinity, see section 6.2.3.
3. Scale invariance: the system has the same behavior at all scales. This property leads to a fractal geometry and means that the system has a specific symmetry (scale invariance itself). The method of renormalization is used to evidentiate this behaviour.
4. The determination of the system is global and no longer local.

These properties are the key motivations for the biological interest of this field of physics. The global "coherence structure" that is usually formed at critical transitions provides a possible understanding, or at least, an analogy for the unity of an organism. In the current terminology, it gives a form of "global determination or causation": the global structure has a causal role on its "components". Also, power laws, so frequent in biology as we discussed in chapter 2, are ubiquitous in critical phenomena. They are mathematically well-behaved functions (e. g. $f(x) = x^\alpha$) with respect to the change of scale [typically, λ is the scale change in $f(\lambda x) = \lambda^\alpha f(x) = \lambda^\alpha x^\alpha$, a power law in α], and they yield *scale symmetries*, see section 2.1.1 or more the mathematically oriented annex A.1. In our example, scale changes just multiply the function f by a constant λ^α. Now, a power law depends on a quantity without physical dimension (α in the notation above). In critical transitions, these quantities are called *critical exponents* and describe how the change of scale occurs. In our terminology, they describe the properties due to the objective determination of a phase transition because they are the invariants associated with the scale symmetry.

Specific analytic methods, called renormalization methods, are used to theoretically establish these quantities [Delamotte, 2004]. These methods, which we will discuss more extensively in section 6.2.2 below for the interested reader, consist in analyzing how scale changes transform a model representing the system, and this analysis is made "asymptotically" toward large scales. One may deduce the critical exponents from the mathematical operator representing the change of scale. The key point is that a variety of models ultimately lead to the same quantities, which means that they have the same behavior at macroscopic scales. Thus, they can be grouped in so-called *universality classes*. This analytic feature is confirmed empirically, both by the robustness of its results for a given critical point and more stunningly by the fact that very different physical systems happen to undergo the same sort of phase transitions; that is, they are associated with the same critical exponents, thus with the same scale symmetries. Finally, there exist fluctuations at all scales, which means, in particular, that small perturbations can lead to very large fluctuations.

The physical situation is that in the disordered phase the system is macroscopically symmetric. Still it has coherent fluctuations limited in size. Now, the spatial extent of fluctuations increases when the system gets closer to the critical point. At this point coherent fluctuations encompass the whole system (whatever its size, that is they to infinity) and when crossing this point, the system is finally dominated by a macroscopic order that arose from these fluctuations (metaphorically we could say that the system is stuck in a fluctuation so large that it gets beyond infinity, which does not prevent smaller fluctuation to occur).

To conclude, the transition through a specific point of the parameters' space, i.e., a transition between two very different kinds of behavior is associated in physics to a change of symmetries. At this point, the system has very peculiar properties and symmetries. Symmetries by dilation (by a coefficient λ as above) yield a scale invariance. This latter invariance is associated to a global determination of the system and the formation of a "structure of coherence". As observed above, this allows to describe a global determination of local phenomena and a unity that goes beyond the idea of understanding the global complexity of a system as the sum of many local behaviors or by adding more and more local, possibly hidden, variables. For some physical phenomena this theoretical framework presents peculiar and very relevant forms of "systemic unity".

A well-known example is given by an Ising spin lattice. This is a mathematical lattice, where the field value of each element of the lattice, Φ_i, can either be 1 or -1. We assume that the Hamiltonian, which determines the system, is symmetric by the *global* permutation of these two directions (namely by the transformation g, with $g(1) = -1$ and $g(-1) = 1$). Physically, this assumption also means that these two signs, at least at this scale, are arbitrary labels (the resulting algebraic structure, that is, the fact that there is two different signs, however, is not arbitrary). Thus, if the system is disordered, it follows globally the symmetry group[4] $\mathcal{O}(1)$, which corresponds to this permutation and identity. However, if this symmetry is broken for the global system, we have a macroscopic distinction between these two orientations. This distinction can be taken into account by the order parameter $\Phi = \langle \Phi_i \rangle$, the mean of the spins. When $\Phi = 0$, the symmetry is macroscopically respected. However, when $\Phi \neq 0$, Φ is either positive or negative, and the symmetry is broken by the state of the system, and replaced by the trivial subgroup which contains

---

[4] $\mathcal{O}(n)$ is the group of symmetry of the sphere in dimension n.

only the identity, {1}. We will provide a further, more mathematical, insight into the mathematics of Ising models, in section 6.2.1.2 below.

As for now, observe that, in this case, we have considered a discrete symmetry. The simplest extension to a continuous symmetry is the $\mathcal{O}(2)$ symmetry, which leads to an order parameter that can be written in the form $\rho e^{\iota\theta}$ and models a circle. In this case, the Hamiltonian does not depend on a global multiplication by a phase factor $e^{\iota\theta}$. In general, the cases of $\mathcal{O}(n)$ symmetries are especially widely studied. Their breaking corresponds to the usual "choice" of an oriented direction in an n dimensional real space, and appears, for example, to the magnetization of magnets. We can, however, easily provide alternative examples. The liquid crystals in LCD screens are nematic, which leads to the determination of a direction but not of an orientation (an angle from 0 to 180°).

This theoretical framework has also been applied to a possible understanding of life phenomena. In the next chapter, we will survey some of these approaches but we will mostly look at biology through a different insight into the symmetries of biological criticality. In particular, we will still refer to critical transitions, but we will describe them as given on a interval or a (dense) subset of an interval and possibly controlled by several parameters. The main aim of these chapter is to prepare the following one on the analysis of biological processes in terms of extended critical transitions: an organism is always in a transition, beginning with each individual mitosis. We claim that this transition is "critical" in the terms specified here. As for now, we thus discuss more closely the mathematical approaches to criticality in physics. Also a qualitative understanding of the technical issues in this chapter may suffice to follow the main discussion in the next one.

6.2 Renormalization and Scale Symmetry in Critical Transitions

We present here the basic mathematical ideas and techniques that have been used in the current analyses of critical phase transitions.

6.2.1 Landau Theory

Landau theory is a remarkably straightforward approach of second order phase transitions[5]. This theory, in spite of its simplicity, provides information on the basic properties of such situations. In particular, Landau theory allows a first account of the relation between symmetry breaking and phase transitions and gives a first account, valid only in certain cases, of the singular behavior at the critical point.

---

[5] In Ehrenfest classification, a transition of order n has its first discontinuity for a derivative of order n of the free energy. In the modern classification, first order transition involve a latent heat, while second order transition do not.

6.2.1.1 Statistical Mechanics

We will first recall basic aspects of statistical mechanics. The point here is mainly to show the general relation between fluctuations, on the one side, and susceptibilities and heat capacities on the other side.

The key function that governs a systems behavior is the partition function, see for example [Sethna, 2006]:

$$Z = \sum_{s \in \text{states}} \exp(-\beta E_s) \tag{6.1}$$

where E_s is the energy of the state s, and β is roughly the inverse of the temperature: $\beta = \frac{1}{k_b T}$. The probability to obtain a state s is then:

$$P(s) = \frac{\exp(-\beta E_s)}{Z} \tag{6.2}$$

The theoretical crucial point, here, is that the states with the same energy have the same probabilities (they are symmetric from a statistical perspective). This is usually justified by an assumption of ergodicity of the microscopic trajectories. As a result, statistical mechanics is related to the geometry of the phase space in high dimensions.

Note that the temperature just "tunes" the impact of energy on the probabilities distribution. At temperature ∞ ($\beta = 0$), all states have the same probability which amounts to ignore the consequences of energy. On the contrary, when tending to 0 temperature ($\beta \to \infty$), only the minimum energy states are possible (the probability of the other states vanish), so that energy directly determines the possible states. At and near 0 temperature (but not exclusively), other approaches are needed because fluctuations in energy are engendered by quantum uncertainty; this leads to quantum phase transitions, which are a very active and promising research field. We will not describe them here, see for example [Belitz et al., 2005].

The equivalents to the usual thermodynamic functions are then obtained from the partition function. We provide next the *extensive* version of these quantities; however, at the thermodynamic limit, it is their intensive version that is actually relevant. It can be obtained by dividing the results by n, the number of elementary objects or in an experimentally more practical way, by V the volume or a mass m.:

ENERGY. The mean energy is obtained as follows:

$$-\frac{\partial \ln Z}{\partial \beta} = -\frac{\partial Z}{\partial \beta} \frac{1}{Z} \tag{6.3}$$

$$= -\sum_{s \in \text{states}} -E_s \exp(-\beta E_s) \frac{1}{Z} \tag{6.4}$$

$$= \sum_{s \in \text{states}} P(s) E_s \tag{6.5}$$

$$= \langle E \rangle \tag{6.6}$$

6.2 Renormalization and Scale Symmetry in Critical Transitions

ENERGY FLUCTUATIONS. A similar reasoning leads to:

$$\langle (\Delta E)^2 \rangle = \langle (E - \langle E \rangle)^2 \rangle = -\frac{\partial^2 \ln Z}{\partial \beta^2} \tag{6.7}$$

HEAT CAPACITY. Heat capacity is the energy needed to increase the temperature of the system.

$$C = \frac{\partial \langle E \rangle}{\partial T} = \frac{1}{k_B T^2} \langle (\Delta E)^2 \rangle \tag{6.8}$$

This relation is particularly interesting; it relates the heat capacity to the energy fluctuations.

HELMHOLTZ FREE ENERGY.

$$\mathscr{F} = -\frac{\ln Z}{\beta} (= \langle E \rangle - TS) \tag{6.9}$$

ENTROPY.

$$S = -\frac{\partial \mathscr{F}}{\partial T} \tag{6.10}$$

If one introduces another parameter B (which can be a vector), for example associated to an external field, then a similar reasoning allows to define the corresponding susceptibility. This can be introduced by $E_s = -\sum_i B.\Phi_{s,i} + E'_s$.

MAGNETIZATION. We consider the magnetization along the u direction of the field ($u.B$ denotes the scalar product along direction u); the reasoning is then the same than the one for the mean energy:

$$\langle \Phi.u \rangle (\beta, B) = -\frac{\partial \mathscr{F}}{\partial (B.u)} \tag{6.11}$$

SUSCEPTIBILITY. The susceptibility corresponds to the ability of an external field to to change the internal field of the object.

$$\chi_i(\beta) = \frac{\partial \langle \Phi.u \rangle}{\partial (B.u)} = -\frac{\partial^2 \mathscr{F}}{\partial (B.u)^2} \tag{6.12}$$

FIELD FLUCTUATIONS. In this case we will detail the calculus which relates the susceptibility to the fluctuations.

$$-\frac{\partial^2 \mathcal{F}}{\partial (B.u)^2} \quad (6.13)$$

$$= \frac{1}{\beta} \sum_{s \in \text{states}} \beta \sum_i u.\Phi_{s,i} \frac{\partial \exp(-\beta(E_s - \sum_i B.\Phi_{s,i}))/Z}{\partial (B.u)} \quad (6.14)$$

$$= \sum_{s \in \text{states}} \beta \left(\sum_i u.\Phi_{s,i}\right)^2 \frac{\exp(-\beta(E_s - \sum_i B.\Phi_{s,i}))}{Z} \quad (6.15)$$

$$- \sum_{r,s \in \text{states}} \beta \left(\sum_i u.\Phi_{s,i}\right) \left(\sum_i u.\Phi_{r,i}\right) \frac{\exp(-\beta(E_s - \sum_i B.\Phi_{s,i} + E_r - \sum_i B.\Phi_{r,i}))}{Z^2} \quad (6.16)$$

$$= \beta \sum_{s \in \text{states}} \left(\sum_i u.\Phi_{s,i}\right)^2 P(s) - \sum_{s \in \text{states}} \sum_{r \in \text{states}} \left(\sum_i u.\Phi_{s,i}\right) P(s) \left(\sum_i u.\Phi_{r,i}\right) P(r) \quad (6.17)$$

$$= \beta \left\langle \left(\sum_i u.\Phi_{s,i}\right)^2 \right\rangle - \beta \left\langle \sum_i u.\Phi_{s,i} \right\rangle^2 \quad (6.18)$$

$$\chi_u(\beta, B) = \beta \left\langle \left(\Delta \sum_i u.\Phi_{s,i}\right)^2 \right\rangle \quad (6.19)$$

6.2.1.2 Some General Remarks on Phase Transitions

As already mentioned, a very classical example for the study of phase transition is the Ising model, which is described by the Hamiltonian:

$$E_s = \sum_{i<j} 2J_{ij}\phi_i\phi_j + \sum B_j\phi_j \quad (6.20)$$

where $\Phi = \pm 1$. Simulations of this model with coupling constant of 1 for neighbours in 2 dimension are given in figure 6.1. This model is particularly important because exact solutions are known in dimension 1 and 2 (without external field).

It is straightforward that the partition function is *analytic* when there is a finite number of spin, in the Ising model. More generally, this result also holds directly, even when an infinite number of states is generated by a symmetry of the Hamiltonian, with a finite number of elementary objects. The general point is then that finite-size systems have analytic partition function, and thus all the above (pre-)thermodynamic functions are analytic (the only "dangerous" operation is the application of the logarithm, but it is applied on a finite sum of exponentials with the same positive coefficient). By contrast, the order parameter, typically one of the $\Phi.u$, has a non-analytic behavior, since,it is different from 0 in the ordered phase and has

6.2 Renormalization and Scale Symmetry in Critical Transitions

non-isolated zeros (the whole ordered phase). As a result, *no phase transition occurs in finite-size systems*, and the transition occurs only *at the thermodynamic limit*[6].

In general, the partition function (and all thermodynamic functions) are analytic except at the critical point. When mathematically extended to the whole complex plane (abstract complex temperatures), the non-analyticity propagates to a set of non-isolated, dense zeros, which touches the real line at the critical point. This structure is a vertical line for isotropic Ising model (Lee-Yang approach), but can be far more complex in the case of anisotropic magnets [van Saarloos & Kurtze, 1984]. This point of view can be used to characterize finite-size phase transition (in the isotropic case), by the density of complex zeros.

Another crucial aspect is that a spontaneous symmetry breaking, in this context, leads to an *ergodicity breaking*. Indeed, without entering in many details, Boltzmann approach to entropy assumes that the most probable states are those that have a maximum entropy (with given constraints); however, the symmetry breaking confines the system in a given direction of the order parameter, whilst other states are symmetric in particular with respect to their entropy (following the above description). This confinement of the microscopic phase space is precisely the situation that we called strong symmetry breaking in section 5.4, and that is described by [Strocchi, 2005]. The description in terms of ergodicity breaking means that the mean time evolution (which is constrained by a given breaking of the symmetry) is not equivalent to the mean on the phase space (which is not constrained by it).

6.2.1.3 Landau Theory

Landau theory allows to straightforwardly approach phase transitions from their thermodynamic description. This approach is based on an almost self-contradictory point of view, since it assumes that a limited development makes sense whilst the considered function is typically non-analytic at the critical point. We will further discuss this question below. Another limitation is analyzed in [Sen, 2010], which tries to distinguish the situation where this approach can be used at least as a heuristic and when it cannot, depending on the properties of the symmetry breaking.

We will suppose that the system depends only on the temperature T and the external field B. We will assume that $B = 0$. We will also suppose that the thermodynamic functions are symmetric with respect to the change of sign of Φ, so that the thermodynamic potential verifies $\mathscr{F}(-\Phi) = \mathscr{F}(\Phi)$ ($\mathscr{O}(1)$ symmetry). In the following, all quantities are considered intensive. Considering this symmetry, the limited expansion in Φ, near $\Phi = 0$, of the thermodynamic potential only has even terms (the odd terms would break the symmetry of the free energy):

$$\mathscr{F} = \mathscr{F}_0 + \frac{1}{2}a_T t \Phi^2 + \frac{1}{4}c_T \Phi^4 \qquad a_T = a \qquad t = \frac{T - T_c}{T_c} \qquad c_T = c > 0 \qquad (6.21)$$

---

[6] At least in the usual sense. For small finite-size systems, other definitions can be used. This field is an active research topic.

The equilibrium state Φ_0 of the system minimizes the thermodynamic potential $\mathscr{F}$, we consider then:

$$\left(\frac{\partial \mathscr{F}}{\partial \Phi}\right)_{\Phi_0} = at\Phi_0 + c\Phi_0^3 = 0 \tag{6.22}$$

This yields:

$$\Phi_0 = \pm\sqrt{\frac{a}{c}}(-t)^{1/2} \qquad T < T_c \tag{6.23}$$

$$\Phi_0 = 0 \qquad T > T_c \tag{6.24}$$

We will now consider that an external field can be present, which will allow us to compute the zero field susceptibility, and therefore the fluctuations of the order parameter.

$$\mathscr{F} = \mathscr{F}_0 - B\Phi + \frac{1}{2}at\Phi^2 + \frac{1}{4}c\Phi^4 \tag{6.25}$$

Thus we have at equilibrium:

$$\left(\frac{\partial \mathscr{F}}{\partial \Phi}\right)_{\Phi_0} = -B + at\Phi_0 + c\Phi_0^3 = 0 \tag{6.26}$$

We differentiate this equality with respect to B at zero external field.

$$0 = -1 + at\frac{\partial \Phi_0}{\partial B} + 3c\frac{\partial \Phi_0}{\partial B}\Phi_0^2 \tag{6.27}$$

$$\frac{\partial \Phi_0}{\partial B} = \frac{1}{at + 3c\Phi_0^2} \tag{6.28}$$

This leads to:

$$\chi(T,0) = \beta\langle\Phi^2\rangle - \beta\langle\Phi\rangle^2 = \begin{matrix}\frac{1}{2a}(-t)^{-1} & T < T_c \\ \frac{1}{a}(t)^{-1} & T > T_c\end{matrix} \tag{6.29}$$

Thus, the susceptibility and the fluctuations diverge when we are getting closer to the critical point.

These results, the divergence of the derivative of the order parameter and of the fluctuations, are at odd with the hypotheses of regularity that we assumed in order to perform the limited expansion, at the first step of this model. Physically, this corresponds also to the issue of only approaching the system by its macroscopic properties, namely by the macroscopic value of the order parameter. Moreover, it is also not obvious that the dependencies on T of the different functions, in particular $\mathscr{F}_0$, are regular (because of the fluctuations at the critical point).

This approach is related to the microscopic description by the mean field theory, which simply assumes that the interaction of one element with a second one, Φ_i say,

6.2 Renormalization and Scale Symmetry in Critical Transitions

can be considered as an interaction with the mean value of this quantity $\langle \Phi_j \rangle$. This approach, by elementary means (*in fine* a consistency equation), allows to predict possible phase transitions from the definition of the Hamiltonian, and then to find some of their properties. However, the arguments we discussed above for Landau's approach are still relevant: this approach does not allow to take the contribution of the fluctuations into account. As a result, its validity is limited and incorrect predictions occur, such as missing phase transitions or wrong behaviour near critical points.

6.2.1.4 First Approximation of the Fluctuations

In order to take the fluctuations into account, we consider that Φ is a function of the position r, which is simply written $\Phi(r)$ but leads also to the consideration of a free energy density $\mathscr{F}'$. Then, we limit ourselves to the second derivatives and use the symmetries of the the system, which leads to:

$$\mathscr{F}' = \mathscr{F}'_0 - B(r)\Phi(r) + \frac{1}{2}at\Phi^2(r) + \frac{1}{4}c\Phi^4(r) + g\left(\frac{\partial\Phi(r)}{\partial r}\right)^2 \quad (6.30)$$

Here, we have assumed that the mixed partial derivatives are irrelevant and that the system is isotropic. Moreover, $g > 0$, so that the situation without fluctuations is favored (they give a lower potential).

As usual, we look at the derivative of the free energy with respect to Φ, which is now a function, at the equilibrium point:

$$\left(\frac{\partial \mathscr{F}'}{\partial \Phi}\right)_{\Phi_0} = -B(r) + at\Phi(r) + c\Phi^3(r) + 2g\nabla^2\Phi(r) = 0 \quad (6.31)$$

Then we differentiate it with respect to the value of the external field at another position $B(r_2)$. We use the generalized susceptibility $\chi_T(r, r_2) = \frac{\delta\Phi(r)}{\delta B(r_2)}$

$$-\delta(r - r_2) + \left(at + c3\Phi^2(r) + 2g\nabla^2\right)\chi_T(r, r_2) = 0 \quad (6.32)$$

The generalized susceptibility is proportional to the correlation function (the computation is similar to the one for the susceptibility and it relation with fluctuations), and we assume translation invariance so that:

$$\left(at + c3\Phi^2(r) + 2g\nabla^2\right) G(r - r_2) = k_b T \delta(r - r_2) \quad (6.33)$$

Assuming $\Phi(r) \simeq \Phi_0$, we obtain:

$$\left(\frac{1}{\xi^2(t)} - \nabla^2\right) G(r - r_2) = \frac{k_b T}{2g} \delta(r - r_2) \quad (6.34)$$

with:

$$\xi(t) = \begin{cases} \sqrt{\frac{g}{2a}}(-t)^{-1/2} & T < T_c \\ \sqrt{\frac{g}{a}}t^{-1/2} & T > T_c \end{cases} \qquad (6.35)$$

We can solve the latter by Fourier transform. It leads to:

$$G(\|r\|) \propto \frac{e^{-\frac{\|r\|}{\xi(t)}}}{\|r\|^{d-2}} \qquad (6.36)$$

The quantity $\xi(t)$ is thus the correlation length of the system. As expected, it diverges at the critical point. More details on this derivation can be found in [Altland & Simons, 2006, Schulte-Frohlinde & Kleinert, 2001]. Technically, this derivation is the first perturbative correction to the Landau theory. Note, that the Φ^4 terms is not taken into account in the fluctuation (except through the equilibrium value)

6.2.1.5 Ginzburg Criterion

The Ginzburg criterion tests the consistency of the Landau theory by verifying that the fluctuations determined in this framework are not strong enough to break its validity. We will follow the reasoning in [Als-Nielsen & Birgeneau, 1977], a more extensive discussion can be found in[Schulte-Frohlinde & Kleinert, 2001], both in the Landau Framework and in the renormalization framework, which allows to provide a better determination of these fluctuations.

The criterion checks what quantitatively determines the value of Φ over the magnitude of the fluctuations, that is an appropriate volume Ω.

$$(\delta\Phi)^2_\Omega \ll \Phi^2_\Omega \qquad (6.37)$$

We will write $t = \frac{T-T_c}{T_c}$ in the following. In order to assess the strength of the fluctuations, the correct volume is given by the characteristic length $\xi(t)$, so that $\Omega = \omega_\xi$. This region is indeed the region where the spins are correlated, so that we have a local order that may depart from the global mean.

$$(\delta\Phi)^2_{\Omega_\xi} \ll \Phi^2_{\Omega_\xi} \qquad (6.38)$$

$$N(\Omega_{\xi(t)})\chi(t)k_bT \ll N(\Omega_\xi)^2 \Phi_0^2(t) \qquad (6.39)$$

$$\chi(t)k_bT \ll N(\Omega_\xi)\Phi_0^2(t) \qquad (6.40)$$

we use d, the dimension of space

$$\chi(t)k_bT \ll \xi(t)^d \Phi_0(t)^2 \qquad (6.41)$$

We use equations 6.35, 6.29 and 6.23 so that:

6.2 Renormalization and Scale Symmetry in Critical Transitions

$$\frac{1}{ta}k_b T \ll \left(\frac{g}{a}(t)^{-1}\right)^{d/2} \frac{a}{c}(-t) \tag{6.42}$$

We are only interested by the behavior at the critical point so for small t

$$1 \ll t^{\frac{4-d}{2}} \tag{6.43}$$

We thus observe that the criterion is met for $d > 4$ and not met for $d < 4$. $d = 4$ is then called the critical dimension. For $d > 4$, the Landau approach is thus consistent, whilst it is inconsistent for $d < 4$. Note that we have made some important assumptions, in particular on the isotropy of the system. Anisotropy can change the situation [Als-Nielsen & Birgeneau, 1977]. We want to emphasize that the consistency of Landau theory is a limited result (other problems may arise); on the opposite its inconsistency for $d < 4$ is a definitive answer (modulo the other hypotheses on the system).

A qualitative and heuristic approach can also be done by dimensional analysis, on the basis of equation 6.30.

$$\int \frac{1}{2} at \Phi^2(r) + \frac{1}{4} c \Phi^4(r) + g \left(\frac{\partial \Phi(r)}{\partial r}\right)^2 dr \tag{6.44}$$

To show the internal structure of this object, we divide by g

$$\int a't \Phi^2(r) + c' \Phi^4(r) + \left(\frac{\partial \Phi(r)}{\partial r}\right)^2 dr \tag{6.45}$$

We can now perform a dimensional analysis with respect to space, assuming this quantity has no spatial dimension.

$$\int \left(\frac{\partial \Phi(r)}{\partial r}\right)^2 dr \rightarrow [\Phi] = L^{\frac{2-d}{2}} \tag{6.46}$$

$$\int a't \Phi^2(r) dr \rightarrow [a't] = L^{-2} \tag{6.47}$$

$$\int c' \Phi^4(r) dr \rightarrow [c'] = L^{d-4} \tag{6.48}$$

- If we set $a't = 0$ and $c' = 0$, we have what is called a free massless field. In this case, the equation is *space-scale invariant*, in association with the exponent $\frac{2-d}{2}$.
- With $a' \neq 0$ and $c' = 0$, we have a free massive field. The above scale invariance is broken by the Φ^2 term. We get then that the characteristic length of the system is proportional to $\sqrt{\frac{1}{a't}}$, which is consistent with the result in equation 6.35. The term $a't$, however, vanishes at the critical point, and the corresponding length diverges.

Looking the system from a greater distance means that the lengths shrink. We thus see that the "mass" term a' gets larger in the process. This means simply that

when we are looking at the system from a long distance, the correlations vanish (except at the critical point).
- With $a' \neq 0$ and $c' \neq 0$, the Φ^4 term also breaks the (naive) scale invariance, when $d \neq 4$.

  Using the same reasoning than above, the coupling "constant" c' gets larger when $d < 4$. Which means that it is not limited to local interactions. On the contrary, when, $d > 4$, this coupling constant vanishes at large scales, which justifies the fact that this term (and the corresponding fluctuations) can be neglected at large scales. This also means that the system does not generate more and more relevant interactions.

We understand also that, for $d < 4$, there is an instability of the behavior of the system associated to the Φ^4 term (at the critical point), since this term increases exponentially with the scale.

As a general conclusion, we can say that the crucial point for the Ginzburg criterion is the interplay between the local terms (derivation) and global terms (integration). The latter leads to the dependence on the spacial dimension of the system.

6.2.2 Some Aspect of Renormalization

In order to go beyond the Landau theory and the perturbative approach that can be used to estimate the fluctuations, an original approach is needed. This approach is called renormalization and, can be seen as a (highly) refined and quantitative version of the dimensional analysis we have just performed.

The basic idea behind renormalization is that, when more and more interactions are relevant (typically when $d < 4$ in the above case) and when the system fluctuates strongly, we can nevertheless understand its behavior by looking how its equational description (the Hamiltonian, or the partition function, but it can also be an evolution function, over time) changes when we integrate a part of the contributions to this description. This could be seen as a reduction of the number of degree of freedom of the system, but this alone would be pointless because the number of degree of freedom is infinite. The point is then more rigorously to observe the change of the equational form, its stability, and crucially its parameters (the coupling constants typically).

Renormalization then succeeds when this operation allows to obtain an equational stability with vanishing quantities (coupling constants), which are then irrelevant, and non-vanishing quantities, which are relevant and correspond to the asymptotic scaling behavior of the system. Here, we will focus on statistical mechanics, but renormalization is also crucial in quantum field theory.

Before going further, let us discuss briefly the situation in quantum field theory. In this theory, Divergences occur in particular with respect to small space-scales (equivalently high energy). The conceptual question is then how can "long" range interactions be approached whilst we cannot compute them (integrate them) from arbitrarily small length scales up. Renormalization allows to answer this question by saying that the equational forms stay the same when we integrate the couplings over

6.2 Renormalization and Scale Symmetry in Critical Transitions 151

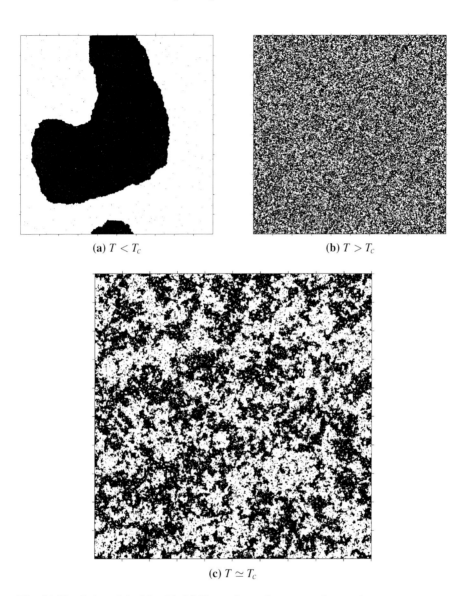

Fig. 6.1 *Simulation of the Ising Model.* From a large distance, we have an homogeneous gray block for $T > T_c$. For $T < T_c$, here, the white spins dominate the system, and the system is ordered. Near the critical point, we have, for all sizes, clusters where the different signs of spin dominate, leading to a fractal distribution. Note that the simulation near the critical point needs an important amount of iterations because the relaxation time diverges, see section 6.2.3.

a larger set of scales, and we mostly need to change the coupling constants by doing so (this is of course a highly simplified discussion). The key here is the stability of the equations with scale changes, this stability removes the need of an explanation at a fundamental scale, as the equations remains coherent from the bottom up. Renormalizability of a quantum field theory is then crucial because its failure leads to the appearance of new parameters when we extend the scale domain of integration, which is then incoherent because the going from arbitrary scales to a given scale would lead to an arbitrary large quantity of parameters for the same scale. The result is then that we need to know at which scale we should start (and what happens at this scale). A crucial example to this issue is quantum gravitation, which typically leads to non-renormalizable theories [Zinn-Justin, 2007], and leads to an attempt of a theorization at a fundamental scale, namely string theory. The other three fundamental interactions have, however, been formalized as renormalizable theory in the standard model.

We will now describe the principle of renormalization in the statistical mechanics context. One should note that there are different versions of this method. Typically, the renormalization can be performed in real space or in momentum space, after a Fourier transform. Moreover, renormalization can start from the upper critical dimension, where the mean field approximation is valid (4, in the situation discussed above) and approach the other dimensions $4 - \varepsilon$ as an expansion over ε, or to be performed more directly, ...

6.2.2.1 Principle of Renormalization

Here, we will follow the presentation of the basic steps of the method which is given in [Lesne, 2003]. Another general account can be found in [Fisher, 1998]; a technical presentation is also given in [Zinn-Justin, 2007].

We first consider a small scale a, corresponding, for example, to the resolution of the measurement apparatus or to an estimated minimal scale of the system (when such a scale is known). This scale corresponds to the *microscopic* description of the system. We will also consider a *macroscopic* scale of observation, L. We will work on a family of models $\mathcal{M}$, indexed by the scale at which they describe the same physical system $\mathcal{S}$. These models have a subjective flavor, since they usually are highly simplified versions of the physical situation[7].

The reductionist stance is not relevant, since the choice of a minimum scale a determines the sub-systems of $\mathcal{S}$ that we regard as elementary, and this choice is arbitrary and can even be based on the resolution of the measurement apparatus. These elementary parts can be described by a small number of quantities, noted s.

---

[7] In the case of quantum field theory, the precise description at a smallest scale is not even available. Some physicists think that this description is epistemologically necessary on the basis of a reductionist stance. However, such a description is not technically necessary, precisely because of the renormalization method which we are exposing here, in the context of statistical mechanics — see, for example, the Ising model of magnets: the magnetic interactions are only local, and the states have only binary possibilities, [Zinn-Justin, 2007]. see also [Longo et al., 2012c] for a qualitative discussion.

6.2 Renormalization and Scale Symmetry in Critical Transitions

These sub-systems typically have a linear extension of magnitude a and will be described as points in the model $\mathcal{M}_a$. In this model, the state of $\mathcal{S}$ will be described by a *configuration* $\bar{s} \equiv (s_1, \ldots, s_N)$, which provides the states of all elementary components. In the example of lattices that we described above, one has $s_i = \Phi_i$. Typically, $N = \left(\frac{L}{a}\right)^d$ in dimension d. We can then define for a model $\mathcal{M}_a$:

- The *phase space* $\mathcal{E} = \{\bar{s}\}$, which is the set of possible configurations (and thus depends on a and N).
- The *evolution or structure function* describing the system (depending on the static or dynamic nature of the problem): $F(\bar{s})$. It is thus a function on $\mathcal{E}$, which determines the behavior of the system (statistical weights in the static case, evolution rules in the non-equilibrium case, probability transitions, ...). $F \in \mathcal{F}$ depends on the scale. In the statistical mechanics framework, F is usually a partition function (and thus indirectly the Hamiltonian).

The renormalization methods are based on three basic operations: a decimation, scale changes and parameter transformations. The principle of renormalization is then to iterate a correct combination of these operations. This combination is called the *transformation of renormalization*. The core idea is then to analyze the effect of this transformation and more precisely of its iterations in the space of models.

- The *decimation* is a procedure which handles the effects of a change of resolution $a \to ka$ on the configurations $\bar{s} \in \mathcal{E}$. This operation is performed by grouping the elementary constituents of scale a (*coarse-graining*) and is represented formally by a function $\bar{s}' = T_k(\bar{s})$.

    This operation allows to reduce the number of degrees of freedom of the system by a factor k^d in dimension d (or equivalently, with a constant N, to increase linearly the scope of observation by a factor k). By performing this transformation, we lose all information on scales smaller than ka. Since T_k is chosen partly arbitrarily, one should try to keep as much information on the smaller scales as possible. This can be obtained by focusing on the elements that can be considered qualitatively as crucial.
- The *scale changes* do not change the minimal scales of the model, the aim of this operation is to keep the same phase space and to highlight the properties of self-similarity. In order to do so, one determines a family (usually unique) $(k, k^{\alpha_2}, \ldots, k^{\alpha_n})_{k \geq 1}$, which verifies the following property: when multiplied to the parameters of the models (including the spatial scale) they lead to a nontrivial limit when $k \to \infty$.

    These exponents can be chosen *a posteriori* in certain cases.
- The *effective parameters* are used to replace the former parameter in order to describe the same physical system in spite of the former transformations. We transform then F in $R_k(F)$ so that $R_k(F)$ describe the statistics or the evolution of $T_k(\bar{s})$. R_k is called the *renormalization operator* and acts on the $\mathcal{F}$. It is crucial here that (a part of) the structure of the interactions is taken into account in this step.

The notion of *covariance* is then essential, because it refers to the conditions on the transformations of the various aspects of the model for it to describe the same physical object. It comes into play, for instance, with respect to the consideration of the systems symmetries.

A particularly strong notion is then the notion of *invariance by renormalization*, which can be written as: $R_k(F^*) = F^*$. It corresponds, therefore, to a fixed point by renormalization, and to an exact scale invariance (asymptotically): the system has a property of *self-similarity*. This notion is very powerful because of the following reasons:

- All models which converge towards the same fixed point by renormalization have the same properties for large scales. This consideration leads to the notion of *universality class*.
- The characteristic scale ξ^*, associated to a fixed point is either 0 or infinite because $k\xi^* = \xi^*$. In the first case, the fixed point corresponds to a limit situation, where there is no coupling between the (sufficiently renormalized) components of the system seen at large scales. In the second case, we have a critical situation.
- In the case of a critical phenomenon, the analysis of R_k near the fixed point F^* determines the asymptotic scaling laws of the different situations which converge towards this fixed point. The critical exponents are, in particular, determined by the eigen values provided by the linearization of R_k in the neighbourhood of F^*.

The latter point leads to the notion of *class of universality*, which regroup models having the same behavior by renormalization. This allows to distinguish among variants of a model, what changes lead to *relevant* or *irrelevant* contributions. Overall it is the relatively strong robustness of classes of university that makes this approach especially powerful, as it allows to show that even simple models can provide a genuine account of the critical behaviour.

An elementary example is the (nearest neighbor) Ising model. We will not develop the corresponding calculation (which are not particularly difficult in dimension 1 or 2). Then, the assumption of a scale invariant minimum of F leads to a unique parameter $K(k)$ with $K(a) = \beta J$. Then, we obtain two degenerate fixed point: $K_0^* = 0$ and $K_1^* = \infty$. They can be interpreted straightforwardly: the first fixed point corresponds to a situation with $\beta = 0$ ($T = \infty$) and, therefore, to no coupling between the different elements of the system. On the contrary, the second fixed point corresponds to $T = 0$, and, therefore, to an ordered situation. Physically, this means that since in both cases the correlations have a limited range, corresponding to the different temperatures; the large scale equational form is not concerned with these scale limited aspects, and the situation collapses to one of these fixed points. In dimension 2, we have a critical point at a finite temperature, which separate the basin of attraction of the two aforementioned fixed points.

It should be clear that the "philosophy" of renormalization basically departs for standard reductionism, as there is no privileged level, there is no reduction to presumed "elementary and simple" theoretical entities at the bottom of reality. The cascades of models may start at any scale, in the intended frame.

6.2 Renormalization and Scale Symmetry in Critical Transitions

6.2.3 Critical Slowing-Down

In a critical situation, a system needs qualitatively more time to stabilize than a normal situation. This phenomenon is called critical slowing-down. It is one of the aspects of the long range or global correlations in critical phase transitions: in a sense, the entire structure is involved in relaxation phenomena and produces slow decays of the effect of perturbations. Once more, this is one of the reasons, both for the autonomous physical interest of these phenomena in physics and for their resemblance with the "global" nature of biological organisms. In the next chapter we will go further than this and other aspects of physical criticality by looking at biological processes, as we said, as extended (permanent, ongoing) critical transitions.

We will illustrate the phenomenon of critical slowing-down with an elementary mathematical example in the context of bifurcation theory, following the presentation in [Scheffer et al., 2009]. This context is simple and sufficient to understand the basic ideas underlying critical slowing down in general. Let us consider a dynamical system with a pitchfork bifurcation:

$$\frac{dx}{dt} = -x(\alpha + x^2) \tag{6.49}$$

When $\alpha < 0$, the system has three equilibrium points: $x_0 = 0$ and $x_{\pm 1} = \pm\sqrt{-\alpha}$. Otherwise, it has only one equilibrium point: $x_0 = 0$. We can obtain the stability of the system by looking at its second derivative:

$$\frac{d^2x}{dt^2} = -\frac{dx}{dt}\left((\alpha + x^2) + 2x^2\right) \tag{6.50}$$

Thus, at the point x_0, we have $\frac{d^2x}{dt^2} = -\frac{dx}{dt}\alpha$; therefore, a perturbation is amplified when $\alpha < 0$ and stabilized when $\alpha > 0$. At the point x_1, when defined, we have similarly $\frac{d^2x}{dt^2} = \frac{dx}{dt}2\alpha$, so that x_1 is stable when $\alpha < 0$. When $\alpha > 0$ it is anyway not defined. The same analysis applies, *mutadis mutandis*, for x_{-1}. As a result, when x_0 is stable, the other points are not defined, and when they are defined $x_{\pm 1}$ are stable and x_0 is unstable.

Assuming that x_i is stable and that we are near the equilibrium, we can write $x(t) = x_i + \varepsilon(t)$ where ε remains small. In general, a linearization near the equilibrium leads to:

$$\frac{dx_i + \varepsilon}{dt} = f(x_i + \varepsilon) \tag{6.51}$$

$$\frac{d\varepsilon}{dt} = f(x_i) + \varepsilon \frac{\partial f}{\partial x}(x_i) \tag{6.52}$$

which in our example yields (we have $b - a > 0$):

$$\frac{d\varepsilon_0}{dt} = -\alpha\varepsilon_0 \qquad \frac{\varepsilon_{\pm 1}}{dt} = 2\alpha\varepsilon_{\pm 1} \qquad (6.53)$$

The latter equations defines exponential decreases of ε over time, characterizing a fast return to equilibrium.

When approaching the critical point, however, the characteristic times of recovery, $\frac{1}{\alpha}$ and $\frac{1}{2\alpha}$ tend to ∞. At the bifurcation point we have:

$$\frac{dx}{dt} = -x^3 \qquad (6.54)$$

So that the dynamic starting from a perturbation a at time $t = 0$ is

$$\frac{2}{x^2} - \frac{2}{a^2} = t \qquad (6.55)$$

Thus, we get a relaxation of the form $t^{-1/2}$ and the return to the equilibrium follow a power law, which is slower than any exponential decrease.

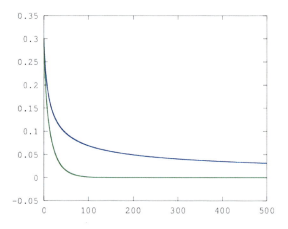

Fig. 6.2 *Bifurcation and critical slowdown.* In green (below), the relaxation of the system discussed at a generic point. In blue (above), we represent the relaxation at the bifurcation point.

We will now consider, as an example, that this dynamical system is subject to additive noise (which leads to Langevin equation). We will show that the slowdown has dramatic consequences. We rewrite the dynamics:

$$\frac{dx}{dt} = -x(\alpha + x^2) + \zeta(t) \qquad (6.56)$$

Then, we get near "normal" stable equilibrium, for $x = x_i + \varepsilon$

6.2 Renormalization and Scale Symmetry in Critical Transitions

$$\frac{dx_i + \varepsilon}{dt} = \frac{\partial f}{\partial x}(x_i)\varepsilon + \zeta(t) \tag{6.57}$$

We write $-\frac{1}{\tau_i} = \frac{\partial f}{\partial x}(x_i)$. This yields:

$$\frac{d\varepsilon}{dt} = -\frac{1}{\tau_i}\varepsilon + \zeta(t) \tag{6.58}$$

$$\varepsilon(t) = \exp\left(-\frac{t}{\tau_i}\right)\left[\int_{-\infty}^{t}\zeta(t')\exp\left(\frac{t'}{\tau_i}\right)dt'\right] \tag{6.59}$$

$$\varepsilon(t) = \int_{-\infty}^{t}\zeta(t')\exp\left(\frac{t'-t}{\tau_i}\right)dt' \tag{6.60}$$

We assume that ζ is in fact a random process, following a white Gaussian noise distribution with standard deviation σ:

$$\varepsilon(t)^2 = \int_{-\infty}^{t}\int_{-\infty}^{t}\zeta(t')\zeta(t'')\exp\left(\frac{t'-t}{\tau_i}\right)\exp\left(\frac{t''-t}{\tau_i}\right)dt'dt'' \tag{6.61}$$

$$\langle\varepsilon(t)^2\rangle = \int_{-\infty}^{t}\int_{-\infty}^{t}\langle\zeta(t')\zeta(t'')\rangle\exp\left(\frac{t'-t}{\tau_i}\right)\exp\left(\frac{t''-t}{\tau_i}\right)dt'dt'' \tag{6.62}$$

Our white noise assumption leads to no correlation between different time points

$$\langle\varepsilon(t)^2\rangle = \int_{-\infty}^{t}\langle\zeta(t')^2\rangle\exp\left(2\frac{t'-t}{\tau_i}\right)dt' \tag{6.63}$$

$$\langle\varepsilon(t)^2\rangle = \tau_i\frac{\sigma^2}{2} \tag{6.64}$$

Thus, the variance of ε is proportional to τ_i. It therefore tends towards ∞ when $\frac{1}{\tau_i}$ tends to 0 (that is to say when the parameter tends to the bifurcation point).

Critical slowing-down is a fundamental property; it makes the convergence of straightforward Monte-Carlo simulations[8] of critical phenomena exceedingly slow. Moreover, [Scheffer et al., 2009] argue that the slow decay of the effect of perturbation can be used to detect the vicinity of a critical point in a complex system. This approach has been used to study the dynamic of molecular networks in cells, see section 2.6.2. Qualitatively, in statistical physics, critical slowdown corresponds to the fact that the system has to be understood as global so that the relaxation involve destabilization and restabilization at long ranges.

Note also that self-organized criticality usually relies (implicitly or explicitly) on a feedback equation, based on the value of the order parameter [Sornette et al., 1995]. The critical slowing-down gets in the way of this feedback and, therefore, justifies the necessary slow input needed for criticality to be observed in these systems.

---

[8] i.e. using pseudo randomness and the probabilities given by the partition function.

6.2.4 Self-tuned Criticality

A situation that illustrates well the high level of sensitivity of a critical system is the model of hair cells described by [Camalet et al., 2000] that we will discuss again latter in a more biological perspective, in section 7.2.2.2. The basic idea is that the hair bundles of these cells oscillates spontaneously because their dynamic is close to a Hopf bifurcation. This situation, however, cannot be understood as spontaneously appearing (it is a point in the parameter space), so a second dynamical system is needed which tunes the system to the bifurcation point.

Let us consider a dynamical system $x(t)$ controlled by a parameter C. When $C > C_c$ the system has a stable fixed point, however for $C < C_c$ the system oscillates spontaneously. We are interested in the systems output with respect to a stimulus which has a frequency $v = \frac{\omega}{2\pi}$. Therefore, we will consider the Fourier decomposition of the output: $x(t) = \sum x_n e^{in\omega t}$. Near the Hopf bifurcation, the mode $n = \pm 1$ is prevalent. We can then write, for a stimulus $f(t) = f_1 e^{i\omega t} + f_{-1} e^{-i\omega t}$:

$$f_1 = \mathscr{A}(\omega,C)x_1 + \mathscr{B}(\omega,C)|x_1|^2 x_1 + \ldots \tag{6.65}$$

where $\mathscr{A}(\omega,C)$ and $\mathscr{B}(\omega,C)$ are functions with complex values.

For $C < C_c$, the system oscillates spontaneously with an amplitude $|x_1|^2 \simeq \Delta^2 \frac{(C_c-C)}{C_c}$ where Δ is a characteristic magnitude. Moreover, we have $\mathscr{A}(\omega_c, C_c) = 0$, for $C = C_c$ and $\omega = \omega_c$, the form of the output is then:

$$|x_1| \simeq |\mathscr{B}|^{-\frac{1}{3}} |f_1|^{\frac{1}{3}} \tag{6.66}$$

The gain is then:

$$r = \frac{|x_1|}{|f_1|} \sim |f_1|^{-\frac{2}{3}} \tag{6.67}$$

We see that the gain becomes arbitrary large for small stimuli. The output has thus a linear component $\mathscr{A}(\omega, C_c) \simeq A_1(\omega - \omega_c)$, at first order. It, however, remains predominantly non-linear when the linear component remains small with respect to the cubic component.

When this situation is not verified, that is when:

$$|\omega - \omega_c| \gg |f_1|^{\frac{2}{3}} \frac{|\mathscr{B}|^{\frac{1}{3}}}{|A_1|} \tag{6.68}$$

The output has the form:

$$|x_1| \simeq \frac{|f_1|}{|(\omega - \omega_c)A_1|} \tag{6.69}$$

and the gain, $|x_1|/|f_1|$, no longer depends on f.

6.2 Renormalization and Scale Symmetry in Critical Transitions

A system near a Hopf bifurcation behaves thus as a highly selective filter with an important gain for weak stimuli. For stronger stimuli, the system is less selective but has also a weaker gain.

We introduce now a second equation in order to maintain C near C_c, but without an explicit dependence on the latter (because C_c depend of the critical frequencies and thus is different for different cells):

$$\frac{1}{C}\frac{\partial C}{\partial t} = \frac{1}{\tau}\left(\frac{x^2}{\delta^2} - 1\right), \text{ where } \delta \text{ is a typical amplitude} \qquad (6.70)$$

When no external force is applied, this equation leads to C_δ, where the system oscillates spontaneously with an amplitude $|x_1| \simeq \delta$. If δ is small in comparison with Δ, we are near the bifurcation since:

$$\frac{C_\delta - C_c}{C_c} \simeq \left(\frac{\delta}{\Delta}\right)^2 \qquad (6.71)$$

As a result there are two kinds of responses: for short stimuli in comparison with τ, the system stays at C_δ and leads to the above discussed non-linear response. For long stimuli, it maintains $|x_1| \simeq \delta$.

6.2.4.1 Remarks on Reductionism and Renormalization

In critical transitions, by renormalization, the intelligibility of the phenomenon has an "upward" flavor since renormalization is based on the stability of the equational determination when one considers only a part of the interactions occurring in the system. That is, the understanding of the global situation may seem to be given in terms of its (elementary) components. Now, the "locus of the objectivity" is not in the description of the parts but in the stability of the equational determination when taking more and more interactions into account. This is also true for those critical phenomena, where the parts, atoms for example, can be objectivized extrinsically to the renormalization and have a characteristic scale. In general, though, only scale invariance matters and the contingent choice of a fundamental (atomic) scale is irrelevant. Even worse, in quantum fields theories, the parts are not really separable from the whole, as this would mean to separate an electron from the field it generates. Thus, there is no relevant elementary scale which would allow one to get rid of the infinities. Moreover, this would be rather arbitrary, since the objectivity needs the inter-scale relationship, see for example [Zinn-Justin, 2007].

In short, even in physics there are situations where the whole is not the sum of the parts because the parts cannot be summed on. This is not specific to quantum fields as it is also relevant for classical fields, in principle. In these situations, the intelligibility is obtained by the scale symmetry. This is why fundamental scale choices are arbitrary with respect to these phenomena, see [Longo et al., 2012c] for further discussions.

6.3 Conclusion

We have seen that symmetry and symmetry breaking have fundamental consequences on the determination of the behavior of objects. Theoretical symmetries (of the Lagrangian, here) correspond to conserved quantities, which are the properties of physical objects and allow their theoretical determination.

On the contrary, at a spontaneous symmetry breaking critical point, the loss of the determination of *both* phases behaviors leads to a particular determination, which is associated to the non-analyticity of the partition function. More precisely, the critical point constitutes a singularity in the determination of the system because it is right between two different behaviors, characterized by different relevant macroscopic phase spaces.

The strength of these singularities can be of different magnitudes; because of this, an original method, renormalization, can be required, depending on the Ginzburg criterion. This qualitatively corresponds to the bigger averaging nature of models in higher spatial dimensions since the higher the dimension of space, the more neighbors a point has. When this averaging is insufficient, renormalization methods are necessary to take into account the global structure of determination of the system that results from the coupling between fluctuations and local averages.

In the following chapter, we will go back to biology and discuss the role played by symmetries and criticality in our theoretical proposal for understanding the phenomena that this field aims to understand.

Chapter 7
From Physics to Biology by Extending Criticality and Symmetry Breakings

> The artificial products do not have any molecular dissymmetry; and I could not indicate the existence of a more profound separation between the products born under the influence of life and all the others.
>
> L. Pasteur

Abstract. In comparison to modern physics, symmetries play a radically different role in biology. By arguing on the relation between symmetries and conservation and stability properties in physics, we posit that the dynamics of biological organisms, in their various levels of organization, are not "just" processes, but permanent (extended, in our terminology) critical transitions and, thus, symmetry changes. Within the limits of a relative structural stability (or interval of viability), variability is at the core of these transitions. And biological adaptivity and diversity are a consequence of it[1].

Keywords: symmetries, systems biology, critical transitions, levels of organization, coherent structures, downward causation, variability.

7.1 Introduction and Summary

A synthetic understanding of the notion of organism requires drawing strong correlations between different levels of organization as well as between the global structure and the local phenomena within the organism. These issues should govern any systemic view on biology. Here, we sketch an approach in which the living state of matter is interpreted as a permanent "transition", conceived as an ongoing or *extended* and *critical* transition. A large amount of very relevant work pertaining to the Theories of Criticality in physics, that we discuss in the previous chapter, has been successfully applied to biology, as we hint below. The mathematical core of these theories rests upon the idea that a "phase transition," which can be either critical

---

[1] This chapter discusses and expands ideas first presented in [Longo & Montévil, 2011a].

or not, may be described as a *point* along the line where the intended control parameter runs. For example, the ferromagnetic / paramagnetic transition takes place for a precise value of the temperature, the Curie temperature. Mathematically, this is expressed by the "point-wise" value of this temperature, i.e., one mathematical point in this parameter's space. When the temperature decreases and passes through that point, the magnetic orientation organizes along one direction and magnetism appears. When the temperature increases through that point, disorder prevails and magnetism disappears. We call a (phase) transition truly critical when it forms a global system, where the Ginsburg criterion is typically not met, see 6.2.1.5 and [Longo et al., 2012c]. This corresponds to the appearance of a "coherent structure", that is to say space and/or time correlations at all scales. These correlations at the transition point give a "global" aspect to the new physical object. As we already observed, the physics of criticality constructed new objects of knowledge, by delimiting and singling out some familiar structures, never or badly analyzed before: the coherent structure of percolation, of a ferromagnetic material, a ...snow flake after the transition from suspended, homogeneous, water to the new symmetries of coherently linked ice crystals....

In contrast to known critical transitions in physics, biological entities, in our view, should not be analyzed just as transient over a point of a phase change; instead, they permanently sustain criticality over a non-zero interval and this with respect to many control parameters (time, temperature, pressure ...). This represents a crucial change of perspective. First, the mathematical tools used in physics for the analysis of criticality, i.e, the renormalization methods, essentially use the point-wise nature of the critical transitions. Secondly, *symmetries* and *symmetry breakings* radically change when enlarging the mathematical locus of criticality from one point to a non-zero interval. These symmetry changes make a key theoretical difference with respect to the few cases in physics where the transition seems extended (see footnote 12, below). Our approach may be seen as a move from physics to biology by an analysis of the radically different symmetries and symmetry breakings at play in their respective theoretical frames. In particular, we will focus on physical vs biological criticality in terms of symmetries and then apply this method to the analysis of the difference between physical and biological "objects" as well as of physical vs biological "trajectories".

In short, in our perspective, living entities are not "just" processes, but something more: they are lasting, *extended critical transitions*, always transient toward a continually renewed structure. In general, physical processes do not change fundamental symmetries: to the contrary, they are mostly meant to preserve them. Typically, conservation properties (of energy, of momentum) are symmetries in the equations of movement. Critical transitions are an exception to the preservation of symmetries in physics; their "extension" radically changes the understanding of what biological processes are. This perspective also proposes a possible way of overcoming a key issue in the analysis of the complexity of the living state of matter.

As for the construction of physico-mathematical or computational models, it is difficult to take the global structure of an organism into consideration, with its correlations between all levels of organization and in all lengths, including the many forms of integration and regulation. Thus, the complexity of the living unity is often modeled by the stacking of many but *simple* elementary processes. Typically, these formal systems deal with many observables and parameters. Since the framework is classical in a physical sense, these variables are local, i.e. they depend on point-wise values of the intended phase space. Instead, conceptual and mathematical dependencies in biology should be dealt with as "global" ones, where variables may depend on systemic or *non-local* effects. In physics, these dependencies are a relevant aspect of critical transitions, and they are even more so in biology, where criticality is extended.

7.1.1 Hidden Variables in Biology?

In classical and relativistic physics, once the suitable "phase space" and the equations that mathematically determine the system are given, the knowledge of the point-wise position-momentum of the intended object of analysis allows to describe *in principle* the subsequent dynamics. This is "in principle" since physical measurement, which is always approximated, may produce the phenomenon of *deterministic unpredictability*, in particular in the presence of non-linear mathematical determination[2]. Moreover, not all "forces" in the game may be known and there may be "hidden variables" (like the frictions along the trajectory of bouncing dice). Yet, these theories are deterministic and, once all pertinent variables and forces are assumed to be known, it is the *epistemic* lack of knowledge which yields classical randomness. In other words and *per se*, a dice follows a "geodesic". This is a unique, optimal and "critical" path, completely determined by the Hamiltonian and may be computed as an optimum of a Lagrangian functional[3]. This very beautiful paradigm, which may be summarized as the "geodesic principle", may be further grounded on *symmetries* by an analysis of conservation principles, as we hinted above, see section 5.3.

In order to compare this situation with other fields of physics and subsequently to biology, we refer to the point-wise or local nature of the mathematical variables. Cantorian (and Euclidean) points are *limit* conceptual constructions; that is, they are the limit of a physical access to space and time by an always approximated measurement, i.e., an "arbitrarily small" interval. Yet, their perfect theoretical "locality" makes all classical dynamics intelligible (in principle). So, if something is unknown, one expects that by adding enough observables and/or more variables with definite values at any given time, one could increase knowledge, since the values of these observables are intrinsic and independent of the context.

The situation is rather different in Quantum Mechanics. The simultaneous, perfect, point-wise knowledge of position *and* momentum (or energy *and* time) are, in principle, forbidden because indeterminacy is intrinsic to the theory. Moreover,

---

[2] More generally, unpredictability may appear when the dynamics is determined by an evolution function or equations that mathematically represent "rich" interactions. Non-linearity is a possible mathematical way to express them.

[3] These are mathematical operators, that is, functions acting on functions that contain all known physical information concerning the energy state of the system.

suppose that two quanta interact and form one system and that they later separate in space. Then, acquiring knowledge regarding an observable quantity by performing a measurement on one of these quanta produces an instantaneous knowledge of the value of the measurement made on the other, i.e., the two quanta are "entangled"[4]. This feature of the theory has several consequences: for instance, variables cannot always be associated to separated points and quantum randomness is intrinsic. That is, under the form of Schrödinger equation, the "determination" gives the *probability* to obtain a value by measurement. Within this theoretical framework, quantum randomness differs from the classical one: two interacting dice which later separate obeying independent statistics, while the probability values of an observable of two previously interacting quanta are correlated. This is the so called "violation of Bell inequalities", which has been empirically verified repeatedly since the experiments described in [Aspect et al., 1982]. Quantum entanglement requires considering some phenomena as being "non-local" and inseparable by any physical measurement ("non-separability").

Since the '30s, some have found this situation unsatisfactory and have searched for "hidden variables" like in the epistemic approach to randomness and determination of classical and relativistic physics — a flipping coin goes along a geodetics, yet we do not know all hidden variables (forces, frictions ...). The idea is that the hidden variables corresponding to quantum mechanical observables have definite (point-wise/local) values at any given time, and that the values of those variables are intrinsic and independent of the device used to measure them. A robust result has instead shown that these assumptions contradict the fundamental fact that quantum mechanical observables need not be commutative [Kochen & Specker, 1967]. Moreover, even when assuming the existence of, or the need for, hidden variables, these would be "non-local" (that is depending on remote, but theoretically inseparable points) and thus, far from the point-wise/local usual mathematical or set-theoretic treatment of variables.

The difference between the classical and quantum frameworks has the following consequence: quantum systems may have a proper systemic unity for at least two reasons. Conjugated observables (position and momentum) are "linked" by joint (in-)determination. Entangled quanta remain a "system", in the sense of their non-separability by measurement[5].

Can this perspective help us in biology? On technical grounds, surely not, or rather not yet. Perhaps, "entangled molecular phenomena" or "tunnel effects ... in the brain" may clarify fundamental issues in the future. However, theoretical ideas in quantum mechanics may at least inspire our attempts in system biology, in particular by considering the methodological role of symmetries and symmetry breakings in this area of physics.

---

[4] This property was considered an inconsistency or incompleteness of the theory, in the seminal paper [Einstein et al., 1935]; it turned out to be a fundamental fact in quantum mechanics, [Aspect et al., 1982], and a key property of quantum vs. classical randomness as we hint next, see also [Longo et al., 2010].

[5] Superposition should also be mentioned, see [Silverman, 2008].

A living organism is a system. And entanglement, non locality, non-separability, superposition, whatever these concepts may mean in biology, may present themselves both at each specific level of organization and in the interactions between levels of organization. In an organism, physiological interactions among cells, tissues, organs do not simply sum each other up: they are "entangled", "non-local", "non-separable" ... they are "superposed" (see the examples described by [Noble, 2006, Soto et al., 2008]). Thus, the theoretical and mathematical approaches to biology cannot be based only on a continual enrichment of "local" views: mathematical models cannot work just by assuming the need for more and more variables, possibly hidden to the previous models. A global view of the system and of its symmetries is required. In the previous chapters, the physics of criticality suggested to us a beautiful frame where the symmetry changes, at the point-wise limit of the critical transitions, allow to construct and analyze new theoretical objects. In the present context, the differences in the role of symmetries and their breakings will help in clarifying and facilitating the passage from physics to biology.

7.2 Biological Systems "Poised" at Criticality

Before progressing to our theoretical development, we give here some precise background on the approaches of biological phenomena in terms of physical criticality.

The title of this section refers to a recent survey [Mora & Bialek, 2011], which emphasizes the omnipresence of criticality in biological systems[6]. The term poised stands for the fact that criticality is typically a non-generic behavior and, in this case, it describes point-wise transitions, as usual in physics where a trajectory is given by a one dimensional parameter, but where the system seems to "stay at or near" the critical transition. As a result, this situation is not observed spontaneously, in its classical form proper to physics; the surprising result is then that criticality is observed, as if the biological system was "poised" at these specific points. Criticality is then obtained experimentally by an original methodology, that we will describe before providing some biological examples.

7.2.1 *Principle*

Usually, statistical mechanics is used to determine macroscopic behaviors (thermodynamic equations) by knowing the symmetries of the interactions of elementary components (or in other words the microscopic Hamiltonian). Let us recall that, in the case of critical phenomena, the macroscopic behavior is not thoroughly described by classical thermodynamics and is better described as a scale symmetry (or more generally a conformal symmetry), in particular determined by the critical exponents. These observable quantities are obtained by renormalization in the asymptotic limit of large scales, see chapter 6.

---
[6] An introduction to the theoretical aspects of physical criticality have been provided in the previous chapter, chapter 6.

Here, however, the strategy is different. Since the "microscopic" objects are usually bigger than in condensed matter physics (cells, birds, ...)[7] and are in a smaller number, thus their individual trajectory can be observed. From this observation, the statistical structure of a collective phenomenon can then be inferred, and, from the latter, a technical transfer in terms of statistical mechanics can then be performed. From this transfer, an abstract "Hamiltonian" and "temperature" are defined. These objects, here, have only a statistical meaning, in particular there is no particular reason for them to have the same physical dimensionality than their usual physical counterparts. In particular, heat transfers cannot be performed in the usual sense, so that this temperature is not measurable by a thermometer — even one abstracted from the physical dimension of temperature. The formal and mathematical meaning of these objects corresponds, however and *mutatis mutandis*, to their statistical mechanical counterpart.

The mathematics, behind what we have just described, are the following simple equations, for a microscopic state σ:

$$P(\sigma) = \frac{1}{Z}\exp\left(-\frac{\mathcal{H}(\sigma)}{kT}\right) \qquad Z = \sum_{\sigma}\exp\left(-\frac{\mathcal{H}(\sigma)}{kT}\right) \qquad (7.1)$$

More details can be found in section 6.2.1.1.

Thus, assuming that the measurement is performed at a fixed temperature, by defining k so that $kT = 1$, and conventionally assuming $Z = 1$[8], we get:

$$\mathcal{H}(\sigma) = -\log(P(\sigma)) \qquad (7.2)$$

Of course, additional assumptions are required to actually determine a "Hamiltonian" from data. These assumptions essentially concern the symmetries of this functional and depend of the specific features of the system under study. General assumptions, however, are that $\mathcal{H}$ is stationary (does not change with time) and that only means and pairwise contributions should be taken into account in the Hamiltonian. These symmetries allow then to go from the data to an estimated Hamiltonian.

With such a Hamiltonian and the equations 7.1, a "natural" macroscopic degree of freedom is then the "temperature". The theoretical collective behavior can then be mathematically analyzed, depending on this parameter, and quantities such as susceptibilities can be evaluated. Finally, the *observed* statistical distributions can be localized in the parameter space. It is then in this sense that biological systems are surprisingly found at or near critical points, or "poised" at criticality in the words of [Mora & Bialek, 2011].

Other aspects of criticality are of course observable, such as the effects of perturbations, the spatial distribution of states (or better of energy in the above formal sense), ...

---

[7] Note that statistical mechanics is also used for nebula and galaxies, thus for even bigger objects.

[8] All these assumptions, except the first one, do not lead to a loss of generality because the situation is not related to physical dimensionalities.

7.2.1.1 Some Examples

We will now describe some biological situations, where this methodology has been applied. These examples correspond to recent studies, because the simultaneous observation of a large number of states is required, and such a feat has been only recently made possible.

Flocks of Birds

The collective behavior in flocks of birds and similar systems such as schools of fish have recently raised considerable interest. Indeed, such situations are not dominated by the behavior of a single individual but are nevertheless able to adapt swiftly to environmental conditions, such as the presence of a predator. This interest originates also from the recent finding, in [Cavagna et al., 2010], of signs of criticality in their spatio-temporal statistical structure.

This kind of system is understood as described by states that are the position and velocities of each bird, in 3d. The statistical mechanical interpretation of the situation is that birds tend to mimic their neighbors' speed, in a way similar to the tendency of elementary magnets to align with their neighbors in an Ising spin glass (because of the magnetic field)[9]. In this sense, the "temperature", as described above, is the propensity of birds to ignore their neighbors (a temperature of 0 means that they follow strictly their neighbors while temperature ∞ means that they are completely independent).

In [Cavagna et al., 2010], it is mainly the correlation of birds velocities that is studied. More precisely, both the correlations of the absolute values of speed and the orientations are evaluated, in function of the spatial distance of birds. Two important results are to be mentioned:

- First, there are strong correlations at small distances and anti-correlations at longer distances. Let us recall that anti-correlations are not the contrary of correlations (which is uncorrelated behavior or in other words independence). Anti-correlation is the opposite of correlation in the algebraic sense; but also implies a high level of coherence. In the case of birds, it intuitively means that if a large *spatially structured* subgroup has its speed that varies coherently (with respect to the group) in a direction, then an other spatially structured subgroup has a variation in the opposite direction (these considerations are not limited to combinations of two subgroups; we have considered this case for illustrative purposes).
- The other key result of this study is that the correlation functions, for both speeds and orientations, follow power laws, so that $C(r) \propto r^{\gamma}$ in the infinite size limit. However, in this case, the situation is even more remarkable than usual scale-free behaviors, since the evaluation of γ is remarkably close to 0. This means that there is almost no decrease of correlations with distance. From another perspective, the correlation function does not depend (much) on the size of the flock,

---

[9] This situation is, however, somewhat more complicated than spin lattices, because the collective structure concerns velocities, so that it changes also the positions of the objects in space, whilst in spin lattices these positions are fixed, and only the fields vary.

if written in the form $C(\frac{r}{\xi})$. In this equation, ξ is the characteristic length and is found to be proportional to L, which is the size of the flock. This relation holds for both orientations and magnitudes of velocities.

These results are found for flocks of sizes spanning approximately one order of magnitudes, which is limited; however, as we said, the situation has very strong signatures of criticality in this range. These results strongly suggest that the system should be considered critical.

Neural Network of the Retina

We do not discuss this case with much details, but we nevertheless provide a basic description of its properties, following [Mora & Bialek, 2011]. The (salamander) retina is used as a paradigmatic case since it allows the recording of a relatively high number of neurons ($\simeq 40$ neurons in current experiments) during a large interval of time. The basic modeling technique of the situation is then the following. The activity of each neuron, i is recorded, so that its state σ_i for time windows of a given length $\Delta\tau$ is defined to be either 1 if a spike (or more) is recorded or -1 if no spike is recorded.

The collective, statistical behavior of the system is then approached as equivalent to the Ising model, which allows to fit the means and the two points correlation functions of the model on the observed data.

$$\mathcal{H}(\sigma) = -\sum_i h_i \sigma_i - \sum_{i<j} J_{i;j} \sigma_i \sigma_j \qquad (7.3)$$

The observed results exhibit a peak for the susceptibility with respect to the "temperature" parameter. This maximum converges towards the temperature of the observed data (conventionally set to 1), when considering systems of greater sizes. This support the hypothesis that the system is poised at criticality. Another aspect of the situation which is consistent with criticality, is the distribution of probability of global activity, which follows approximately a power law (with an exponent 1). The empirical results are closer to this relationship when the number of neurons observed is larger.

Percolation in Myofibril Mitochondria

In [Aon et al., 2003] and [Aon et al., 2004a], the state of mitochondria (depolarization and concentration of ROS[10]) in a myofibril of a cardiomyocyte is understood as, and found experimentally to be, approximately equivalent to the physical situation of percolation. A clarification, here, is needed; these studies consider a limited depolarization, and, as a result, do not involve the permeability transition pores or intracellular C_a^{2+} overload [Aon et al., 2003]. This point is crucial since it means that we are not considering transitions, usually not reversible, towards a state close to cellular death. Instead, these processes are understood as "regenerative" by the

---

[10] Reactive Oxygen Species, we mentioned these observables earlier, in subsection 2.4.2.

7.2 Biological Systems "Poised" at Criticality

authors, inasmuch they allow to control excessive ROS leaks by decreasing mitochondria activity by depolarization.

Now, the mitochondrial system of a myofibril is approached as an approximately bi-dimensional lattice, where each square of the lattice is occupied by one or two mitochondria. The evaluated quantities are then the ratio of depolarized mitochondria, their spatial distribution, the effect of a perturbations (by a local increase of ROS concentration), and particularly characteristic times. When global depolarization occurs, these observed aspects are consistent with the physical universality class of 2d percolation transition (which thus leaves fractal like clusters of polarized mitochondria, as in percolation). In [Aon et al., 2004a], these features are argued to explain oscillations of mitochondria depolarization observed in [Aon et al., 2003] and discussed also in section 2.4.2.

In conclusion, the critical state is the "ordinary" state for these biological phenomena. They may then be interpreted as being "poised at criticality". In our perspective, though, these cases and more below are better understood as cases of "extended criticality", a notion that we will present in this chapter: the processes is not bouncing around an always identical critical point, but moves from one critical point to another in a (mathematically dense) subset of a extended interval of criticality, with respect to all pertinent parameters.

7.2.2 Other Forms of Criticality

In the literature, there are other forms of criticality that are argued to be relevant in the study of biological systems, and used to describe particular spatio-temporal situations. They usually more or less revolve around the idea of a stabilization with respect the point-wise nature of criticality.

7.2.2.1 Self-organized Criticality

The basic idea behind self-organized criticality is that the critical point is understood as an attractor of the dynamic of the system [Jensen, 1998]. The paradigmatic, first example of this kind of situations is the sand pile model [Bak et al., 1988], which is an automaton. This model is defined by the following basic rules: when there is too much difference between adjacent grains heights, the grain that is too high falls, and the grains are added slowly so that the system has sufficient time to relax between two perturbations. This system spontaneously converges towards a critical situation, where the size of the avalanches follows a power law distribution. Notice that, for large systems, this behavior means that the input has to be infinitely slow.

The pitfall of this approach is that the criticality obtained this way remains valid only for a point. This fact is "hidden" by the association of this point with a conservation property. In this sense, then, the critical behavior remains specific and can be mathematically analyzed as a standard form of criticality [Sornette et al., 1995]. In this analysis, the suitable parameter is associated to a feedback mechanism, based on the value of the order parameter. This reasoning also explains why the system needs to have a slow input in order to exhibit criticality.

From an empirical point of view, which, here, is ours, the concept of self-organized criticality leads to focus on the observation of avalanches. We will now provide some biological examples, where self-organized criticality is useful for the analysis of the situation.

The first example is the detailed analysis of respiration, and of the flow of oxygen in the lungs. During exhalation, peripheral airways, in the lungs, tend to close up. Reinflation then leads to the progressive reopening of these airways. The resulting phenomenon is then a succession of jumps, corresponding to these re-openings, which are caused by the pressure differences. This dynamic can be observed by measuring the airway resistance. The two observed distributions, in [Suki et al., 1994], are then the time intervals between jumps and the magnitude of the jumps. Both of them, according to the results of this study, follow power laws over 2 orders of magnitudes, with exponents 2.5 ± 0.2 and 1.8 ± 0.2, respectively. This dynamic is argued to correspond to a situation where the openings occurs in bursts, the opening of one branch being followed by an avalanche of successive opening. In particular, by modeling this process, the authors also show that it leads also to a power law for the air volume increase of the lungs, which corresponds to an exponent 1.1 ± 0.2. The size of the volume freed by an avalanche can therefore be extremely large.

In [Phillips, 2009a,b], some aspects of protein folding are shown to be equivalent to self-organized systems. In particular, this approach allows to focus on long-range interactions and to understand their consequences on the structure and physical properties of proteins, such as hydrophobicity. Another quantity associated with scaling is the solvent-accessible surface area which follows statistically power laws, depending on the amino acids involved.

7.2.2.2 Self-tuned Criticality

The sensitivity of hair cells to stimuli of the order of magnitude of thermal noise can be understood by the proximity of the system to a Hopf bifurcation [Camalet et al., 2000, Balakrishnan & Ashok, 2010]. However, this situation is not generic; the authors thus introduce a second dynamical system, which determines the behavior of the parameter and leads to a convergence of the first system towards its bifurcation point.

In this situation, the gain is highly non-linear, and diverges for small inputs, see section 6.2.4 for more details on the mathematics involved. It has the following form:

$$r = \frac{|x_1|}{|f_1|} \sim |f_1|^{-\frac{2}{3}} \qquad (7.4)$$

where f_1 is the first coefficient of the input (by Fourier analysis) and x_1 is the corresponding coefficient of the output, the spatial displacement of the "hairs".

This amplification occurs for the critical frequency, which depends on the precise values of the parameters. These parameters differ from cell to cell, so that the cochlea is globally able to react to a wide class of sound frequencies. However, the value of the parameter of the first system at bifurcation point is not a parameter of the second system, so that the self-tuning of the system is generic.

7.2.2.3 "Attached" Criticality

A particularly interesting theoretical and empirical situation is analyzed in [Machta et al., 2011]. Plasma membranes fluctuate in relation with a critical point observed at a temperature of 22 °C, on giant plasma membrane vesicles, which are separated from living cells. In the context of a cell, the membrane has relatively large heterogeneities in particular in the form of lipid rafts, associated to proteins, receptors, and disordered lipid structures .... The third element that comes into play is the cytoskeleton, which is attached to the plasma membrane and seems necessary to observe these heterogeneous structures.

The question then arises as to how these large heterogeneities are sustained. A related aspect is that temperatures below the critical point, in isolated membranes, lead to a phase separation, between the liquid ordered and liquid disordered phases. This transition is not observed in the context of the cell. The idea developed in [Machta et al., 2011], is then that these peculiar, relatively large scale structures are explained by the features of the critical point and the interaction of the critical fluctuations with the cytoskeleton. Indeed, even though the system is not at the critical point, it remains not far from it in physiological temperature, 37 °C (one should recall that the physically relevant quantities are in K). At these temperatures, the critical fluctuations have ranges consistent with observations (for isolated membranes). The cytoskeleton comes into play by forcing the behavior of determined points of the plasma membrane. below the critical temperature, this prevent the system from undergoing the phase transition by limiting the size of homogeneous clusters, whilst at physiological temperatures it stabilizes the fluctuations, and bestow them a larger lifetime. This also leads to an explanation of the confinement observed in the diffusion of the lipid rafts and allows at the same time interactions over large distances. Both feats are possible because of the formation of a pavement of the membrane by relatively thin frontiers (allowing fast diffusion) and bounded surfaces where the lipid rafts are confined.

7.2.3 *Conclusion*

We have seen some examples of biological situations interpreted with the methods of physical criticality. Note that this is not an exhaustive list. We have also seen other examples analyzed in terms of criticality in chapter 2, as in networks in section 2.6.2 or as an interpretation of heart rhythms, in section 2.4.3.1.

When considering biological criticality, the amount of studies is limited for practical reasons. The number of units that one needs to observe is indeed approximately $(L/l)^d \times T/\tau$, where L is the spatial extend of the object, d is the dimension of space (1, 2 or 3) and l is the spatial resolution, T is the duration of the measurement and τ the temporal resolution. This should be put in contrast with the fact that an empirical evidence of scaling requires large objects, precisely because of the collective and multi-scale nature of such structures. These technical limitations prevent us from actually discussing the question of variability in these systems. Note that as for physical criticality no variation is allowed for the critical exponents since they are the specific invariants of these systems, see chapter 6.

Nevertheless, the results we have reported are compelling. They show in particular that collective behaviors occurs in biology, and they look comparable to the physical situation of criticality. In a sense, this logic is reinforced by the very limitation of these approaches. Indeed, they require a certain homogeneity of the objects considered (birds, neurons, membrane, mitochondria ...), but, even in these cases of relative homogeneity of the constituents, the biological cases exhibit an extension of criticality, whereas the physical situations corresponding to strongly collective behaviors are usually met only on a single point of the parameter space (with a one-dimensional parameter) and the chances of being poised at it are mathematically 0 (and the chances of being near it are small).

7.3 Extended Criticality: The Biological Object and Symmetry Breakings

In the previous chapters, we presented some basic ideas concerning the role of symmetries in physics and, in particular, their relevance in understanding critical transitions. In view of the interest of criticality in biology and by looking at the organism as a peculiar "coherent structure", our idea is to understand the biological unity in relation to symmetry changes, along the lines of theories of criticality. In order to do so, it may be worthwhile to look at the symmetries which may be involved in biological theorizing. As it should be clear by now, the concept of symmetry is used here in a more fundamental context than when used in a purely spatial way. By the latter we mean, for instance, discussing the bilateral symmetry of some "bauplans". This kind of use is the main one where the concept of symmetry is explicitly applied in biology. In physics, one mostly deals with *fundamental* or *theoretical* symmetries as typically given by the equations, see chapter 5. Recall, for example, that the already mentioned fundamental principle of energy conservation corresponds to a time translation symmetry in the equations of movement, see section 5.3. This use of symmetries also justifies the soundness of empirical results: Galilean inertia is a special case of conservation of energy and it may be empirically verified. In biology, as in any science, a missing analysis of invariants may give unreliable results and data. For example, early measurements of membrane surfaces gave very different results, since their measure is not a scale invariant property: as in fractal structures, it depends on the scale of observation. More details on this question are given in chapter 2[11]. In other words, in physics, both the generality of equations and the very

---

[11] Let us recall a precise example that we discussed in section 2.3.2. In [Weibel, 1994], a "historical" example is given as for the different results that are obtained according to different experimental scales (microscope magnifications). One team evaluated the surface density of the liver's endoplasmic reticulum at $5.7\,m^2/cm^3$ the other at $10.9\,m^2/cm^3$ (!).

7.3 Extended Criticality: The Biological Object and Symmetry Breakings

objectivity of measures depend on theoretical symmetries and their breakings, such as scale invariants and scale dependencies.

As discussed in chapter 6, critical transitions in physics are mathematically analyzed as isolated points[12].

In our approach to biological processes as "*extended* critical transitions", "extended" means that *every point* of the evolution/development space is near a critical point. More technically, at the mathematical limit, the critical points form a dense subset of the multidimensional space of viability for the biological process[13]. Thus, criticality is extended to the space of all pertinent parameters and observables (or phase space), within the limits of viability (tolerated temperature, pressure and time range, or whatever other parameter, say for a given animal), see [Bailly et al., 1993, Bailly & Longo, 2008, 2011]. In terms of symmetries, such a situation implies that biological objects (cells, multicellular organisms, species) are in a *continual transition between different symmetry groups*; that is, they are in transition between different phases, according to the language of condensed matter[14]. These phases swiftly shift between different critical points and between different *physical determinations* through symmetry changes.

Our perspective approaches the mathematical nature of biological objects as a *limit* or asymptotic case of physical states: the latter may yield the dense structure we attribute to extended criticality only by an asymptotic accumulation of critical points in a non-trivial interval of viability — a situation not considered by current physical theories. In a sense, it is the very principles grounding physical theories that we are modifying through an "actual" limit. In short, biological objects are analysed here in terms of partial but continual changes of symmetry within an interval of viability, as an extended locus of critical transitions. Thus, a biological object is mathematically and fundamentally different from a physical object. In particular, this mathematical view of "partial preservation through symmetry changes" is a way to characterize the joint dynamics of *structural stability* and *variability* proper to life. We then consider this characterization as a tool for the mathematical intelligibility of fundamental biological principles: the global/structural stability is crucially associated with variability.

A first consequence of these permanent symmetry changes is that there are very few invariants in biology. Mathematically, invariants depend on stable symmetries. Structural stability in biology, thus, should be understood more in terms of *correlations of symmetries within an interval of the extended critical transition*, rather than

---

[12] The Kosterlitz-Thouless transition in statistical physics presents a marginally critical interval; that is, it is a limit case between critical and not critical behaviour. It presents correlations at large scales, as critical features, but with no symmetry changes. Thus, this particular situation is not a counter-example to our statement (the essentially point-wise nature of the proper physical transitions), in view of a lack of symmetry changes that are essential to our notion of extended criticality.

[13] Here, in a more practical sense, dense means that for every small volume of the intended phase space being considered, there is a critical point in such volume.

[14] The dense set of symmetry groups may be potentially infinite, but, of course, an organism (or a species) explores only finitely many of them in its life span, and only viable ones.

on their identical preservation. It is clear that the *bauplan* and a few more properties may be "identically" preserved. Yet, in biology, theoretical invariants are continually broken by these symmetry changes. A biological object (a cell, a multicellular organism, a species) continually changes symmetries, with respect to all control parameters, including time.

As a fundamental example of symmetry change, observe that mitosis yields different proteome distributions, differences in DNA or DNA expressions, in membranes or organelles: the symmetries are not preserved. In a multi-cellular organism, each mitosis asymmetrically reconstructs a new coherent structure, also in the sense of the physics of critical transitions: a new tissue matrix, new collagen structure, new cell-to-cell connections . . . . Now, mitosis is one of the fundamental biological processes, associated to Darwin's first principle "descent with modification" or reproduction with variation (the second is selection). And a multi-cellular organism undergoes millions or billions of mitosis each day.

This variability, under the mathematical form of symmetry breaking and constitution of new symmetries, is essential both for evolution and embryogenesis. The interval of criticality is then the "space of viability" or locus of the possible structural stability.

The changes of symmetries in the dense interval of criticality, which provide a mathematical understanding of biological variability, are a major challenge for theorizing. As a matter of fact, we are accustomed to the theoretical stability warranted by the mathematical invariants at the core of physics. These invariants are the result of symmetries in the mathematical (equational) determination of the physical object. This lack of invariants and symmetries corresponds to the difficulties in finding equational determinations in biology[15].

As a further consequence of our approach, phylogenetic or ontogenetic trajectories cannot be defined by the geodesic principle. Indeed, they are not theoretically determined by invariants and their associated symmetries. Trajectories are continually changing in a relatively minor but extended way. Moreover, we expect the rate of these changes themselves not to be regular with respect to physical time, so that some temporal region can be "calm" while others correspond sudden burst of changes.

Biology may be considered to be in an opposite situation with respect to physics: in contrast to physics, in biology, *trajectories* are *generic* whereas *objects* are *specific* [Bailly & Longo, 2011]. That is, a rat, a monkey or an elephant are the *specific* results of *possible* (generic) evolutionary trajectories of a common mammal ancestor — in other words each of these individuals is *specific*. They respectively are the

---

[15] In a rather naive way, some say this by observing that any (mathematized) theory in biology has a "counterexample". This conceptual instability of the determination may be understood by the minor role played by conservation properties, in terms of stable symmetries. Yet, it goes together with the "structural stability" of biological entities, which is largely due to the stabilizing role of integration and regulation effects between different levels of organization. The autonomous functionality of organisms is another way to approach their structural stability, see [Moreno & Mossio, 2013]

7.3 Extended Criticality: The Biological Object and Symmetry Breakings

result of a unique constitutive history, yet a possible or *generic* one [Bailly et al., 1993, Bailly & Longo, 2011].

The evolutionary or ontogenetic trajectory of a cell, a multicellular organism or a species is just a *possible* or *compatible* path within the ecosystem. The genericity of the biological trajectories implies that, in contrast to what is common in physics, we cannot mathematically and *a priori* determine the ontogenetic and phylogenetic trajectory of a living entity be it an individual or a species. In other words, in biology, we should consider *generic* trajectories (or possible paths) whose only constraints are to remain compatible with the survival of the intended biological system. Thus, phylogenesis and embryogenesis are *possible* paths subject to various constraints, including of course the inherited structure of the DNA, of the cell and the ecosystem. The *specificity* of the biological object, instead, is the result of critical points and of symmetry *changes* of the system considered *along its past history* (evolutive and ontogenetic). These constitute the specific "properties" of this object, which allow to define it. A rat, a monkey or an elephant or their species are *specific* and cannot be interchanged either as individuals nor as species. A living entity is the result of its history and cannot be defined "generically" in terms of invariants and symmetries as it is done for physical objects.

This situation has a particular meaning when we consider time translation and time reversal symmetries. In physics, time symmetries correspond to the maintaining of the system's invariant quantities that define the geodesics, that is mainly the conservation of energy. In biology both symmetries are broken. In particular, evolutionary and ontogenetic paths are both irreversible and non-iterable; there is no way to identically "rewind" nor "restart" evolution or ontogenesis. This corresponds to the breaking of time translation and reversal symmetries. In particular, this lack of time symmetries is associated with the process of *individuation*, understood here as the specificity of cells, organisms and species (as much as this latter notion is well defined). It is crucial to understand that time plays a particular role in this approach, since the *history* of all the changes in symmetry is not reducible to a specific trajectory in a given space. Thus,

The sequence of symmetry changes defines the historical contingency of a living object's phylogenetic or ontogenetic trajectory.

Biological processes are more "history based" than physical processes. Usual physical processes preserve invariants, whereas extended critical transitions are a permanent reconstruction of organization and symmetries, i.e., of invariants. This situation also points to a lack of symmetry by permutation. For example, even in a clonal population of bacteria, different bacteria are not generic, because they are in general not interchangeable, i.e., they cannot be permuted. This allows to understand biological variability in a deeper way than the usual (Gaussian or combination of Gaussians) random distribution for a set of observables.

Consider, say, organs (and organelles). Some organs have a functional role that can be expressed in a physical framework, particularly when energy or matter transfers are concerned. This functional role can lead to restrictions on the variability of the cells that constitute the organ. At least for certain aspects of their behavior and on average, these restrictions make cells behave symmetrically. In other words, the cells in an organ or in a tissue of an organ behave, in part and approximately, like generic objects with specific trajectory (geodesics). They may approximately be interchangeable, like physical objects.

The simple case of cells secreting a protein such as erythropoietin (EPO) under specific conditions indicate that on average, a sufficient amount of the protein must be produced, independently of the individual contribution of each cell (which become "relatively" generic). Since the result of these cells' production is additive (linear), their regulation does not need to be sharp. Even if some cells do not produce EPO there is no functional problem as long as a sufficient quantity of this protein is secreted at the tissue level. However, when cells contribute to a non-linear framework as part of an organ, the regulation may need to be sharper. This is the case, for example, for neuronal networks or for cell proliferation where non-linear effects may be very important. In the latter case, regulation by the tissue and the organism seems to hold back pathological developments, like cancer, see [Sonnenschein & Soto, 1999]. This point of view can possibly be generalized in order to understand the robustness of development.

The role of physical processes in shaping organs is crucial; for example, exchanges of energy (or matter) force/determine the optimal (geodesic) fractal structure of lungs and vascular systems. Organs in an organism may even be replaced by man-made artifacts (as for kidneys, heart, limbs, etc.). As biological entities, organisms and even cells are specific or, at most, weakly generic given that they can be interchanged only within a given population or tissue and only sometimes. In general, they are not generic, and by their specificity they cannot be replaced, even not by an artifact — structurally. In other words, they are not an invariant of the theory nor of empiricity. This is why, empirical evidence requires constraining conditions that "symmetrize" as much as possible the organism and its context [Montévil, 2013].

In summary, in critical transitions one may consider variables depending on global processes because of the formation of coherent structures. For example, there may be functional dependencies on a network of interactions, which cannot be split into a sum of many local dependencies (local variables). Thus, the search for more variables would not take into account this fundamental property of biological systems, the global dependency of many processes, which is instead considered from the perspective of extended critical transitions. Moreover, symmetries in physics allow to define generic objects which follow specific trajectories (the latter allowing to find invariants in terms of symmetries, which are robust with respect to measurement). On the contrary, in biology, the continual symmetry changes lead to generic trajectories that remain compatible with the survival of the system. The generic/specific duality with respect to physics helped us understand this key issue, in relation to extended criticality — which is a form of "relatively stable instability." In other words, it is stability under changes of symmetries in an interval of viability. In a sense, the biological object is also defined by its symmetries but in a very different way: as we said, it is the *specific* result of a history, where its dynamics is punctuated by symmetry changes. This makes it "historical" and *contingent*.

7.4 Additional Characteristics of Extended Criticality

In physics, criticality implies more than a symmetry change; in our perspective, it also leads to peculiar behaviors at the critical point that is relevant to biology. The first property observed at this point is a global determination, instead of a simply local one. More precisely, the singularities involved in criticality lead to a change of the level of organization in a very strong sense. Also in physics, in view of the mathematical divergence of some observables, the singularities break the ability of the "down level" to provide a causal account of the phenomena. Thus, they lead to the need for a "top level" of description to overcome this difficulty. In mathematical physics, this upper level can be found in the renormalization operator (it is the abstract level of *changing scale*). In biology, instead, the upper level is the functional unity of an organism. As a result, the existence of different levels of organization is a component of our notion of extended critical transition. "Downward causation" may find the right frame of analysis in this theoretical context.

The permanent reconstruction of these levels of organization is mathematically represented by the density of the critical points and by the continual changes of determination (symmetry changes) in the passage between these points within the interval of extended criticality.

The second property is the presence of power laws which seem to be ubiquitous in biology. They appear regularly especially when regulation is concerned, see chapter 2 for a comprehensive review of those laws, and on the limitation of these approximate symmetries.

Extended critical transitions also concern the relevant lengths of local and global exchanges, the temporalities mobilized for such exchanges and biological rhythms. To summarize, the extended critical situation, in biology, has at least the following characteristics [Bailly & Longo, 2008, 2011]:

1. A spatial volume enclosed within a semi-permeable membrane;
2. Correlation lengths of the order of magnitude of the greatest length of the above referred volume;
3. A metabolic activity that is far from equilibrium and irreversible, involving exchanges of energy, of matter and of entropy with the environment, as well as the production of entropy due to all these irreversible processes, see [Bailly & Longo, 2009];
4. An anatomo-functional structuration in levels of organization that can be autonomous but also coupled to each other. They are "entangled" in the sense defined by [Bailly & Longo, 2009, Soto et al., 2008]. These levels are likely to be distinguished by the existence of fractal geometries (membranous or arborescent), where the fractal geometries can be considered as the trace (or a "model") of effective passages to the infinite limit of an intensive magnitude of the system

(for example, local exchanges of energy[16]). The different levels of organization induce, and are a consequence of, the alternation of "organs" and "organisms", such as organelles in cells, which, in turn, make up the organs in multicellular organisms. Organisms stay in an extended critical transition, while organs (or, at least, those exchanging energy or matter) are partially "optimally shaped" by the physical exchanges. Note also that fractal geometries essentially manifest in organs that are also the privileged loci of endogenous rhythms (see below). Correlation lengths are manifested both *in* and *between* these levels[17]. Likewise, the various biological "clocks" are coupled, and in some cases even synchronized, within and between these levels.

With the purpose of providing biological temporality with a structuring of the mathematical type, we will consider two other aspects as being specific to extended criticality.

- The two-dimensionality of time, discussed in chapter 3:
  1. One dimension is classical and is parameterized according to the line of real numbers limited by fertilization on one side, and death on the other. This dimension is linked to the bio-physicochemical evolution of the organism in relation to an environment.
  2. The other dimension is compactified, i. e. it is parameterized on a circle. This second dimension is linked to the organism's endogenous physiological rhythm that is manifested through *numeric quantities without dimension* such as the mean total number of heartbeats and respirations during the lifetime of mammals. These are the interesting interspecific invariants and they are "pure" numbers, *not frequencies* (they have no dimension; they are the "total number of ..."). They become frequencies (with the inverse of time as a dimension), according to the average lifespan. The extra dimension is needed exactly because the invariant phenomenon is not defined by a period which has the dimension of time, but by this new invariant observable. Recall, for example, that, on average, the identical (invariant) number of total heartbeats give different frequencies according to the different lifespans of an elephant or of a mouse.

Moreover, the temporality of extended criticality involve protention (i.e. pre-conscious expectation) and retention (i. e. pre-conscious memory) as described

---

[16] The fractal dimension of some organs may be calculated by optimizing the purely physical exchanges within the intended topological dimension (for example, the maximization, within a volume, of surfaces for lungs, or of volumes for the vascular system, [West et al., 1997]), and it may be subjected to constraints in terms of stericity and homogeneity, as in the cases mentioned (lung, vascular system, kidney, etc).

[17] The term "entanglement" in [Soto et al., 2008] does not correspond, of course, to the physical meaning of "quantum entanglement" as expressed by Schrödinger's treatment of the state function and the inseparability of quantum measure, yet it may be appropriate because there is no way to isolate one of the organs mentioned above (e.g. put a brain in a flowerpot) and perform any reasonable physiological measure on it.

7.4 Additional Characteristics of Extended Criticality

in chapter 4, which can be seen as breaking of conservation of information in cognition: pretension uses but also modifies memory.
- The confinement within a volume of a parameter space (such as temperature, pressure, etc) of n dimensions of which 3 are spatial and 2 temporal and whose measure is different from 0 (see above).

7.4.1 Remarks on Randomness and Time Irreversibility

The "arrow of time", that is the idea that time is oriented and irreversible, has been definitively introduced in physics by thermodynamics. Entropy growth as energy dispersal (see chapter 9) is omnipresent in physics, yet fully understood only in thermodynamics and statistical mechanics. The equations of movement, from classical to relativistic physics, are time reversible: planets may go in the current direction or the inverse one, nothing would change in the related theories. Conservation principles and Noether's theorem, based on time and space translation symmetries, guaranty the scientific soundness of classical and relativistic reversible dynamics. Even Schrödinger equation is time reversible: irreversibility steps in by measurement, in the occasion of a symmetry breaking.

As for biology, it is fair to say that no properly biological process may be conceived as reversible: if your preferred theory allows to see evolution or embryogenesis played backwards in time, then forget it[18].

In order to stress the relevance of extended criticality also in the understanding of biological time, we propose here a "conjecture", concerning both physics and biology.

We claim first that, in all the theories of the inert we mentioned, irreversibility appears in presence of random events.

Let's try to express this by a case list. Irreversibility of time is associated to random events, in physics:

1. Classical dynamics: unpredictability with respect to the theory (randomness) sets an epistemic orientation to time. It is true that we cannot retrodict either, yet the situation introduces a fundamental asymmetry, as we can keep records of the past. See also the next point.
2. Thermodynamics:
   Macroscopically: entropy production,
   Microscopically: diffusion as random paths, or ergodic trajectories.
3. Quantum Mechanics: projection of the state vector (measurement); non-commutativity of measurement; creation of a particle; tunneling effects ...
4. Critical transitions: the point-wise symmetry change is associated to fluctuations; the transition may be reversed, but not the symmetry changes (or, more precisely, they cannot be recovered).

---

[18] The state of a cell, even a cancer cell, may be "reversed", by changing its environment. Similarly, one may recover from a disease, but the cell and the organism have got older in this process.

All these phenomena are of course present in biology. In particular, energy exchange and transformation are omnipresent in organisms, with the inevitably associated energy dispersal, thus entropy production (see chapter 9). The associated irreversibility of time is then of physical nature and may be seen a first form of "observable time", an irreversible, thermodynamical time. Yet, biological processes are enriched by another "observable of time", the time due to the proper irreversibility of these very processes: phylogenesis, embryogenesis, ontogenesis[19]. Let's be more specific. We see two reasons to admit such a second observable, in our theory.

First, the setting up of organization, in evolution and embryogenesis, but its maintenance in ontogenesis as well, is always associated to a slight increase of disorder, thus of entropy. At each mitosis, at each construction/reconstruction of an organism, growth and change are always slightly disordered, all the while increasing or maintaining order (by anti-entropy, an issue to be discussed at length in chapter 9). Moreover, as for (pre-)conscious activities, protection and retention are asymmetric and impose an arrow to time, see chapter 4.

Second, following our remark in section 5.5, the extended criticality of organisms forces cascades of random events, associated to symmetry changes. Thus, if we are right, it produces a cascades of locally irreversible events and a global irreversibility of time in the extended time-interval of criticality.

In the next chapter, we will give another reason to analyze time as "doubly irreversible" in biology: randomness shows up at the very construction of the space of possibilities (the phase space of observables and parameters). This means that the theory must embed in itself a further form of randomness, as irreversible multifurcations.

7.5 Compactified Time and Autonomy

Beyond the dimension of usual, irreversible, physical time, we proposed, in chapter 3, a second temporal dimension, to be added to physical time, and which allows to accommodate biological rhythms. This dimension is compactified (i. e. the added dimension has the topology of a circle) and is associated with the allometry of internal rhythms ($\tau \propto W^{1/4}$). As a result, this supplementary time dimension is first justified by an allometric scale symmetry. We examine here the interplay of stabilities and instabilities of (some) biological rhythms, in reference to biological extended criticality.

This symmetry, however, takes only into account the broad tendencies among species sharing similar physiological traits. From a closer point of view, we can look at the time series generated by the iterations of the compactified time and parameterized by physical time, for a specific organism. This approach, in our 2-dimensional framework, is described, in particular, in section 3.4.4. Considering their variability, this kind of precise trajectories, for a given internal rhythm, is not determined by

---

[19] These two observables of time may sit in the same dimension, yet differ. In physics, the dimension of energy contains several different observables: potential, kinetic energy and more.

7.5 Compactified Time and Autonomy

the allometric symmetry alone. The slight, but continual symmetry changes, in an organism, that is its extended criticality, contribute to the determination of actual biological rhythms, as we explain next.

In order to better understand the situation, we will first recall how periodic behaviors (oscillators) are determined in physics and we will then analyze the situation in biology. The core point that we want to emphasize is the relationship between the periodicity of physical phenomena and energy (and its conservation). The other reason why we provide these examples is that the form of elementary repetitive phenomena provides a paradigmatic and usually ubiquitous example of the consequences and of the nature of our theoretical framework.

7.5.1 Simple Harmonic Oscillators in Physics

Oscillatory behaviors in physics have multiple forms. However, the simplest of them, the simple harmonic oscillator, is usually a key and paradigmatic model since it regularly appears, at least as a first approximation, in a wide range of situations.

A reason, why such behaviors regularly appear, is that it is a general linear approximation of a behavior near equilibrium, meaning that it is valid near equilibrium as soon as that the latter is well defined and the associated functions are sufficiently regular.

7.5.1.1 Classical Oscillator

If we consider a classical conservative system near a local stable equilibrium at $x = 0$, we obtain:

$$m\frac{d^2x}{dt^2} = f(x) \simeq f(0) + \frac{\partial f}{\partial x}(0)x \qquad (7.5)$$

$$\simeq -kx \qquad (7.6)$$

Indeed, the condition for stable equilibrium leads to $f(0) = 0$ and to $k > 0$. The general solution for such a system is $x(t) = a\cos(\sqrt{\frac{k}{m}}t + \phi)$, where a depends on the initial energy of the system and ϕ depends on the direction in phase space of the initial state. The corresponding energy is

$$E = \mathcal{H} = \frac{1}{2}m\left(\frac{dx}{dt}\right)^2 + \frac{1}{2}kx^2 \qquad (7.7)$$

as a result the simple harmonic oscillator is a system which continuously transfers its energy from kinetic energy to potential energy and then the other way round. The time an iteration takes depends only on the relative "strength" of kinetic energy (the inertial mass, m) and of potential energy (the coefficient k).

7.5.1.2 Quantum Oscillator

The quantum harmonic oscillator is somewhat different. Its behavior is still determined by energy; however, the structure of determination and, in particular, the nature of the objects used are different. The Hamiltonian is given by:

$$\hat{\mathcal{H}} = \frac{1}{2m}\hat{p}^2 + \frac{1}{2}k\hat{x}^2 \qquad (7.8)$$

$$\text{with} \qquad \hat{x}\psi(x) = x.\psi(x); \qquad \hat{p} = -i\hbar\frac{\partial}{\partial x} \qquad (7.9)$$

$$\text{Possible measures verify} \qquad \hat{\mathcal{H}}\psi = E\psi \qquad (7.10)$$

Solving this system leads to discrete energy levels: $E_n = \hbar\sqrt{\frac{k}{m}}\left(n+\frac{1}{2}\right)$ and correspondingly a set of frequencies $f_n = \frac{1}{2\pi}\sqrt{\frac{k}{m}}\left(n+\frac{1}{2}\right)$. Again, the behavior of the system is determined by energy, but in this context energy is quantized and has a minimum which is different from zero, meaning that the system is never at the corresponding classical equilibrium point[20]. More generally, quantum physics introduces wave-particle duality, in particular through the De Broglie relation: $f = \frac{E}{\hbar}$.

7.5.1.3 Concluding Remarks on Physical Oscillators

We saw that in both cases, the simple harmonic oscillator is characterized by its symmetric form, for energy, of the momentum and the position contribution. Their respective weight in energy leads to the determination of the period of the phenomenon. This example shows how, in fundamental physics, the "repetitive behaviors" have their properties determined by energy, with its underlying symmetries as conservation properties.

The result of these accounts is that the period of such physical phenomena is determined by the form of energy. More precisely, these periods are determined by the ratio of the coefficients, in the Hamiltonian, corresponding to the contributions of position and momentum. In both cases, the period depends then on parameters characterizing the properties of the objects (the mass and the coefficient k). The consequence of this theoretical origin of the period is that its values are robust and rooted in the properties of objects. The further consequence of this situation is then that these phenomena have an intrinsically regular relationship with physical time (which have been conceptualized and mathematized precisely for this purpose).

Of course, physical oscillators are not limited to these cases, which are the most elementary ones. More complex situation can occur, in particular by introducing non-linearities (linearity almost directly leads to a superposition of harmonic oscillators).

---

[20] This kind of quantum behavior leads to quantum fluctuations, even at 0 K. temperature.

7.5.2 *Biological Oscillators: Symmetries and Compactified Time*

In the case of biological internal rhythms, however, energy conservation does not seem to be able to determine the trajectory of the iterative process[21], since there is no underlying symmetry corresponding to a possible energetic determination; that is, energy conservation properties do not suffice to determine the process. Let us consider, for example, the case of the heart rhythm. This rhythm is neither *fully* determined by the properties of the heart (insofar they are defined) nor by direct response to a well defined requirement in "energy" of the organism. On the contrary, the determination of the heart rate involves the full complexity of *both* the heart organ and the regulation associated to the behavior of the whole organism. For example, the various activities of organs are related to the history of the organism at different time scales. As a result, the determination of the heart rate depends also on the environment of an organism. More precisely it depends on the complex relationship of the organism with its environment, including protection and retention, as described in chapter 4 — protensive activities are crucial for heart rhythms. More globally, it entirely reflects the extended critical processes which are at the core of its dynamics.

In other words, the theoretical determination of the dynamics of an organ, such as the heart, involves the activity of the whole organism[22]. As a result, the symmetry changes in the organism break any geodetic determination of the heart trajectory (beat to beat interval, here). This situation leads to the lack of a complete theoretical determination of this trajectory with respect to physical time. From a slightly shifted point of view, the generic trajectories of the organism influence the trajectories of the heart, breaking possible regularities with respect to physical time. Our point here is not to say that there would be some violation of energy conservation in an organism, but to say that there is no invariant form (symmetry) in the determination of something like a heart rate, at least insofar this determination involves the whole organism.

The symmetry we have proposed in chapter 3, to accommodate such a rhythm, including allometry, is based on the iterative nature of the organ considered and on the pure number of their total iterations (an interspecific approximate invariant). This further motivates the strategy we proposed to approach this kind of fundamental "biological oscillator" as based on the introduction of a supplementary temporal structure (generated by the second time dimension), irreducible to the usual physical time. We did so, precisely because of the lacking energetic determination. Now, this conceptual change, from the physical oscillators to the "biological oscillator" — or in more precise terms, biological iterator — allows to understand the peculiar features of heart rate variability. This features are associated to an instability of the beat to beat (physical time) interval during wake, instability which can have different forms depending on age, activity, possible pathologies, ... (see section 3.4.4.2).

---

[21] Understood as a trajectory parameterized by the usual physical time.

[22] In general, the precise aspects of the activity of the organism that are directly relevant for a particular organ depend of course on the nature of the organ.

As a result, our mathematical approach in chapter 3 should be understood as different from the usual physical approaches. The framework is based on very general tendencies (symmetries), such as allometry and the regularity of the physical, external rhythms (based on a regularity among mammals, birds, ...), but the engendered mathematical structure, suggested by these basic symmetries, does not suffice to provide objective, specific trajectories. On the contrary, the observed trajectory of a specific organism, with its crucial instabilities, can be embedded in our framework in a "natural" way, and is therefore a "projection" of (a part of) the extended critical interval of an organism life-span. This operation allows to show the specificity of the biological object by geometrically emphasizing the qualitative features of a trajectory, which can also be seen as the structure(s) of its variability. In other words, our approach stressed the the various shapes of its relationship with physical time and its complex interplay between regularities (basic symmetries) and irregularities (critical transitions), within a global, organismal, structural stability (by regulation and integration).

7.5.3 Conclusion

The aspects we described provide a supplementary justification to the introduction of a second temporal dimension. Indeed, the points we raised and our more general framework leads to understand the inadequacy of the theoretical determination of biological trajectories with respect to the physical time only. Physical time is the parameter where energy conservation, which governs physical oscillators, is described as an invariant quantity and it is largely insufficient to analyze biological time and "oscillators".

Moreover, we pointed out the role of a time trajectory associated to a specific biological object, since our 2-dimensional framework allows to represent and study the complex and diverse relationships between physical time and biological proper rhythms. This kind of trajectory is associated to the individuation of the organism at various time scales; it includes both "spontaneous" variations and association with various factors. It soundly takes place in an extended interval of criticality, i. e. of (regulated and integrated) symmetry changes.

7.6 Conclusion

Since ancient Greece (Archimedes' principle on equilibria) up to Relativity Theory (Noether's and Weyl's work) and Quantum Mechanics (from Weyl's groups to the time-charge-parity symmetry), symmetries have provided a unified view of the principles of theoretical intelligibility in physics. We claimed here that some major challenges for the proposal of mathematical and theoretical ideas in biology depend, in principle, on the very different roles that symmetries play in biology when compared to physics. The unifying theoretical framework in biology is neither associated to invariants nor to transformations preserving invariants like in (mathematical/theoretical) physics. It focuses, instead, on the permanent change of symmetries

7.6 Conclusion

that *per se* modify the analysis of the internal and external processes of life, both in ontogenesis and evolution.

In a sense, variability may be considered as the main invariant of the living state of matter (yet it is not the only one!). In order to explain it, we proposed to consider the role played by local and global symmetry changes along extended critical transitions. In extended criticality, dynamically changing coherent structures as global entities provide an understanding of variability within a global, extended stability. The coherent structure of critical phenomena also justifies the use of variables depending on non-local effects. Thus, an explicitly systemic approach may help in avoiding the accumulation of models and hidden variables. In conclusion, the notion of extended criticality provides a conceptual framework, to be further mathematized, where the dynamics of symmetries and symmetry breakings provide a new, crucial role for symmetries in biology with respect to physics.

It should be clear that by focusing on symmetry changes and variability, as core notions for understanding life adaptivity and diversity, we do not forget biological structural stability and autonomy, under ecosystemic and internal constraints. No extended criticality would ever be possible without the integrating and regulating activities proper to an organism and its relation to the ecosystem. Actually, the main motivation we provided to look at criticality, was derived from the role of the coherent structures that are witnessed at critical transitions in physics. These structures, though changing along all control parameters in a biological organism, are the mathematical representation of the organismal (changing) stability, its internal and external coherence, as a whole.

This chapter is a major turning point for our perspective. Its conclusion directly concerns the nature of the mathematical accounts that we can provide on biological organization, given the instability of biological symmetries.

In the following chapter, we will explore the consequences of our analysis on the notion of phase spaces, discuss causality and introduce the notion of enablement.

Chapter 8
Biological Phase Spaces and Enablement

Abstract. This chapter analyzes, in terms of critical transitions, the phase spaces of biological dynamics. The phase space is the space where the scientific description and determination of a phenomenon is given. We first hint to the historical path that lead physics to give a central role to the construction of a sound notion of phase space, as a condition of possibility for physico-mathematical analyses to be developed. We then argue that one major aspect of biological evolution is the continual change of the pertinent phase space and the unpredictability of these changes. This analysis will be based on the role of theoretical symmetries in biology and on their critical instability along evolution. Our hypothesis deeply modifies the tools and concepts used in physical theorizing, when adapted to biology. In particular, we argue that causality has to be understood differently, and we discuss two notions to do so: differential causality and enablement. In this context constraints play a key role: on one side, they restrict possibilities, on the other, they also enable biological systems to integrate changing constraints in their organization, by correlated variations, in un-prestatable ways. This corresponds to the formation of new phenotypes and organisms.

Keywords: Conservation properties, symmetries, biological causality, phase space, unpredictability, enablement.

8.1 Introduction

As a major guideline for our work, we used the perspective proposed by H. Weyl and B. van Fraassen: XXth century physics has been substituting to the concept of law that of symmetry. Thus, the concept of symmetry may be "considered the principal means of access to the world we create in theories", [Van Fraassen, 1989].

In this chapter, we will discuss the question of biological phase spaces in relation to critical transitions and symmetries[1].

---

[1] We revise extensively and develop here early joint work with Stuart Kauffman [Longo et al., 2012b] and a subsequent paper [Longo & Montévil, 2013].

In order to understand the peculiarities of biological theorizing, we will first shortly recall the sense of "phase spaces" in physics. A phase space is the space of the pertinent observables and parameters in which the theoretical determination of the system takes place[2]. As a result, to one point of the phase space corresponds a complete determination of the intended object and of the features that are relevant for the analysis.

Aristotle and Aristotelians, Galileo and Kepler closely analyzed trajectories of physical bodies, but without a mathematical theory of a "background space". In a sense, they had the same attitude as Greek geometers: Euclid's geometry is a geometry of figures with no space. It is fair to say that modern mathematical physics (Newton) begun by the "embedding" of Kepler and Galileo's Euclidean trajectories in Descartes' spaces. More precisely, the conjunction of these spaces with Galileo's inertia gave the early relativistic spaces and their invariant properties, as a frame for all possible trajectories — from falling bodies to revolving planets[3]. In modern terms, Galileo's symmetry group describes the transformations that preserve the equational form of physical laws, as invariants, when changing the reference system.

Along these lines, one of the major challenges for a (theoretical) physicist is to invent the pertinent space or, more precisely, to construct a mathematical space which contains all the required ingredients for describing the phenomena and to understand the determination of its trajectory, if any. So, Newton's analysis of trajectories was embedded in a Cartesian space, a "condition of possibility", Kant will explain, for physics to be done. By this, Newton unified (he did not reduce) Galileo's analysis of falling bodies, including apples, to planetary orbits: Newton derived Kepler's ellipsis of a planet around the Sun from his equations. This is the astonishing birth of modern mathematical-physics as capable of describing exactly (and predicting, many hoped) the theoretical trajectory, once given the right space and the exact boundary conditions. But, since Poincaré, we know that if the planets around the Sun are two or more, prediction is impossible due to deterministic chaos. Even though the planets trajectories are fully determined by Newton-Laplace equations, the non-linearity of this equations yields the absence almost everywhere of analytic solutions and forbids predictability, even along well determined trajectories at equilibrium.

As a matter of fact, Poincaré's analysis of chaotic dynamics was essentially based on his invention of the so-called Poincaré section (analyze planetary orbits only by their crossing a given plane) and by the use of momentum as a key observable. In his analysis of this early and fundamental case of deterministic chaos, as it will be later called, stable and unstable trajectories in the *position-momentum* phase space, nearly intersect infinitely often, in "infinitely tight meshes" and are also "folded upon themselves without ever intersecting themselves", (1892). Since then,

---

[2] Note that this definition is a little more inclusive than the usual meaning of the expression in physics, as physicists mostly refer to the space given by momentum and position, or by energy and time.

[3] The Italian Renaissance painters invented the mathematical "background" space by the perspective, later turned into mathematics by Descartes and Desargues, see [Longo, 2011b].

8.1 Introduction

in physics, the phase space is mostly given by all possible values of momentum and position, or energy and time. In Hamiltonian classical mechanics and in Quantum Physics, these observables and variables happen to be "conjugated", a mathematical expression of their pertinence and tight relation[4]. These mathematical spaces are the spaces in which the trajectories are determined. Even in Quantum Physics, when taking Hilbert's spaces as phase spaces for the wave function, Schrödinger's equation *determines* the dynamics of a probability density. The indeterministic aspect of quantum mechanics appears when quantum measurement projects the state vector — and gives a probability, as a real number value.

It is then possible to give a broader sense to the notion of phase space. For thermodynamics, say, Boyle, Carnot and Gay-Lussac decided to focus on pressure, volume and temperature, as the relevant observables: the phase space for the thermodynamic cycle (the interesting "trajectory") was chosen in view of its pertinence, totally disregarding the fact that gases are made out of particles. Boltzmann later unified the principles of thermodynamics to a particle's viewpoint and later to Newtonian trajectories by adding the ergodic hypothesis. Statistical mechanics thus, is not a reduction of thermodynamics to Newtonian trajectories, rather, as we said already, an "asymptotic" unification, at the infinite time limit of the thermodynamic integral, under the novel assumption of "molecular chaos" (and ergodicity). In statistical mechanics, ensembles of random objects are considered as the pertinent objects, and observables are derived as aspects of their (parameterized) statistics.

It should be clear that, while the term phase space is often restricted to a position/momentum space, we use it here in the general sense of the suitable or intended space of the mathematical and/or theoretical description of the system. In this sense the very abstract Hilbert space of complex probability densities is a phase space for the state function in Quantum Mechanics, very far form ordinary space-time.

Now, in biology, the situation poses several new challenges. Along the lines of [Kauffman, 2002, Bailly & Longo, 2008, Longo et al., 2012b], we will argue the following: in contrast to existing physical theories, where phase spaces are pre-given, if one takes organisms and phenotypes as observables in biology, the intended phase spaces need to be analyzed as changing in unpredictable ways through evolution. Our approach allows to understand this phenomenon by stressing the peculiar biological relevance of critical transitions and the related role of symmetry changes. It then adds a new form of proper biological randomness on top of the two main physical treatments of randomness that we will also discuss.

A further and major challenge is then posed, or made explicit, to the study of biological phenomena. We will motivate it by different levels of analysis. Of course, our result is a "negative result", but negative results may open the way to new scientific thinking, in particular by the very tools proposed to obtain them, [Longo, 2012]. Our tools are based on the role of symmetries and criticality, which will also suggest some possible ways out. In particular, we will add to this frame the notion of "enablement", as a further conceptual tool for the analysis of biological processes.

---

[4] One is the position and the other takes into account the mass and the change of position.

8.2 Phase Spaces and Symmetries in Physics

We understand the historically robust "structure of determination of physics" (which includes unpredictability, thus randomness) by recalling that, since Noether and Weyl, physical laws may be described in terms of theoretical symmetries in the intended equations (of the "dynamics", in a general sense, see chapter 5 and below). These symmetries in particular express the fundamental conservation laws of the physical observables (energy, momentum, charges ...), both in classical and quantum physics. And the conservation properties allow us to compute the trajectories of physical objects as geodetics, by extremizing the pertinent functionals (Hamilton principle applied to the Lagrangian functionals). It is the case even in Quantum Mechanics, as they allow to derive the trajectory of the state function in a suitable mathematical space, by Schrödinger equation.

As we said, only with the invention of an (analytic) geometry of space (Descartes), could trajectories be placed in a mathematically pre-given space, which later became the absolute space of Newtonian laws. The proposal of the more general notion of "phase space" dates of the late XIX century. Then momentum was added to spatial position, or energy to time, as an integral component of the analysis of a trajectory. This allowed to apply the corresponding conservation properties, thus the corresponding theoretical symmetries. In general, the phase spaces are the right spaces of description in the sense that they allow one to soundly and completely specify "trajectories": if one considers a smaller space, processes would not have a determined trajectory and could behave arbitrarily with respect to the elements of the description (for example, ignoring the mass or the initial speed in classical mechanics). Adding more quantities would be redundant or superfluous (for example, considering the color or flavor, in the usual sense, in classical mechanics).

In other words, in physics, the observables (and parameters), which form the phase space, derive from the choice of the (pertinent/interesting) invariants / symmetries, possibly by a suitable mathematization of trajectories. That is, they derive from the invariants and the invariant preserving transformations in the intended physical theory. So, Poincaré's momentum is preserved in the dynamics of an isolated system, similarly as Carnot's product pV is preserved at constant temperature while p and V may vary. Again, one uses these invariants in order to construct the "background space" where the phenomena under analysis can be accommodated. Thus, the conceptual construction of the phase space *follows* the choice of the relevant observables and invariants (symmetries) in the physico-mathematical analysis.

In summary, the historical and conceptual development of physics went as follows:

- analyze trajectories
- pull-out the key observables as (relative) invariants (as given by the symmetries)
- construct out of them the intended phase space.

8.2 Phase Spaces and Symmetries in Physics

Thus, physical (phase) spaces are not "already there", as absolutes underlying phenomena: they are our remarkable and very effective invention in order to make physical phenomena intelligible [Weyl, 1983, Bailly & Longo, 2011].

As H. Weyl puts it, the main lesson we learn from XX century physics is that the construction of scientific objectivity (and even of the pertinent objects of science) begins when one gives explicitly the reference system (or the phase space with its symmetries) and the metric (the measurement) on it. We do not consider anymore ether or phlogiston as pertinent observables nor parameters, thus they have been excluded from our phase spaces. The role of symmetries is also exemplified by the passage form the Galileo group to Lorentz-Poincaré group that frame Relativity Theory, as it characterizes the relevant physical invariants (in particular the new one, the speed of light) and invariant preserving transformations (Poincaré group) in the phase space.

In summary, the modern work of the theoretical physicist begins by setting the phase space and the measure in it, on the grounds of the observables he/she considers to be essential for a complete description of the intended dynamics — in the broadest sense, like in Quantum Physics, where quanta do not go along trajectories in ordinary space-time, but the wave or state function does, in a Hilbert space.

As for the formal foundation, from Descartes' spaces up to the later more general phase spaces (Hilbert spaces or alike), all these spaces are finitistically (axiomatically) describable, because of their symmetries. That is, their regularities, as invariants and invariant preserving transformations in the intended spaces (thus their symmetries), allow a finite description, even if they are infinite. Consider, say, a tri- (or more) dimensional Cartesian space, since Newton our preferred space for physics. It is infinite, but the three straight lines are given by symmetries (they are axes of rotations) and their right angles as well (right angles, says Euclid, are defined from the most symmetric figure you obtain when crossing two straight lines)[5]. When adding the different groups of transformations (the symmetries) that allow to relativize the intended spaces, one obtains the various physical theories that beautifully organize the inert matter, up to today.

Hilbert and Fock's spaces require a more complex but conceptually similar definition, in terms of invariants and their associated transformations. These invariants (symmetries) allow to handle infinity formally, possibly in the terms of Category Theory. Note that symmetries, in mathematics, have the peculiar status of being both invariant (structural invariants, say) and invariant preserving transformations (as symmetry groups).

In summary, symmetries thus allow to describe infinite spaces and mathematical structures, even of infinite dimension, in a very synthetic way, by the finitely many words of a formal definition and of a few axioms. We will argue for the intrinsic incompressibility of the phase space of intended observables in biology: no way to present it a priori, as a time invariant system, by finitely many pre-given words.

---

[5] More generally, modern Category Theory defines Cartesian products in terms of a symmetric commuting diagrams.

8.2.1 More Lessons from Quantum and Statistical Mechanics

As we observed, quantum mechanics takes as state function a probability density in possibly infinite dimensional Hilbert or Fock spaces. More generally, in quantum mechanics, the density matrix allows to deal also with phase spaces which are known only in part. In such cases, physicists work with the part of the state space that is known and the density matrix takes into account that the system can end up in an unknown region of the state space, by a component called "leakage term". The point is that this term interferes with the rest of the dynamics in a determined way, which allows us to capture theoretically the situation in spite of the leakage term.

In Quantum Field Theory (QFT) it is even more challenging: particles and anti-particles may be created spontaneously. And so one uses infinite dimensional Hilbert's spaces and Fock spaces to accommodate them. Of course, quanta are all identical in their different classes: a new electron is an electron ... they all have the same observable properties and underlying symmetries. Also, the analysis by Feynman diagrams allows us to provide the participation in the quantum state of each possible spontaneous creation and annihilation of particles (and, basically, the more complex a diagram is, the smaller its weight). The underlying principle is that everything that can happen, for a quantum system, happens, but only a limited number of possibilities are quantitatively relevant.

In statistical mechanics one may work with a randomly varying number n of particles. Thus, the dimension of the state space *stricto sensu*, which is usually $6n$, is not pre-defined. This situation does not, however, lead to particular difficulties because the possibilities are known (the particles have a known nature, that is relevant observables and equational determination) and the probabilities of each phase space are given[6]. In other terms, even if the exact finite dimension of the space may be unknown, it has a known probability — we know the probability it will grow by 1, 2 or more dimensions, and, most importantly, they are formally symmetric. The possible extra particles have perfectly known properties and possible states: the pertinent observables and parameters are known, one just misses: how many? And this becomes a new parameter ... (see for example [Sethna, 2006], for an introduction).

In these cases as well, the analysis of trajectories or the choice of the object to study (recall the role given to momentum or the case of the thermodynamic cycle or the probability density for QM) lead to the construction of the pertinent phase space, which contains the proper observables and parameters for the trajectories of the intended object. Then, as mentioned above, the symmetries of the theories allowed synthetic, even axiomatic, definitions of these infinite spaces, even with infinite or fluctuating dimensions. In other words, the finite description of these spaces of possibly infinite dimension, from Descartes to Quantum spaces, is made possible by their regularities: they are given in terms of mathematical symmetries.

---

[6] In general, n changes either because of chemical reactions, and it is then their rate which is relevant, or because the system is open, in which case the flow of particles is similar to an energetic flow, that is the number of particles plays the same role than energy: they are both fluctuating quantities obeying conservation laws.

8.2 Phase Spaces and Symmetries in Physics

And, since Newton and Kant, physicists consider the construction of the (phase) space as an "*a priori*" of the very intelligibility of any physical process.

8.2.2 Criticality and Symmetries

Critical transitions are particularly interesting with respect to phase spaces and their symmetries. As we know, they are characterized by a change of global behavior of a system, which is largely understood in terms of symmetry changes. More precisely there is two different symmetry changes for critical transitions: the change from the ordered versus disordered phase and the symmetry at the critical point.

Recall, for example, from 6, that spin lattices phase transitions are understood, from a purely macroscopic point of view, as a change of phase space: a parameter (the order parameter which is the global field in this example) shifts from being degenerate (uniformly null) to finite, non zero quantities. In other words, a new quantity becomes relevant. From a microscopic point of view, this quantity, however, is not exactly new: it corresponds to aspects used for the description of the microscopic elements of the system (a field orientation, for example). In the equational determination of the system as a composition of microscopic elements, there is no privileged directions for this observable. In the disordered phase (homogeneous, in terms of symmetries), the order parameter, that is the average of the field, is 0. However, in the ordered phase, the state of the system has a global field direction and its average departs from 0.

The appearance of this observable at the macroscopic level is understood thanks to an already valid observable at the microscopic level, and by changing macroscopic symmetries. That is, at the critical point, the point of transition, we have a collapse of the symmetry of the macroscopic orientations of the field (the symmetry is verified when the field is null). This change corresponds to the formation of a coherence structure which allows microscopic fluctuations to extend to the whole system and *in fine* to lead to a non null order parameter, the global field, after the transition. The system at the transition has a specific determination, associated to this coherence structure. Depending in particular on the dimension of space see section 6.2.1.5, this physical process can require a specific mathematical approach, the renormalization method, which allows to analyze the characteristic multi-scale structure of coherence, dominated by fluctuations at all scales, proper to critical situations. In all cases, this situation is associated to a singularity in the determination of the system, which stems from the order parameter changing from a constant to a non-zero value.

As a matter of fact, the concept of *extended critical transition* in 7 has been proposed initially to account for the specific coherence of biological systems, with their different levels of organization. The notion of different levels of organization is rather polysemic. It refers usually to the epistemic structuring of an organism by different forms of intelligibility, thus, *a fortiori* and if mathematically possible, by different levels of determination or mathematical description (molecular cascades, cells' activities and interactions, tissues' structures, organs, organisms ...). In the context of extended criticality, however, we propose to objectivize the levels of organization and especially the change of level by the mathematical breaking of the determination at one fixed level, by singularities. This approach sheds an original light on the notion of level of organization, as the new level corresponds to a coupling between scales and not simply to a higher scale, see [Longo et al., 2012c], an issue that we will not discuss here.

Recall that the core hypothesis of extended criticality is that, while physical systems have a mainly point-wise criticality, organisms have ubiquitous critical points (dense in a viability space, for example). Note that, in physics, the critical *point* can be an attractor: this is the paradigm of self-organized criticality. In our approach, the interval of extended criticality may be given with respect to any pertinent parameter. Its main properties along this line are given in chapter 7 (see also [Bailly & Longo, 2008, 2011]). In this context, the different levels of organization are presented by fractal or fractal-like structures and dynamics, as proposed by Werner and others, see [West, 2006, Werner, 2010, Longo et al., 2012c]. More recent applications of this concept may be found in [Lovecchio et al., 2012]. Note also that criticality enables a multi-scale heterogeneity to take place, which is usually impossible under the constraints of "normal" physical states. This heterogeneity is of interest for biological symmetry changes, as hinted in [Werner, 2010, Machta et al., 2011].

As discussed in 7, a crucial aspect of extended criticality is given by the role of symmetries and symmetry changes in biological dynamics. The density of critical points leads to omnipresent symmetry changes. Now, this has consequences for the very constitution of the scientific object. Physical objects are *generic* inasmuch different objects with the same equational determination will behave in the same way, and this way is determined by the *specific* trajectory provided precisely by the equations, a geodetic in the intended phase space. These specific trajectories, possibly after some transformations, allow to state that objects behave the same, both in the theory and in experiments (i.e. they have invariant properties). The trajectory is thus obtained by using theoretical symmetries (conservation principles, see above) and *in fine* it allows to define physical objects as generic, because they are symmetric (interchangeable, or they behave the same way).

In contrast to this core perspective in physics, variability, adaptability and diversity are at the core of biological objects and their dynamics. In this book, we proposed to capture this, by suggesting that biological objects do not have sufficiently stable *theoretical* symmetries, and, thus, that their trajectories are not specific. That is, there are no sufficiently stable symmetries and corresponding invariants, as for phenotypes, which would allow to determine the evolutionary dynamics of the object. On the contrary, and as we claimed in several places above, the biological object follows a *possible* evolutionary trajectory, which may be considered generic, or a "possible" one. Conversely, the living entity is not generic but specific, since it is determined by a *historical* cascade of symmetry changes (see chapter 7 and [Longo & Montévil, 2011a]). In our approach, the inversion of generic vs. specific is a core conceptual duality of biological theorizing vs. physical one. It deeply modifies the status of the object.

The starting assumption in this approach to evolutionary trajectories is based on Darwin's first principle (and default state for biology, [Sonnenschein & Soto, 1999]: *Descent with modification*. Darwin's other principle, *selection*, would make little sense without the first[7].

Notice that Darwin's first principle, descent with modification, may be understood as a *non-conservation* principle as for phenotypes (see 8.5): any reproduction yields (some) changes. It is crucial for us that this applies at each individual cellular mitosis. As a matter of fact, each mitosis may be seen as a critical transition. In a multicellular organism, in particular, it is a bifurcation that yields the reconstruction of a whole coherence structure: the tissue matrix, the collagen's tensegrity structure, the cells' networks and dialogue in general. And this besides the symmetry breakings due to proteome and DNA variations, which we will further discuss. In short, in view of the "density" of mitoses in the life interval of an organism, we may already consider this phenomenon at the core of biological analysis in terms of extended criticality. And mitoses are the fundamental processes of life, both uni- and multicellular life.

In this context, the mathematical un-predefinability of biological phase space we discuss below will follow by comparing the physico-mathematical constructions of trajectories and phase spaces to the needs of biology, where theoretical symmetries are not preserved. Let us recall that we work in a Darwinian frame and consider organisms and phenotypes as the pertinent observables.

8.3 Non-ergodicity and Quantum/Classical Randomness in Biology

We will discuss here the issue of "ergodicity" as well as the combination of quantum and classical random phenomena in biology. By ergodicity, we broadly refer to Boltzmann's assumption in the 1870's that, in the course of time, the trajectory of a closed system passes arbitrarily close to every point of a constant-energy surface in the given phase space. This assumption allows to understand a system without taking into account the details of its dynamics.

From the molecular viewpoint, the question is the following: are (complex) phenotypes the result of a random exploration of *all* possible molecular combinations and aggregations, along a path that would (eventually) explore all molecular possibilities, most later excluded by selection?

In physics, an easy combinatorial argument shows that at levels of complexity above the atom, for example for molecules, the universe is grossly non-ergodic, that is it does not explore all possible paths or configurations. Following an example in

---

[7] Note that some physicists claim to use a Darwinian scheme in order to understand physical systems where modification are purely quantitative. Our discussion in chapter 7 allows to understand the gap between the physical cases and actual biological evolution where modification involve changes of the theoretical symmetries. Of course, the notion of critical transition, where the symmetries of a physical object may change, provides a conceptual connection (or a point of "*conceptual* critical transition") ....

[Kauffman, 2002], the universe will not make all possible proteins length 200 amino acids in 10 to the 39th times its lifetime, even were all 10 to the 80th particles making such proteins on the Planck time scale. So, their "composition" in a new organ, function or organism (thus, in a phenotype) cannot be the result of the ergodicity of physical dynamics[8].

The point is that the lack of ergodicity presents an immediate difficulty for the (naive) reductionist approach to the construction of a phase space for biological dynamics, as given in purely molecular terms. In order to understand this, let's consider the role of ergodicity in statistical mechanics. A basic assumption of statistical mechanics is a symmetry between states with the same energetic level, which allows to analyze their probabilities (on the relevant time scales). This assumption is grounded on a hypothesis of ergodicity as for the dynamics of the particles: at the infinite time limit, they "go everywhere" in the intended phase space, and they do so homogeneously (with a regular frequency). In this case, the situation is described on the basis of energetic considerations (energy conservation properties, typically), without having to take into account the Newtonian trajectory or the history of the system.

In biology, non-ergodicity in the molecular phase space allows to argue that the dynamic cannot be described without historical considerations, even when taking only into account molecular aspects of biological systems. *A fortiori*, this holds when considering morphological and other higher scale biological aspects (the phenotypes in the broadest sense). In other terms, non-ergodicity in biology means that the relevant symmetries depend on a history even in a tentative phase space for molecules, which is in contrast with (equilibrium) statistical mechanics.

To sum the situation up, non-ergodicity prevents us to symmetrize the possible dynamics. With respect to a Darwinian phase space, most complex things will never exist and don't play a role, [Kauffman, 2002]. The history of the system enters into play and canalizes evolution.

Note that some cases of non-ergodicity are well studied in physics. Symmetry breaking phase transitions is a simple example: a crystal does not explore all its possible configurations because it has some privileged directions and it "sticks" to them. The situation is similar for the magnetization of a magnet, see [Strocchi, 2005] for a mathematical analysis. A more complex case is given by glasses. Depending on the models, the actual non-ergodicity is valid either for infinite time or is only transitory, yet relevant at the human time scales. Crucially, non-ergodicity corresponds to a variety of possible states, which depend on the paths in the energetic landscape that are taken (or not taken) during the cooling. This can be analyzed as an entropic distance to thermodynamic equilibrium and corresponds to a wide variety of "choices". However, the various states are very similar and their differences are relatively well described by the introduction of a time dependence for the usual thermodynamic quantities. This corresponds to the so-called "aging dynamics"

---

[8] Notice here that this argument only states that ergodicity in the molecular phase space does not help to understand the biological dynamics of phenotypes. The argument does not preclude the trajectories from being ergodic *in infinite time*. We can then say that infinite time ergodicity is biologically irrelevant and can take this irrelevance as a principle.

8.3 Non-ergodicity and Quantum/Classical Randomness in Biology

[Jensen & Sibani, 2007]. The example of glassy dynamics shows that the absence of a relevant ergodicity is not sufficient in order to obtain phase space changes in the sense we will describe, because in this example the various states can be understood in an *a priori* well-defined phase space and are not qualitatively different.

Note, finally, that an ergodic trajectory is a "random", yet complete, exploration of the phase space. However, ergodicity does not coincide with randomness, per se: a step-wise random trajectory (i.e. each step at finite time is random), does not need to be ergodic, since ergodicity, in mathematical physics, is an asymptotic notion.

Now, biological dynamics are a complex blend of contingency (randomness), history and constraints. Our thesis here is that biological (constrained) randomness is essential to variability, thus to diversity, thus to life.

The most familiar example is provided by meiosis, as gametes randomly inherit chromosomes pairs from the parents. Moreover, chromosomes of a given pair may exchange homologous portions and, so far, this is analyzed in purely probabilistic terms. It is a well established fact that DNA recombinations are a major contribution to diversity. However, all aspects of meiosis depend on a common history of the mixing DNA's and viable diversity is restricted by this history, starting with the common history as organism of the same "species".

A finer analysis can be carried on, in terms of randomness. In a cell, classical and quantum randomness both play a role and "superpose". Recall first that, in physics, classical and quantum randomness differ: different probability values (thus measures of randomness) may be associated to classical events vs. (entangled) quantum events. Bell inequalities distinguishes them (see [Aspect et al., 1982]).

Some examples of biologically relevant quantum phenomena are electron tunneling in cellular respiration [Gray & Winkler, 2003], electron transport along DNA [Winkler et al., 2005], quantum coherence in photosynthesis [Engel et al., 2007, Collini et al., 2010]. Moreover, it has been shown that double proton transfer affects spontaneous mutation in RNA duplexes [Ceron-Carrasco et al., 2009]. The enthalpic chaotic oscillations of macro-molecules instead have a classical nature, in physical terms, and are essential to the interaction of and with DNA and RNA. Quantum randomness in a mutation is typically amplified by classical dynamics (including classical randomness), in the interaction between DNA, RNA and the proteome (see [Buiatti & Longo, 2013] for a discussion). This kind of amplification is necessary in order to understand that changes at the nanometer scale impact the phenotype of the cell or of the organism. Moreover, it may be sound to consider the cell-to-cell interactions and, more generally, ecosystem's interactions as classical, at least as for their physical aspects, yet affecting the biological observables, jointly with quantum phenomena.

Poincaré discovered the destabilizing effects of planetary mutual interactions, in particular due to *gravitational resonance* (planets attract each other, which cumulates when aligned with the Sun); by this, in spite of the deterministic nature of their dynamics, they go along unstable trajectories and show random behavior, in astronomical times (see [Laskar J., 1994]). In [Buiatti & Longo, 2013], by analogy, the notion of "bio-resonance" is proposed. Different levels of organization, in an organism, affect each other, in a stabilizing (regulating and integrating), but also in a destabilizing way.

A minor change in the hormonal cascade may seriously damage a tissue's coherence and, years later, cause or enable cancer. A quantum event at the molecular level may be amplified by cell to cell interaction and affect the organism, whose changes may downwards affect tissues, cells, metabolism. Note that Poincaré's resonance and randomness are given at a unique and homogeneous level of organization (actually, of mathematical determination). Bio-resonance instead concerns different epistemic levels of organization, thus, a fortiori and if mathematically possible, different levels, yet interacting, of determination or mathematical description (molecular cascades, cells, tissues, organs, organisms ...).

In evolution, when a (random) quantum event at the molecular level (DNA or RNA-DNA or RNA-protein or protein-protein) happens to have consequences at the level of the phenotype, the somatic effects may persist if they are inherited and compatible both with the ever changing ecosystem and the "coherence structure" of the organism, that is, when they yield viable Darwin's correlated variations. In particular, this may allow the formation of a new function, organ or tool or different use of an existing tool, thus to the formation of a new properly relevant biological observable (a new phenotype or organism). This new observable has at least the same level of unpredictability as the quantum event, but it does not belong to the quantum phase space: it is typically subject to Darwinian selection at the level of the organisms in a population, thus it interacts with the ecosystem as such. Recall that this is the pertinent level of observability, the level of phenotypes, where biological randomness and unpredictability is now to be analyzed.

We stress again that the effects of the classical / quantum blend may show up at different levels of observability and may induce retroactions. First, as we said, a mutation or a random difference or expression in the genome, may contribute to the formation of a new phenotype. Second, this phenotype may retroact downwards, to the molecular (or quantum) level. A molecular activity may be excluded, as appearing in cells (organs / organisms) which turn out to be unfit — selection acts at the level of organisms, and may then exclude molecular activities associated to the unfit organism. Moreover, methylation and de-methylation downwards modify the expression of "genes". These upwards and downwards activities contribute to the integration and regulation of and by the whole and the parts. They both contribute to and constrain the biological dynamics and, thus, they do not allow to split the different epistemic levels of organization into independent phase spaces.

We recall that our choice of the biologically pertinent observables is based on the widely accepted fact that nothing makes sense in biology, if not analyzed in terms of evolution. We summarized the observables as the "phenotype", that is, as the various (epistemic) components of an organism (organs, tissues, functions, internal and ecosystemic interactions ...).

Thus, evolution is both the result of random events at all levels of organization of life and of constraints that canalize it, in particular by excluding, by selection, incompatible paths — where selection is due both to the interaction with the ecosystem and the maintenance of a possibly renewed internal coherent structure of the organism, constructed through its history. So, ergodic explorations are restricted or prevented both by selection and by the history of the organism (and of the ecosystem). For example, the presence and the structure of a membrane, or a nucleus, in a cell canalizes also the whole cellular activities along a restricted form of possible dynamics[9].

In conclusion, the "canalizing" role of history and selection, which excludes what is incompatible with the ecosystem and/or with the internal coherence of the organism, coexists with the various forms of randomness we mentioned. We find it critical that neither quantum mechanics alone, nor classical physics alone, account for evolution. Both seem to work together. Mutations and other molecular phenomena may depend on random, *acausal*, indeterminate quantum events. Thus they may interfere or happen simultaneously to or be amplified by classical dynamics, as well as by phenotype - phenotype interaction. In this amplification, evolution is also not completely random, as seen in the similarity of the octopus and vertebrates' camera eye, independently evolved (see below). Thus, evolution is both strongly canalized (or far from ergodic) and yet indeterminate, random and acausal. Our key point is then that random events, in biology, do not "just" modify the (numerical) values of an observable in a pregiven phase space, like in physics (even in the broad sense mentioned above, as in statistical physics). They modify the very phase space, or space of pertinent biological (evolutionary) observables, the phenotypes.

8.4 Randomness and Phase Spaces in Biology

Recall that we understand randomness in full generality as *unpredictability with respect to the intended theory*. As the practice of physics shows, this is a relativized notion, for example in the quantum vs. classical randomness debate. In either case, randomness is "measurable" and its measure is given by a probability. In pre-given spaces of possibilities (the pertinent phase spaces), modern probability theory is usually treated in terms of Lebesgue Measure Theory. More precisely, the measure (the probabilities) is given in terms of (relative) probabilities defined by symmetries with respect to the observable in a prestated phase space, as for the 6 symmetric faces of a fair dice. A more sophisticated example is the microcanonical ensemble of statistical mechanics, where the microstates with the same energy have the same probability (are symmetric or interchangeable), on the grounds of the ergodic hypothesis. In either case, the random event results in a symmetry breaking: one out of the six possible (symmetric) outcomes for a dice, the random exploration of a specific microstate in statistical mechanics (see section 5.5).

Recall that, by "theoretical symmetries", in biology, we refer both to the phenomenal symmetries in the phenotype and to the "coherence structure" of an organism, a niche, an ecosystem, in the broadest sense. In some cases, these symmetries may be possibly expressed by balance equations, at equilibrium or far from equilibrium, like in physics, or just by the informal description of its working unity as balanced

---

[9] See [Machta et al., 2011] for an analysis of the molecular spatial heterogeneity in the membrane as enabled by the coupling of phase transition fluctuations and the cytoskeleton.

processes of functions, organs and global autopoietic dynamics [Varela et al., 1974, Mossio & Moreno, 2010]. Under all circumstances, a permanent exploration and change is at the core of biology, or, as Heraclitus and Stuart Kauffman like to say: "Life bubbles forth". Yet, it does so while struggling to preserve its relative stability and coherence. We gave a more exact sense of this by the correlations between symmetry breackings and randomness in 5.5.

We need to understand this rich and fascinating interplay of stabilities and instabilities. Extended critical transitions in intervals of viability, the associated symmetry changes and bio-resonance may be a core tool for this: they yield coherence structures and change them continually, through epistemic levels of organization. Bio-resonance integrates and regulates the different levels within an organisms, while amplifying random effects due to transitions at one given level. At other levels of organization, these random events may yield radical changes of symmetries, coherent structures and, eventually, observable phenotypes.

In biology, randomness enhances variability and diversity. It is thus at the core of evolution: it permanently gives diverging evolutionary paths, as theoretical bifurcations in the formation of phenotypes. We also stressed that variability and diversity are key components of the structural stability of organisms, species and ecosystems, alone and together. Differentiation and variability within an organism, a species and an ecosystem contribute to their diversity and robustness, which, in biology, intrinsically includes *adaptiveness* — and, thus, it should be better called "resilience", see [Lesne, 2008]. Thus, robustness or resilience depend also on randomness and this by low numbers: the diversity in a population, or in an organ, which is essential to their resilience by adaptiveness, typically, may be given by few individuals (organisms, cells). This is in contrast to physics, where robustness by statistical effects inside a system is based on huge numbers of elementary components, like in thermodynamics, in statistical physics and in quantum field theory [Lesne, 2008]. Actually, even at the molecular level, the vast majority of cell proteins are present in very low copy numbers, so the variability due to proteome (random) differences after a mitosis, yields new structural stabilities (the new cells, possibly their differentiation in an organism) based on low but differing numbers.

Moreover, there exists a theoretical trend of increasing relevance that considers gene expression as a stochastic phenomenon. The theory of stochastic gene expression, usually described within a classical frame, is perfectly compatible, or it actually enhances our stress on randomness and variability, from cell differentiation to evolution[10]. In these approaches, gene expression must be given in probabilities and these probabilities may depend on the context (e. g. even the pressure on an embryo, see [Brouzés & Farge, 2004]). This enhances variability even in presence of a stable DNA.

Besides the increasingly evident stochasticity of gene expression, contextual differences may also force very different uses of the same (physical) structure. For example, the crystalline in a vertebrate eye and the kidney and their functions use the

---

[10] A pioneering paper on this perspective is [Kupiec, 1983]: recent surveys may be found in [Paldi, 2003, Arjun & van Oudenaarden, 2008, Heams, 2013].

8.4 Randomness and Phase Spaces in Biology

same protein [Michl et al., 2006], with different uses in these different context. Thus, if we consider the proper biological observable (crystalline, kidney), each phenotypic consequence or set of consequences of a chemical (enzymatic) activity has an *a priori* indefinite set of potential biological uses: when, in evolution, that protein was first formed, there was no need for life to build an eye with a crystalline. There are plenty of other way to see, and animals do not need to see. Similarly, a membrane bound small protein, by Darwinian pre-adaptation or Gould's exaptation, may become part of the flagellar motor of a bacterium, while originally it had various, unrelated, functions [Liu & Ochman, 2007]. Or, consider the bones of the double jaw of some vertebrates that evolved into the bones of the middle ears of mammals (one of Gould's preferred examples of exaptation), see [Allin, 1975]. A new function, hearing, emerged as the "bricolage" (tinkering) of old structures. There was no mathematical necessity for the phenotype nor for the function, "listening", in the physical world. Indeed, most complex things do not exist in the Universe, as we said.

Evolution may also give divergent answers to the same or to similar physical constraints. That is, the same function, moving, for example, or breathing, may be biologically implemented in very different ways. Trachea in insects versus vertebrates' lungs (combined with the vascular system), are due both to different contexts (different biological internal and external constraints) and to random symmetry changes in evolutionary paths. Thus, very different biological answers to the "same" physical context make phenotypes incomparable, in terms of physical optima: production of energy or even exchanging oxygen may be dealt with in very different ways, by organisms in the "same" ecosystem.

Conversely, major phenomena of convergent evolution shape similarly organs and organisms. Borrowing the examples in [Longo et al., 2012b], the convergent evolutions of the octopus and vertebrate eye follow, on one side, random, possibly quantum based acausal and indeterminate mutations, which contributed to very different phylogenetic paths. On the other, it is also "not-so-random" as both eyes converge to analogous physiological structures, probably due to physical and biological similar constraints — acting as co-constituted borders or as selection. The convergent evolution of marsupial and mammalian forms, like the Tasmanian wolf (a marsupial) and mammalian wolf are other examples of convergent, not-so-random components of evolution, in the limited sense above.

In conclusion, randomness, in physics, is "constrained" or mathematically handled by probabilities, in general with little or no relevance of history, and analyzed possibly by decorrelating events from contexts. In biology, *histories* and *contexts* (sometimes strongly) canalize and constraint random evolutions.

That is, *randomness may be theoretically constrained, in physics, by probability values in a pre-given list of possible future events; in biology, it is constrained by the past history and the context of an event.*

8.4.1 Non-optimality

Given the lack of ordered or orderable phase spaces, where numbers associated to observables would allow comparisons, it is hard to detect optimality in biology, except for some local organ construction. In terms of physical or also biological observables, the front legs of an elephant are not better nor worst than those of a Kangaroo or a bear: front podia of tetrapodes diverged (broke symmetries differently) in different biological niches and internal milieu. And none of the issuing paths is "better" than the other, nor follows physical optimality criteria, even less biological ones: each is just a possible variation on an original common theme, just compatible with the internal coherence and the co-constituted ecosystem that enabled them.

In general, thus, there is no way to define a real valued (Lagrangian) functional to be extremized as for phenotypes, as this would require an ordered space (ordered by a real valued functional), where "this phenotype" could be said to be "better" than "that phenotype". The exclusion of the incompatible, in a given evolutionary context, in no ways produces the "fittest" or "best", in any physico-mathematical rigorous sense. Even Lamarckian effects, if they apply, may contribute to fitness, not to "fitter", even less "fittest". Only *a posteriori* can one say that "this is better than that" — and never "best" in an unspecified partial upper semi-lattice: the *a posteriori* trivial evidence of survival and successful reproduction is not an *a priori* judgment, but an historical one. Dinosaurs dominated the Earth for more than 100 millions years, leaving little ecological space to mammals. A meteor changed evolution by excluding dinosaurs from fitness: only *a posteriori*, after the specific consequences of that random event, mammals may seem better — but do not mention this to the mammals then living in Yucatan. The blind cavefish, an "hopeful monster" in the sense of Goldschmidt, *a posteriori* seems better than the ascendant with the eyes, once it adapted to dark caverns by increasing peripheral sensitivity to water vibrations, a new or strongly enhanced phenotype. This incomparability *a priori* corresponds to the absence of a pre-given partial order among phenotypes, thus of optimizing paths, simply because their space is not pregiven. At most, sometimes, one can make a pair-wise *a posteriori* comparisons, which may be often associated to experimental situation, with controlled, simple conditions. This incomparability is also due to the relative independence of niches, which are co-constituted by organisms.

More generally, conservation or optimality properties of physical observables (the various forms of physical energy, for example) cannot help to determine the evolutionary trajectory of an organism. No principle of "least free energy" (or "least time consumption of free energy", if it applies) can help to predict or understand completely the evolution of a proper and specific biological observable, nor of an organism as object of selection. At most, some organs, where exchanges of energy are the key functions, may be partially analyzed in terms of an optimizing physical dynamics: morphogenesis as a sometimes mathematically beautiful contribution to phyllotaxis and organogenesis. In other words, the analysis of physical forces may help to determine, by "optimality" principles, the dynamics only locally, typically the form of some organs, where exchanges of matter or energy dominate (lungs, vascular system, phyllotaxis, ...). Their forms *partly* follow optimality principles (dynamical branching, sprouting or fractal structures or alike, see [Jean, 1994, Fleury, 2000, Bailly et al., 1988]). In these cases, physical forces (the pushing of the embryonic heart, respiration ..., tissue matrix frictions ...) must be understood as fundamental *dynamical constraints* to biology's default state: proliferation with variation and motility. Then selection applies at the level of phenotypes and organisms. Thus, the resulting form is *incompletely* understood by looking only at the physical dynamical constraints, since those dynamics depend also on the integration and regulation in and by the organism, including its DNA. Moreover, variability and diversity (the irregularity of lungs, of plants organs in phyllotaxis...) contribute to robustness in an essential way. They are not "noise" as in crystals' formation, but they are at the core of adaptivity and biological resilience.

Moreover, a given physical ecosystem may yield very different organisms and phenotypes, by variability and adaptivity, as reproduction with variation and selection. As Darwin says, descent implies modification, even without being prompted by the environment. This does not forbid to think that, in some cases, modifications may be also prompted by the environment; the simplest example is given by the accelerating (bacterial) mutations under ecosystemic stress.

8.5 A Non-conservation Principle

The phylogenetic change underlying evolution may be understood in terms of a "non-conservation principle" of biological observables. Darwin proposed it as a principle, to which we extensively referred: *descent with modification*, on which selection acts. This is the exact opposite of the symmetries and conservation properties that govern physics and the related equational and causal approaches. There is of course structural stability, in biology, which implies similar, but never identical iteration of a morphogenetic process. As for organisms, this has been extensively described as autonomy under constraints, autopoiesis and alike — yet, without change, the early autopoietic systems would still be at the first bacterium, see [Moreno & Mossio, 2013] for a recent insight and account . That is, evolution requires also and intrinsically this non-conservation principle for phenotypes in order to be made intelligible. In particular, one needs to integrate randomness, variability and diversity in the theory in order to understand phylogenetic and ontogenetic adaptability and the permanent exploration and construction of new niches.

In a sense, we need, in biology, a similar enrichment of the perspective as the one quantum physicists dared to propose in the '20th: intrinsic indetermination was introduced in the theory by formalizing the non-commutativity of measurement (Heisenberg non-commutative algebra of matrices) and by Schrödinger equation (the deterministic dynamics of a probability law). We propose here an analysis of indetermination at the level of the very formation of the phase space, or spaces of evolutionary possibilities, by integrating Darwin's principle of reproduction with modification and, thus, of variability, in the intended structure of determination.

As a further consequence, the concept of randomness in biology that we are constructing mathematically differs from physical forms of randomness. Indeed, we cannot apply a probability measure to it because there is no pregiven space of possible phenotypes in evolution (nor, we should say, in ontogenesis, where monsters appear, sometimes hopeful from the point of view of evolution). The lack of probability measures may resemble the "do not care" principle in algorithmic concurrency, over computer networks, mentioned above (and networks are fundamental structures for biology as well). However, the possible computational paths are pregiven and, moreover, processes are described on discrete data types, which are totally inadequate to describe the many biological dynamics that are better analyzed by mathematics of continua. Indeed, the sequential computers, in each node, are Laplacian Discrete State machine, as Turing first observed [Longo et al., 2012a], far away from organisms.

In summary, in biology, the superposition of quantum and classical physics, bioresonance, the coexistence of indeterminate acausal quantum molecular events, with somatic effects, and of non-random historical and contextual convergences do not allow to invent, as physicists do, a mathematically stable, pre-given phase space, as a "background" space for all possible evolutionary dynamics.

Random events break symmetries of biological trajectories in a constitutive way. A new phenotype, a new function, organ ... organism, is a change (a breaking and a reconstruction) of the coherence structure, thus a change of the symmetries in the earlier organism. Like in physics, symmetry changes (thus breakings) and randomness seem to coexist also in life dynamics, but they affect the dynamics of the very phase space.

Our approach to the biological processes as extended critical transitions fits with this understanding of biological trajectories as cascades of symmetry changes 7. Of course, this instability goes together with structural stability and is even an essential component of it: each critical transition is a symmetry change and it provides variability, diversity, thus adaptivity which is at the core of biological viability. Even an individual organism is adaptive to a changing ecosystem, thus biologically robust, by the ever different re-generation and remodeling of its parts. The sensitivity to minor fluctuations close to transition, which a signature of critical phase transitions, enhances adaptivity of organisms (DNA methylation may affect even adaptive behavior, [Kucharski et al., 2008]).

Recall now the historical and conceptual path that lead physics to invent modern mathematization, by first focusing on some key invariants of trajectories, Galileo's inertia as momentum conservation, typically, then by considering them as fundamental observables for the deterministic/determining phase space, with respect to the intended parameter, space in this case. In view of the remarks above, it seems impossible to extract relevant invariants concerning the specific structure of phenotypes or organisms and construct with them a space of all possible phenotypes. It may be even inadequate as variability is *one of* the main theoretical invariants in biology, beginning with individual mitoses. This does not forbid to propose some general invariants and symmetries, yet not referring to the *specific* aspects of the phenotype, as form and function. This is the path we followed when conceptualizing sufficiently stable properties, such as biological rhythms (3), extended criticality (7) and anti-entropy (9).

We follow by this physics' historical experience of "objectivizing" by sufficiently stable *concepts*. In biology, these must encompass change and diversity. As a matter of fact, our investigations of biological rhythms, extended criticality and anti-entropy are grounded also on variability. In a long term perspective, these concepts should be turned all into more precisely quantified (and correlated) mathematical invariants and symmetries, in *abstract* spaces. This is what we hinted as for the two dimensional time of rhythms and as for anti-entropy (see next chapter, 9), by imitating the way Schrödinger defined his equation in Hilbert spaces. That is an abstract phase space for the quantum state function far away from ordinary momentum or energy, parameterized over space or time. Abstract properties such as extended criticality and anti-entropy do not refer to the invariance of specific phenotypes, but they are *themselves* relatively stable, as they seem to refer to the few invariant properties of organisms. Their analysis, in an eventually quantified space of extended criticality, may give us a better understanding of objects and trajectories within the ever changing space of phenotypes.

8.6 Causes and Enablement

We better specify now the notion of *enablement*, proposed in [Longo et al., 2012b] and already used above. This notion may help to understand the role played by ecosystemic dynamics in the formation of a new observable (mathematically, a new dimension) of the phase space. Examples are given below and we will refine this notion throughout the rest of this chapter.

In short, a niche *enables* the survival of an otherwise incompatible/impossible form of life, it *does not cause* it. More generally, niches enable what evolves, while evolving with it. At most, a cause may be found in the "difference" (a mutation, say) that induced the phenotypic variation at stake, as spelled out next.

This new perspective is motivated, on one side, by our understanding of physical "causes and determinations" in terms of symmetries, along the lines above of contemporary physics, and, on the other side, by our analysis of biological "trajectories" in phylogenesis (and ontogenesis), as continual symmetry changes. Note that, in spite of its replacement by the language of symmetries, the causal vocabulary still makes sense in physics: gravitation, for example, *causes* a body to fall (of course, Einstein's understanding in terms of geodetics in curved spaces, unifies gravitation and inertia in terms of symmetries and conversations properties, it is thus more general). Gravitation instead, in embryogenesis, say, is a (fundamental) constraint, not a cause[11].

In biology, without sufficiently stable invariances and symmetries at the level of organisms, thus without (possibly equational) laws for evolutionary dynamics, "causes" positively and completely entailing, at least in principle, these dynamics

---
[11] In microgravity, the less constrained cell's reproduction generates more variability in the cytoscheleton, ongoing work by the ESA groupe in Rome, lead by Bizzarri.

cannot be defined. As part of this understanding, we will discuss causal relations in a restricted sense, that is, in terms of "differential causes". In other words, since symmetries are unstable, causality in biology cannot be understood as "entailing causality" as in classical and relativistic physics and this will lead us to the proposal that *in biology, causal relations are only differential causes*. If a bacterium causes pneumonia, or a mutation causes a monogenetic diseases (anemia falciformis, say), this is a *cause* and it is differential, i.e. it is a difference with respect to what is fairly considered "normal", "healthy" or "wild" as biologist say as for the genome, and it causes a " pathology" or an abnormal phenotype.

A classical mistake is to say: this mutation causes a mentally retarded child (a famous genetic disorder, phenylketonuria), thus ... the gene affected by the mutation is the gene of intelligence, or ... here is the gene that causes/determines the intelligence, [Weiss, 1992, Stewart, 2004], or that encodes for (part of) the brain. In logical terms, this consists in deducing from "notA implies notB" that "A implies B" (or from "not normal A implies not normal B", that "normal A implies normal B"): an amazing logical mistake. All that we know is a causal correlation of differences[12].

We then propose to consider things differently. The observed or induced difference, a mutation with a somatic effect, say, or a stone bumping on someone's head, or a carcinogen (asbestos), does *cause* a problem; that is, the causal dictionary is suitable to describe a *differential cause - effect relation*. The differential cause modifies the space of possibilities, that is the compatibility of the organism with the ecosystem. In other terms, it modifies the "enablement relations". This is for us the way an organism, a niche, an ecosystem may accommodate a phenotype, i.e. when the modified frame becomes viable for a new or different phenotype (a new organ or function, a differentiated organism).

We are forced to do so by the radical change of the default state in biology. Inertial movement, or, more generally, conservation principles in physics, need a force or an efficient cause to change[13]. In biology, in contrast to physics, the default state guaranties change: reproduction with variation and motility, [Longo et al., 2013]. Differential causes "only" affect the intrinsic (the default) dynamics of organisms, which are a priori "active". More precisely, in our view, the differential causes modify the always reconstructed coherence structure of an organism, a niche, an ecosystem. So enablement is modified: a niche may be no longer suitable for an organism, or an organism for the niche, or a new niche and organism may be formed, by a difference. That is, a change in a niche, due to a differential physical cause (a climate

---

[12] Schrödinger, in his 1944 book, was well aware of the limits of the differential analyses of the chromosomes and their consequences: "What we locate in the chromosome is the seat of this difference. (We call it, in technical language, a 'locus', or, if we think of the hypothetical material structure underlying it, a 'gene'.) Difference of property, to my view, is really the fundamental concept rather than property itself.", p.28.

[13] See the revitalization of the Aristotelian distinction efficient vs. material cause, in [Bailly & Longo, 2011]: following the terminology of quantum physics, the first may change states, the second affects properties.

8.6 Causes and Enablement

change, for example), may negatively select existing organisms while enabling the adaptive ones, since the enablement relations differ.

Differential analysis are crucial in the understanding of existing niches. Short descriptions of niches may be given from a specific perspective (they are strictly epistemic): they depend on the "purpose" one is looking at, say. And one usually finds out a feature in a niche by a difference, that is, by observing that, if a given feature is modified, the intended organism dies. Then, as long as niches are compared by differences, one may not be able to prove that two niches are identical or equivalent (in enabling life), but one may show that two niches are different. Once more, there are no symmetries organizing over time these spaces and their internal relations.

In summary, while gradually spelling out our notion of enablement, we claim that only the differential relations may be soundly considered causal. Moreover, they acquire a biological meaning only in presence of enablement. In other words:

1. In physics, in presence of an explicit equational determination, causes may be seen as a formal symmetry breaking of the equations. Typically, $f = ma$, a symmetric relation, means, for Newton, that a force, f, causes an acceleration a, asymmetrically. Thus, one may consider the application of a Newtonian force as a differential cause[14]. This is so, because the inertial movement is the "default" state in physics ("nothing happens" if no force is applied). This analysis cannot be globally transferred to biology, inasmuch the default state is activity, symmetries are not stable and, thus, one cannot write equations for phylogenetic trajectories (nor break their symmetries).
2. As recalled several times, the default state in physics is inertia. In biology instead, the default state is "activity", as proliferation with variation and motility. As a consequence, an organism, a population, a species, *does not need a cause* to be active, e.g. to reproduce with modifications and move, and possibly occupy a new niche[15]. That is, an organisms only needs to be enabled in order to survive by changing. Moreover, in our terms, this default state involves continual critical transitions, thus symmetry changes, up to phase space changes.

Consider for example an adjacent possible empty niche, for instance Kauffman's example of the swim bladder (see for example [Kauffman, 2002, 2012, Longo et al., 2012b]), formed by Gould's exaptation from the lung of some fishes. Is it a boundary condition? Not in the sense this term has in physics, since the swim bladder may enable a (mutated) worm or a bacterium to live and evolve, according to *unpredictable enabling relations*. That is, the observable features of the swim bladder to be used by the new organism to achieve functional closure in its

---

[14] In an informal/naive way, one may say that Einstein reversed the causal implication: a space curvature "causes" an acceleration that "causes" a field, thus a force (yet, the situation is slightly more complicated and the language of symmetries and geodetics is the only rigorous one, in particular in reference to Lorentz-Poincaré group of symmetries).

[15] Energy or matter, of course, is needed in order to reproduce, but it is not a cause. As we spell out in 9, in biology energy is a parameter, like in allometric equations, it is not an "operator", like in physics.

environment may be radically new, possibly originating for both in a quantum based acausal/indeterminate molecular event and by correlated variations: the niche and the bacterium functionally shape each other. As discussed above, the combination of various forms of (physical and biological) randomness modify the set of observables (the new organ, the new bacterium), not just the values of some observables.

Once more, in physics, energy conservation properties allow us to derive the equations of the action/reaction system proper to the physical phenomenon in a pre-given phase space. Random event may modify the *value* of one of the pertinent observable, not the very set of observables. Typically, a river does co-constitute its borders by frictions, yet the observables and invariants to be preserved are well-know (energy and/or momentum), the game of forces as well. It may be difficult to write all the equations of the dynamics and some non-linear effects (frictions ...) may give the unpredictability of the trajectory. Yet, we know that the river will go along a unique perfectly determined geodetics, however difficult it may be to calculate it exactly (to calculate the exact numerical values of the dynamics of the observables). Yet, a river never goes wrong and we know why: it will follow a geodetics. An onto- or phylogenetic trajectory may go wrong, actually most of the time it goes wrong. We are trying to theoretically understand "how it goes", between causes and enablement.

In summary, enablement and proliferation with variation and motility as default state are at the core of the intelligibility of life dynamics. They conceptually frame the development of life in absence of a pre-definable phase space.

As we recalled, niches and phenotypes are co-constituted observables. Typically, the organism adjusting to / constructing a new niche may be a hopeful monster, that is the result of a "pathology" [Dietrich, 2003, Gould, 1977]. Now, notions of "normal" and "pathological" makes no sense in physics. They are contextual and historical in biology; they are contingent yet fundamental.

These differing notions may also help to distinguish between enablement and causality, as the latter may be understood as a causal difference in the "normal" web of interactions. In evolution, a difference (a mutation) may cause a "pathology", as hopeful monster. This monster, which is such with respect to the normal or wild phenotype, may be killed by selection or may be enabled to survive by and in a new co-constituted niche. A dark cavern may be modified, also as a niche for other forms of life, by the presence of the blind fish. And the contingent monster becomes the healthy origin of a speciation.

Thus, besides the centrality of enablement, we may maintain the notion of cause — and it would be a mistake to exclude it from the biological dictionary. As a matter of fact, one goes to the doctor and rightly asks for the cause of pneumonia — not only what enabled it: find and kill the bacterium, please, that is the cause. Yet, that bacterium has been enabled to grow excessively by a weak lung, a defective immune system or bad life habits .... Therefore, the therapy should not only concern the differential cause, the incoming bacteria, but investigate enablement as well [Noble, 2009]. And good doctors do so, without necessarily naming it.

Finally, following [Sonnenschein & Soto, 1999], by our approach we understand cancer as being enabled by a modified "society of cells" (the concerned tissue, organ, organism). A carcinogen affecting the organism (typically, the epithelial stroma, [Sonnenschein & Soto, 1999, Maffini et al., 2004]) deferentially modifies the "normal" tissue-niche for the cells and its coherence structure. The less controlled cells' default state, proliferation with variation, may then lead to the abnormal proliferation, possibly with increasing variation (as an elementary example, a teratoma has a larger number of cell types than a normal tissue).

8.7 Structural Stability, Autonomy and Constraints

Organisms withstand the intrinsic instability / unpredictability of the changing phase space, by the relative autonomy of their structural stability. They have an internal, permanently reconstructed autonomous coherent structure, Kantian wholes (in Kant's sense, see [Kant, 1781, Longo & Perret, 2013]), or Varela's autopoiesis, that gives them an ever changing, yet "inertial" structural stability. We proposed to understand a component of this inertia for organisms in terms of biological protention, in chapter 4. They achieve a closure in a functional space by which they reproduce, evolve and adapt by changing alone or together out of the indefinite and unorderable set of functions, or by finding new uses of pre-existing components to sustain their activity in the ongoing co-evolution in the ecosystem.

The niche is indefinite in features prior to proliferation with variation and selection revealing what will co-constitute "task closure" for the organism. The niche allows the tasks' closure by which an organism survives and reproduces.

Organisms and ecosystems are structurally stable, also because of their *constrained autonomy*, as they permanently and non-identically reconstruct themselves, their internal and external constraints. They do it in an always different, thus adaptive, way. They change the coherence structure, thus its symmetries. This reconstruction is random, but not completely as it heavily depends on constraints, such as the proteins types imposed by the DNA, the relative geometric distribution of cells in embryogenesis, interactions in an organism, in a niche. Yet, the autopoietic activity is based also on the opposite of constraints: the relative autonomy of organisms. In other words, organisms transform the ecosystem while transforming themselves and they can stand this continual changes because they also have an internal preserved coherent structure (Bernard's "milieu intérieur"). Its stability is maintained also by slightly, yet constantly changing internal symmetries, which enhance adaptivity, beginning with individual cellular mitosis in a multicellular organisms.

As we said, autonomy is integrated in and regulated by constraints, within an organism itself and of an organism within an ecosystem. Autonomy makes no sense without constraints and constraints apply to an autonomous unity. So constraints shape autonomy, which in turn modifies constraints, within the margin of viability, i.e. within the limits of the interval of extended criticality.

A way to understand the impossibility of a complete *a priori* description of actual and potential biological organisms and niches may be the following. Recall first the role of observable invariants and conservation properties in establishing physical phase spaces, since Galileo's inertia and the corresponding symmetry group, in chapter 5. Then, recall how this allowed finite definitions, in terms of symmetries, of abstract, possibly infinite, phase spaces. As a consequence of our analysis in terms of symmetry breakings, any given, possibly complete description of an ecosystem is *incompressible*, in the sense that any linguistic description may require new names and meanings for the new unprestatable functions. These functions and their names make only sense in the newly co-constructed biological and historical (even linguistic) environment. There is no way to define them *a priori* with finitely many words. The issue then is not infinity, but incompressibility by the lack of invariant symmetries, which we described in relation to extended criticality.

8.8 Conclusion

We recalled here the role of invariance, symmetries and conservation properties in physical theories, as also hinted in chapter 5. Our preliminary aim, here and in chapter 7 has been to show that the powerful methods of physics that allowed to pre-define phase spaces on the grounds of the observables and the invariants in the "trajectories" (the symmetries in the equations) do not apply in biology.

In biology, symmetries at the phenotypic level are continually changed, beginning with the least mitosis, up to the "structural bifurcations" which yield speciations in evolution. Thus, there are no biological symmetries that are *a priori* preserved, except and for some time, some basic structures such as bauplans (still more or less deeply modified during evolution). There are no sufficiently stable mathematical regularities and transformations to allow an equational and law like description entailing the phylogenetic and ontogenetic trajectories. These are cascades of symmetry changes and thus just cumulative historical dynamics. And each symmetry change is associated to a random event (quantum, classical or due to bio-resonance), at least for the breaking of symmetries, while the global shaping of the trajectory, by selection say, is also due to non-random events. In this sense biological trajectories are generic: they are just possible ones and yield a historical result, that is an individuated, specific organism (see chapter 7, [Bailly & Longo, 2011, Longo & Montévil, 2011a]).

In other words, this sum of individuals and individualizing histories, co-constituted within an ever changing ecosystem, does not allow a compressed, finite or formal description of the space of possibilities, that is, of the actual biological phase space (functions, phenotypes, organisms): these possibilities are each the result of an unpredictable sequence of symmetry breakings. This situation is in contrast to the invariant (conservation) properties which characterize physical "trajectories", in the broad sense (extended to Hilbert's spaces, in Quantum Mechanics).

An immense literature has been tackling "emergence" in life phenomena. Yet, in the technical analyses, the strong and dominating theoretical frames inherited from mathematical physics (or even computing) do not seem to have been abandoned. In approaches from Artificial Life to Cellular Automata and various very rich analysis of dynamical systems, the frame for intelligibility is *a priori* given under the form, often implicitly, of one or more pre-defined phase spaces, possibly to be combined

8.8 Conclusion

by adequate mathematical forms of products (Cartesian, tensorial products ...). A very rich and motivated frame for these perspectives is summarized in [Drake et al., 2007]. Well beyond the many analysis which deal with equilibrium systems, an inadequate frame for biology, these authors analyze interactions between multiple attractors in dissipative dynamical systems, possibly given in two or more phase spaces (the notion of attractor is a beautiful mathematical notion, which requires explicit equations or evolution functions — solutions with no equations — in pertinent phase spaces in order to be soundly presented). Then, two or more deterministic, yet highly unpredictable and independent systems, which interact in the attractor space, may "produce persistent attractors that are offsprings of the parents.... Emergence in this case is absolute because no trajectories exist linking the child to either parent (p. 158) ... [The] source [of emergence] is the creation, evolution, destruction, and interaction of dynamical attractors (p. 179)".

This analysis is compatible with ours and it may enrich it by a further component, in pre-given interacting phase spaces. Yet, we go somewhat beyond pre-given phase spaces, by a critical perspective, which, per se, is a tool for intelligibility. Below, we will hint again to further possible (and positive) work, besides negating the possibility of an a priori and compressed mathematical description of (combined) spaces of evolution.

In summary, in our approach, the intrinsic unpredictability of the very *Phase Space* of phylogenetic (and ontogenetic) dynamics is due to:

1. physical and properly biological randomness, including bio-resonance, due to interacting levels of organization, as a component both of integration and regulation, in an organism, as well as of amplification of random fluctuations in one level of organization through the others;
2. extended criticality, as a locus for the correlation between symmetry breaking and randomness;
3. cascades of symmetry changes in (onto-) phylogenetic trajectories;
4. enablement, or the co-constitution of niches and phenotypes, a notion to be added to physical determination.

These phenomena are also crucial in order to understand life persistence, as they are at the origin of variability, thus of diversity and adaptability, which are an integral part of life stability. Our theoretical frame, in particular, is based on reproduction with variation and motility as proper default state for the analysis of phylo- and ontogenesis. Selection shapes the bubbling forth of life by excluding the incompatible.

By the lack of mathematically stable invariants (stable symmetries), there are no laws that entail, as in physics, the biological observables in the becoming of the biosphere. In physics, the geodetic principle mathematically forces objects never to go wrong. A falling stone follows exactly the gravitational arrow. A river goes along the shortest path to the sea, it may adjust adjust its path by nonlinear well definable interactions as mentioned above, but it will never go wrong. These are all geodetics. Even though it may be very hard or impossible to compute them, they are unique, by principle, in physics. Living entities, instead, may follow many possible paths, and they go wrong most of the time: most organisms are extinct, almost half of fecundations in mammals do not lead to a birth, an amoeba does not follows, exactly, a curving gradient — by retention it would first go along the former tangent, then correct the trajectory, in a protensive action. In short, life goes wrong most of the time, but it "adjusts" to the environment and changes the environment, if possible: it is adaptive. It maintains itself, always in a critical transition, that is within an extend critical interval, whose limits are the edge of death. It does so by changing the observables, the phenotypes and its niche — in the sense of Darwinian correlated variations of organisms and ecosystems. Thus, it is the very nature and phase space of the living object that changes, in contrast to physics.

We must ask new scientific questions and invent new tools, for this co-constitution by organisms as they co-evolve and make their worlds together. This must be seen as a central component of the biosphere's dynamics. The instability of theoretical symmetries in biology is not, of course, the end of science, but it sets the limits of the transfer of physico-mathematical methods, as taught us from Newton onward, to biology. Kant already doubted of this, [Kant, 1781]. In biological evolution we cannot use the same very rich interaction with mathematics as it has been constructed at the core of physical theories. However, mathematics is a human adaptive construction: an intense dialogue with biology may shape for it new scientific paths, concepts, structures, as it did with physics since Newton.

By providing some theoretical arguments that yield this "negative result", in terms of symmetries and critical transitions, we hope to have provided also some tools for a new opening. Negative results marked the beginning of new sciences in several occasions: the thermodynamic limit to energy transformation (increasing entropy), Poincaré's negative result (as he called his Three Body Theorem), Gödel's theorem (which set a new start to Recursion Theory and Proof Theory) all opened new ways of thinking, [Longo, 2012]. Limits clarify the feasible and the non feasible with the existing tools and may show new directions by their very nature, if these limits have a sufficiently precise, scientific content.

The scientific answer we propose to this end of the physicalist certitudes, is based on our analysis of symmetry changes in extended critical transitions and on the notion of "enablement" in evolution (and ontogenesis). Enablement concerns how organisms co-create their worlds, with their changing symmetries and coherence structures, such that they can exist in a non-ergodic universe.

Our thesis is that evolution, as a "diachronic process" of becoming (but ontogenesis as well), "enables", but does not cause, unless differentially, the forthcoming state of affairs. Moreover, Galileo and Newton's entailed trajectories mathematized Aristotle's "efficient cause" only. Instead, in our view, in biological processes, such *entailed causal relations must be enriched by "enablement" relations*, plus differential, physical, often quantum indeterminate, causes.

Life is caught in a causal web, but lives also in a web of enablement and radical emergence of life from life, whose intelligibility may be largely given in terms of symmetry changes and their association to random events at all levels of organization.

8.8 Conclusion

As hinted in 8.5, a long term project would be to better quantify our approaches to two dimensional time for rhythms, to extended criticality and to anti-entropy (see next chapter 9), in order to construct from them an abstract phase space based on these mathematically stable properties. The dynamical analysis should follow the nature of Darwin's evolution, which is an historical science, not meant to "predict", yet giving a remarkable understanding of the living. Thus, the dynamics of extended criticality or anti-entropy should just provide the evolution of these state functions, or how these abstract observables may develop with respect to the intended parameters, including time. And this, without being "projectable" on specific phenotypes, even not in probabilities, as it is instead possible for Schrödinger's state functions in Quantum Mechanics. To this purpose, one should give a biologically interesting measure for extended criticality and describe in a quantitative way, in the abstract space of extended critical transitions, the qualitative evolution of live. In a preliminary way, we have been able to do so, by following Gould's analysis of increasing biological complexity, that is in the analysis of the evolutionary dynamics of a global observable we will call anti-entropy, 9.

Chapter 9
Biological Order as a Consequence of Randomness: Anti-entropy and Symmetry Changes

Abstract. In this chapter, we introduce the notion and the analysis of phenotypic complexity, as anti-entropy, proposed in [Bailly & Longo, 2009] and develop further theoretical consequences. In particular, we analyze how randomness, an essential component of biological variability, is associated to the growth of biological organization, both in evolution and in ontogenesis. Our approach, in particular, will focus on the role of global entropy production and will provide a tool for a mathematical understanding of some fundamental observations by S.J. Gould on how phenotypic complexity increases, on average, along random evolutionary paths, without a bias towards an increase. We also propose a preliminary analysis of biological regenerative processes, which allows to associate entropy production of adults to anti-entropy, by considering "collisions" between entropy and anti-entropy. Lastly, we analyze the situation in terms of theoretical symmetries, in order to further specify the biological meaning of anti-entropy as well as its strong correlations to randomness[1].

Keywords: entropy production, macroevolution, metabolism, regeneration, variability, randomness, anti-entropy.

9.1 Introduction

Notions of entropy are present in different branches of physics, but also in information theory, biology ... even economics. Sometimes, they are equivalent under suitable transformations from one (more or less mathematized) domain to another. Sometimes, the relation is very mild, or may be at most due to a similar mathematical expression. For example, one often finds formulas describing a linear dependence of entropy from a quantity formalized as $-\sum_i p_i \log(p_i)$, where the p_i are a measure of the probability of the system to be in the i-th (micro-)states. Yet, different theoretical frames may give very different physical meanings to these formulas: somehow like a wave equation describing water movement has a similar mathematical formulation

---
[1] Part of these ideas have been presented in [Longo & Montévil, 2012].

as Schrödinger's wave equation (besides some crucial coefficients), yet water waves and quantum state/wave functions have nothing to do with each other.

Besides the formula, another element seems to be shared by the different meanings given to entropy. The production of entropy is strictly linked to irreversible processes.

But ... what is entropy? The notion originated in thermodynamics. The first law of thermodynamics is a conservation principle for energy. The second law states that the total entropy of a system will not decrease other than by increasing the entropy of some other system. Hence, in a system isolated from its environment, the entropy of that system will not decrease.

More generally, in physics, increasing entropy corresponds to *energy dispersion* (or diffusion). And here we have the other feature shared by the different views on entropy: in all of its instances, entropy is linked to randomness, since diffusion, in physics, is based on *random walks*. Thus, energy, while being globally preserved, diffuses. In particular, heat flows from a hotter body to a colder body, never the inverse, and this by random particles' walks. Only the application of work (the imposition of order) may reverse this flow. As a matter of fact, entropy may be locally reversed, in some cases, by pumping energy. For example, a centrifuge may separate two gazes, which mixed up by diffusion. This separation reduces the ergodicity (the amount of randomness, so to say) of the system, as well as its entropy.

Living beings construct order by absorbing energy. In Schrödinger's audacious little book, *What is life?* [Schrödinger, 2000], it is suggested that organisms *also* use order to produce order, which he calls *negentropy* in the second part of his book, that is entropy with a negative sign. And this order is produced by using the order of the chromosomes' aperiodic structure (his first conjecture) *and* by absorbing organized nutrients (don't we, the animal, eat mostly organized fibers?). Of course, a lot can be said, today, against these tentative theorizations by the great physicist, yet they suggest interesting paths for thought — in particular the second part.

But is really entropy the same as disorder? There is a long lasting and sound critique, in physics, of the "myth" of entropy as disorder. F. L. Lambert (see http://entropysite.oxy.edu/, especially [Lambert, 2007]) is a firm advocate of this critical attitude. This is perfectly fair since entropy is "just" energy dispersal in physics, regardless of whether the system is open or closed[2]. Yet, as explained in [Hayflick, 2007], *"in physics, a lowered energy state is not necessarily disorder, because it simply results in the identical molecule with a lowered energy state. The fact that such a molecule might be biologically inactive may not concern the physicist, but it definitely does concern the biologist ...."* In this perspective, it is then sound to relate entropy also to disorder in biological dynamics: a lesser activity of a molecule may mean metabolic instability, or, more generally, less coherent chemical activities of all sorts. As a consequence, this may result in less bio-chemical and biological order.

---

[2] However, the argument that disorder is an epistemic notion, not suitable to physics, is less convincing, since classical randomness, at the core of entropy, is also epistemic (see above and [Bailly & Longo, 2007]).

In either case, though, and by definition, entropy has to be related to energy dispersal. As a matter of fact, the analysis of heat diffusion in animals and humans has a long history that dates back to the '30s, [Hardy, 1934]. Since then, several approaches tried to bridge the conceptual gap between the purely physical perspective and the biologist's concern with organization and with its opposite, disorder, in particular when increasing, in aging typically [Aoki, 1994, Hayflick, 2007, Marineo & Marotta, 2005, Pezard et al., 1998].

Let us now summarize the perspective of this chapter in a very synthetic way: phylogenetic and ontogenetic processes may be globally understood as the "never identical iteration of a morphogenetic process". The conjunction of inheritance and randomness is at the core of that "*never identical* iteration". By adding selection and following Gould's remarkable insight, we will in particular understand below the increasing complexity of organisms along evolution, as the result of a purely random diffusion in a suitable phase space (and its definition is the crucial issue). A short analysis of development, though, will first stress the role of entropy in ontogenesis.

9.2 Preliminary Remarks on Entropy in Ontogenesis

In an organism, the internal entropy production has *in primis* a physical nature, related to all thermodynamic processes, that is to the transformation and exchange of matter and energy. Yet, we will add to this a properly biological production for entropy: the production due to *all irreversible processes*, including biological (re-)construction. In other words, we also consider both embryogenesis and cell replacement and repair (ontogenesis, globally) from the point of view also of entropy production as they constitute irreversible processes: that is, while producing or reproducing organization, an organism also produces entropy, as "disorganization" — this is one of our key points in this section.

Observe first that, in a unicellular organism, entropy is mostly released in the exterior environment and there are less signs of increasing disorder within the cell. Yet, changes in proteome and membranes are recorded and may be assimilated to aging, see [Lindner et al., 2008, Nyström, 2007]. In a metazoan, however, *the entropy produced, under all of its forms, is also and inevitably transferred to the environing cells, to the tissue, to the organism,* [Bailly & Longo, 2009]. Thus, besides the internal forms of entropy (or disorder) production, a cell in a tissue, the structure of the tissue itself ... the organism, is affected by this dispersal of energy, as increasing disorder, received from the (other) cells composing the tissue , thus the organism. Aging, then, is also (or mostly) a tissular and organismic process: in an organism, it is the network of interactions that is affected by entropy growth, while, conversely, this may have a fall-out also in the intra-cellular activities (such as metabolism, oxidative effects ..., see below).

Moreover, the effect of the accumulation of entropy during life contributes, mathematically, to its *exponential increase* over time. Thus, with aging, this increase of entropy exceeds the reconstructive activities, which oppose global entropy growth in earlier stages of life (this theory, articulated in four major life periods, is proposed in [Bailly & Longo, 2009]). Now, we insist, entropy production, in all its forms, implies increasing disorganization of cells, tissues, and the organism. This, in turn, may be physically and biologically implemented by increasing metabolic instability, oxidative stress, affecting cells' activities as well as the structure and coherence of tissues (matrix, collagen's links, tensegrity ...) and many more forms of progressive disorganization, see [Demetrius, 2004, d'Alessio, 2004, Sohal & Weindruch, 1996, Olshansky & Rattan, 2005]. Of course, there may be other causes of aging, but the entropic component should not be disregarded and may also help in proposing a unified understanding of different processes that may contribute to aging.

Thus, our second observation is that entropy production is due to *all irreversible processes*, both the thermodynamic ones and the permanent, irreversible, (re-)construction of the organism itself. This generating and re-generating activity, from embryogenesis to repair and turnover, is typically biological and it has been mathematically defined as "anti-entropy" (see [Bailly & Longo, 2009] and below[3]). In other words, irreversibility in biology is not only due to thermodynamic effects, related to the use and transformation of energy, typically, but also to all processes that establish and maintain biological organization — that is, it is concomitantly due to entropy production and its biological opposite, anti-entropy production: embryogenesis, for example, is an organizing and highly irreversible process "per se". And it produces entropy not only by the thermodynamic effects due to energy dispersion, but also, in our view, by the very biological constructive activities, that is while the organism increases or reconstructs its complex organization. Let's see this more closely.

Cell mitosis is *never an identical "reproduction"*, including the non-identity of proteomes and membranes. Thus, it induces an *unequal diffusion of energy* by largely random effects: typically, the never identical bipartition of the proteome, organelles, ... That is, biological reproduction, as morphogenesis, is *intrinsically associated to variability* and, thus, *it produces entropy also by lack of (perfect) symmetries*. By this, it induces *its proper irreversibility*, beyond (and in addition to) thermodynamic irreversibility.

As a comparison, consider an industrial construction of computers. The aim is to produce, in the same production chain, identical computers. Any time a computer is duplicated, an identical one is produced (identical up to observable use of the machine). Organization then (locally) grows, at the expenses of energy (a computer is an highly complex and structured machine, made out of less complex components). Entropy is then produced, in principle, only by the required use and inevitable

---

[3] The word anti-entropy has already been used, apparently only once and in physics, as the mathematical dual of entropy: its minimum coincides with the entropy maximum at the equilibrium, in mixture of gases at constant temperature and volume [Duffin & Zener, 1969]. This is a specific and a very different context from ours. Our anti-entropy is a new concept and observable with respect to both negentropy and this mathematical dual of entropy: typically, it does not add to an equal quantity of entropy to give 0 (as negentropy), nor satisfies minimax equations, but it refers to the quantitative approach to "biological complexity" (see below), as opposing entropy by the various forms of biological morphogenesis, replacement and repair.

9.2 Preliminary Remarks on Entropy in Ontogenesis

dispersal of energy, while the construction *per se* just increases organization, along the production chain. Moreover, if, in the construction chain of computers, one destroys the second computer, you are back with one computer and you can iterate identically the production of the second. The process is both reversible (destroy one computer) and iterable (produce again an identical machine), by importing a suitable amount of energy, of course. Imperfection must be (and, for 99% or so of the machines, they are) below observability and functionality: they are negligible errors and "noise". Moreover, in general, in computers' and software's increasing complexity, progress or change are not due to errors or noise in construction and design ....

As we said, it is instead a fundamental feature of life that a cell is *never* identical to the "mother" cell. This is at the core of biological variability, thus of diversity, along evolution as well as in embryogenesis (and ontogenesis, as permanent *adaptive* renewal of the organism, never identically). In no epistemic nor objective way this may be considered a result of errors nor noise: variability and diversity are one of the main "invariants" in biology, jointly to structural stability, which is never identity, and, all together, they make life possible.

Thus, while producing new order (anti-entropy), life, as iteration of a never identical and an always *slightly disordered* morphogenetic process, generates also entropy (disorder), by the (somewhat disordered) reproductive process itself. In a metazoan, we insist, each mitosis is a critical transition (see chapter 7) and produces two slightly different cells, both different also from the "mother" cell: the asymmetry is a form of disorder and, thus, of entropy growth, within the locally increasing order. And this, of course, in addition to the entropy due to free energy consumption. It is this variability that gives this further, and even more radical, form of irreversibility to all biological dynamics (in evolution and ontogenesis). There is no way to neither revert nor iterate identically an evolutionary or embryogenetic process: if you kill a cell after mitosis, you are not back to the same original cell and this cell will not iterate its reproduction *identically*[4].

It should be clear that this theoretical frame concerning the overall increase of entropy in biology says nothing about how this disorganization takes place in the various processes, nor anything about its "timetable". The analyses of the detailed phenomena that implement it in ontogenesis are ongoing research projects. So far, we could apply these principles to an analysis of growing complexity in evolution, as summarized next. In the last part of this chapter, we also propose a preliminary analysis of organizational regeneration and its relation to symmetry changes.

---

[4] The incompetent computationalist (incompetent in Theory of Computation), who would say that also computers are not identical, misses the point: the *theory* of programming is based on identical iteration of software processes on reliable hardware, i.e. functionally equivalent hardware (and it works, even in computer networks, see the analysis of primitive recursion and portability of software in [Longo, 2009]). Any biological theory, instead, must deal with variability, *by principle*. As recalled above, variability as never identical iteration, in biology, is not an error: it is an essential component of biological dynamics, diversity and, thus, structural stability, in ontogenesis and phylogenesis.

9.3 Randomness and Complexification in Evolution

Available energy transformation is the unavoidable physical process underlying reproduction and variability. At the origin of life, bacterial reproduction was (relatively) free, as other forms of life did not constraint it. Diversity, even in bacteria, by random differentiation, produced competition and a slow down of the exponential growth (see diagram 9.3). Simultaneously, though, this started the early variety of live, a process never to stop.

S.J. Gould, in several papers and in two books [Gould, 1989, 1997], uses this idea of random diversification in order to understand a blatant but too often denied fact: the increasing "complexification of life. The increasing complexity of biological structures has been often denied in order to oppose finalistic and anthropocentric perspectives, which viewed life as *aiming* at *Homo sapiens* as the "highest" result of the (possibly intelligently designed) evolutionary path.

Yet, it is a fact that, under many reasonable measures, an eukaryotic cell is more "complex" than a bacterium; a metazoan, with its differentiated tissues and its organs, is more "complex" than a cell ... and that, by counting also neurons and connections, cell networks in mammals are more complex that in early triploblast (which have three tissues layers) and these have more complex networks of all sorts than diplobasts (like jellyfish, a very ancient animal). This global, on average, non-linear increase can be quantified by counting tissue differentiations, networks and more, as hinted by Gould and more precisely proposed in [Bailly & Longo, 2009], a text that we will extensively summarize and develop, next. The point is: how to understand this increasing complexity without invoking global aims?

Gould provides a remarkable answer based on the analysis of the *asymmetric* random diffusion of life. Asymmetric because, by principle, life cannot be less complex than bacterial life[5]. So, reproduction by variability, along evolutionary time and space, randomly produces, just as *possible paths, also* more complex individuals. Some happen to be compatible with the environment, resist and proliferate (a few even very successfully) and keep reproducing with further modifications. *Also* more complex individuals, since the random exploration of possibilities may, of course, decrease the complexity, no matter how this is measured. Yet, by principle,
any asymmetric random diffusion propagates, by local interactions, the original symmetry breaking along the diffusion.

Thus there is no need for a global design or aim: the random paths that compose *any* diffusion, also in this case help to understand a random growth of complexity, *on average*. On average, since, as we said, there may be local inversion in complexity; yet, the original asymmetry, from the early bacteria *also* to more complex individuals (life cannot be simpler than those early organisms), randomly forces a drift to the "right" of our figure. This is nicely made visible by figure 9.1, after [Gould, 1989], p. 205. The image explains the difference between a random, but oriented

---

[5] Some may prefer to consider viruses as the least form of life. The issue is controversial, but it would not change at all Gould's and our perspective: we only need a minimum which differs from inert matter.

9.3 Randomness and Complexification in Evolution

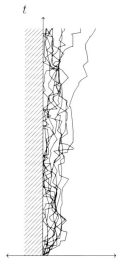 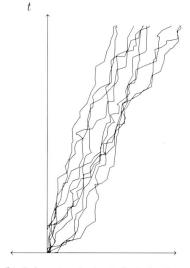

(a) Passive trend, there are more trajectories near 0.

(b) Driven trend, the trajectories have a drift towards an increased mean.

Fig. 9.1 *Passive and driven trends.* In one case the boundary condition, materialized by a left wall (0 or least complexity) is the only reason why the mean increase over time (boundary condition), and this increase is slow. In the case of a driven trend, or biased evolution, the rule of the random walk leads to an increase of the *mean* over time (it shows an intrinsic trend in evolution). Notice that from the neo-Darwinian point of view, a driven trend typically correspond to a selective pressure, here for increased complexity; whilst the passive trend correspond to a neutral theory [Hubbell, 2001], with a domain incompatible with life (the left wall of least complexity). Gould's and our approach are based on passive, random trends, which means that we do not need any intrinsic bias for increasing complexity in the process of evolution. Random changes may include some major critical phase transitions; for example, the formation of eukaryotes or of multicellular organisms, see [Maynard-Smith & Szathmary, 1997].

development (on the right, 9.1b), and the non-biased, purely random diffusive bouncing of life expansion on the left wall, on the left 9.1a.

Of course, time runs on the vertical axis, but ... what is in the horizontal one? Anywhere the random diffusion takes place or the intended phenomenon "diffuses in". In particular, the horizontal axis may quantify "biological complexity" whatever this may mean. The point Gould wants to clarify is in the difference between a fully random vs. a random *and* biased evolution. The biased right image does not apply to evolution: bacteria are still on Earth and very successfully. Any finalistic or selective bias would instead separate the average random complexification from the lower wall (no more bacteria).

We insist that complexity may *locally* decrease and sometimes it may yield compatible organisms, possibly for a new niche: tetrapods may go back to the sea and lose their podia (the number of folding decreases, the overall body structure simplifies). Some cavern fishes may loose their eyes, in their new dark habitat; others, may lose their red blood cells [Ruud, 1954].

Thus, we can understand the increase of *global* complexity on the basis of *the purely random effect of variability* on one side, and of a minimum for the complexity, on the other. In fully general terms, an unbiased diffusion, starting for a pointwise (Dirac) distribution explains more and more explorations of higher (but also lower) complexity: this comes from the increase of variance over time associated to random walks. The left Wall provides a boundary for the exploration of lower complexities, which leads to an overall increase of complexity. This increase is then due to boundary conditions and not to the local dynamic of evolution *per se*.

However, beyond the unbiased, linear diffusion, some "local" interactions can be involved: on average, variation by simplification leads towards a biological niche that has *more chances* to be already occupied, while a more complex organism may have more chances to use or construct a new niche. In mathematical terms, *local selective effects* can be modeled by a term in $-m(t,K)^2$, which would accelerate the increase of the average complexity. Here also, there is no aim towards greater complexity: just the greater chances, for a "simpler" organism, to bump against an already occupied niche and for a more complex one to construct a new way of living. Thus, more complex variants have just slightly more probabilities to survive and reproduce, as they may fit into and/or create new niches.

From another point of view, the biological meaning of an increase of biological complexity involves different forms of organization, including but not exclusively, different ways to inhabit the world. Therefore the exploration of higher levels of complexity implies the exploration and construction of new niches.

Of course, in biology, variability and, thus, diversity are grounded on randomness. No need for finalism nor a priori "global aim" nor "design" at all, just a consequence of an original symmetry breaking in a random diffusion on a very peculiar phase space: biomass times complexity times time (see figure 9.3 for a complete diagram).

Similarly as for embryogenesis, the complexification is a form of local reversal of entropy. The global entropy of the Universe increases (or does not decrease), but locally, by using energy of course, life inverses the entropic trend and creates organisms of increasing complexity. Of course, embryogenesis is a more canalized process, while evolution seems to explore all "possible" paths, within the ecosystem-to-be. In evolution, most paths turn out to be incompatible with the environment, thus they are eliminated by selection, while enablement is at the core of the ever changing dynamics of evolution, see chapter 8. In embryogenesis, increasing complexity seems to follow an expected path and it is partly so: the constraints imposed, at least, by the inherited DNA and zygote, limit the random exploration due to cell mitosis. But only in part, as failures, in mammals say, seem to concern almost 50% of fecundations. Yet, their variability, joint to the many (variable) constraints added to development (first, a major one, a fundamental chemical trace of an history: DNA), is an essential component of cell differentiation. Tissue differentiation involve, for this point of view, a form of (strongly) regulated/canalized variability along cell proliferation.

Thus, by different but correlated effects, biological complexity increases, on average through evolution, and reverts, locally, entropy. We called *anti-entropy*, [Bailly & Longo, 2009], this observable opposing entropy, both in evolution and embryogenesis; its peculiar nature is based on reproduction with random variation, submitted to constraints. As observed in the footnote above, anti-entropy differs from negentropy, which is just entropy with a negative sign, also because, when added to entropy, anti-entropy does not give 0. In our perspective, entropy and anti-entropy, as defined, coexist in a very different singularity (different from 0, the sum of entropy and equal negentropy): they yield a non null interval of extended criticality. In the next section, we will use this notion to provide a mathematical frame for a further insight by Gould.

9.4 (Anti-)Entropy in Evolution

9.4.1 The Diffusion of Bio-mass over Complexity

In yet another apparently naive drawing, Gould proposes a further visualization of the increasing complexity of organisms along evolution. It is just a qualitative image that the paleontologist draws on the grounds of his experience. It contains though a further remarkable idea: it suggests the "phase space" (the space of observables and parameters) where one can analyze complexification. It is *bio-mass density* that diffuses over *complexity*, that is, figure 9.2 qualitatively describes the diffusion of the frequency of occurrences of individual organisms per unity of complexity.

This is just a mathematically naive, global drawing of the paleontologist on the basis of his experience. Yet, it poses major theoretical challenges. The diffusion, here, is not along a spatial dimension. Physical observables usually diffuse over space in time; or, within other physical matter, which also amounts to diffusing in space. Here, diffusion takes place over an abstract dimension, "complexity". But what does biological complexity mean, exactly? Hints are given in [Gould, 1997]: the addition of a cellular nucleus (from bacteria to eukaryotes), the formation of metazoa, the increase in body size, the formation of fractal structures (usually — new — organs) and a few more.... In a sense, complexity is increased by any variation or added novelty provided by the random "bricolage" of evolution. This is often due to "exaptation" (adaptation ex-post of old features to new functions, a key notion by Gould), which happens to be, at least for some time, compatible with the environment. Only a few organisms generate more complex ones over time, but, by the original symmetry breaking mentioned above, this is enough to increase the global complexity.

Of course, the figure above is highly unsatisfactory. It gives two slices over time where the second one is somewhat inconsistent: where are dinosaurs at present time? It is just a sketch, but an audacious one, as we said, if analyzed closely. Mathematics though, may help us to consistently add a more general understanding and the third missing dimension, time.

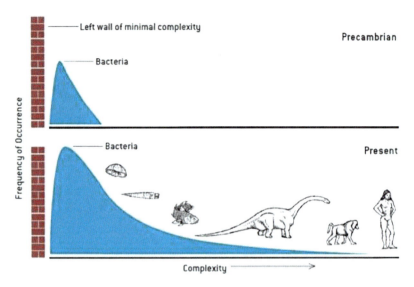

Fig. 9.2 *Evolution of complexity as understood by Gould.* This illustration is borrowed from [Gould, 1997], p.171. This representation is designed on the basis of paleontological observations. The core idea is to explain the biological mean increase of complexity from the left wall of minimal complexity.

A simple form of diffusion equation of a quantity q in time t over space x, goes like:

$$\frac{\partial q}{\partial t} = D\frac{\partial^2 q}{\partial x^2} + Q(t,x) \qquad (9.1)$$

where $Q(t,x)$ is a source term describing growth in mass. The equation gives the variation of q in time ($\partial q/\partial t$) as a function of the variation ($\partial \ldots /\partial x$) of the space gradient ($\partial q/\partial x$).

Yet, in our case, the diffusion of this strange quantity, m, a *bio-mass density*, takes place over an even more unusual "space", biological complexity, whatever the latter may mean. In [Bailly & Longo, 2009], we dared to further specify Gould's hints for biological complexity, as a quantity $K = \alpha K_c + \beta K_m + \gamma K_f$ where α, β, and γ are the respective "weights" of the different types of complexity within the total complexity (we take $\alpha + \beta + \gamma = 1$). We do not get into the details given in that paper and just summarize the basic ideas. The challenge is to quantify a well-defined observable, measuring the complexity of an organism. The measure summarized below is based on rather arbitrary but reasonable choices. The idea is to take an *instantaneous picture* of an organism or to consider just an "anatomical" evaluation of complexity. First, the components of the list below are far from complete: they form a backbone of a possible measure of phenotypic complexity. Second, they are a priori non-positively correlated with "biological organization", which should include functions, interactions, contexts .... That is, we care to distinguish between

9.4 (Anti-)Entropy in Evolution

complexity (an anatomic, static picture) and organization (based on the dynamic functionality of a living organism). Clearly, organization requires complexity, but the anatomic description of a dead organism does not require functions. This distinction between complexity and organization is just our instrumental proposal for a qualitative analysis of complexity growth. Yet, as a side effect, it produced some insight into developmental dynamics as hinted below, see 9.5.1.

In short, phenotypic complexity K is composed by three elements:

1. K_c ("combinatorial" complexity) corresponds to the possible cellular combinatoric (which includes cellular differentiation);
2. K_m ("morphological" complexity) is associated to the topological forms and structures which arise (connexity and fractal structures);
3. K_f ("functional" complexity) is associated to the relational structures supporting, but not identified with, biological functions (cellular and neuronal networks, other interaction networks).

$K = \alpha K_c + \beta K_m + \gamma K_f$ will be more closely defined in section 9.6[6].

K will be used below as a tentative quantification of complexity as *anti-entropy*, in particular in biological evolution: the increase of each of its components (more cellular differentiation, more or higher dimensional fractal structures, richer networks ... yield a more "complex" individual). As we already hinted and will further explain below, we called it anti-entropy as it opposes entropy (as biological disorder). Anti-entropy has the same mathematical dimension as entropy and negentropy — that is, they can face each other in an equation, a fundamental technical property for our mathematical developments.

Of course, many more observables and parameters may be taken into account in order to evaluate the complexity of an organism: [Bailly & Longo, 2009] provides just a mathematical basis and a biological core for a preliminary analysis (an application to ontogenesis as an analysis of *C. Elegans* development is also presented there). They suffice though for a qualitative (geometric) reconstruction of Gould's curve, with a sound extension to the time dimension.

As mentioned above, anti-entropy opposes, locally, to entropy: it has the same dimension, yet it differs from negentropy, since it does not sum up to 0, in presence of an equal quantity of entropy. It differs also from information theoretic frames, where negentropy has been largely used, as a measure of information. Information, as an observable on discrete data types and discrete codings, is *independent from coding and Cartesian dimensions*. That is, as we know since Turing (and Shannon), discrete data bases and their informational elaboration (or transmission, Shannon) do

---

[6] As for now, just notice that, say, humans have some 600 muscles, while a cow or a horse, some 400 — these are connected components of an organs' system; neural networks are far richer in humans than in all other animals — elephant and dolphins have about the same number of neurons as humans, but far less synaptic connections. In no way by this we happen to be "better" than the other animals: bacteria, say, have been the most successful organisms in evolution, as they still compose about half of the biomass. And they will surely overcome the ongoing destruction of our ecosystem, largely caused by our too big, too stupid, randomly produced brains.

not depend on the Cartesian dimension of the space for elaboration or transmission. This is crucial for Shannon-Brillouin as well as for Turing-Kolmogorov-Chaitin information theories, see [Longo et al., 2012a]. Anti-entropy, instead, as defined above, depends on foldings, singularities, fractality .... Thus, it is a *geometric* and *dimensional* notion, and, therefore, by definition, it is *sensitive to codings* (and to dimension), in contrast to information theoretical notions and their correlate, negentropy. Of course, we consider that geometric structuring and physico-mathematical dimensions are crucial for life and its analysis.

The first step now is to adapt equation 9.1 to this new phase space, given by Gould's observables and parameters: the biomass m diffusing over the complexity K. Then write:

$$\frac{\partial m}{\partial t} = D\frac{\partial^2 m}{\partial K^2} + Q(t,K) \tag{9.2}$$

But what is here $Q(t,K)$, the source term? In order to instantiate Q by a specific function, but also in order to see the biological system from a different perspective (and get to the equation also by a different procedure, an "operatorial approach"), we now give a central role, as an observable, to the "global entropy production".

Now, in physics, energy, E, is the "main" observable, since Galileo inertia, a principle of energy/momentum conservation, to Noether's theorems and Schrödinger's equation. Equilibria, geodetic principles etc. directly or indirectly refer to energy and are understood in terms of symmetry principles, as we extensively stressed in this book. Moreover, at least since Schrödinger and his equation in (quantum) physics, one may view energy as an operator and time as a parameter[7].

As stressed above, in biology, also constitutive processes, such as anti-entropy growth (the construction and reconstruction of phenotypic complexity), *produce entropy*, since they also produce some (new) disorder (recall: at least the proteome, after a mitosis, is non-uniformly and partly randomly distributed in the new cells). In these far form equilibrium, dissipative processes – possibly even non-stationary, as the energy flow may be non constant – such as evolution and ontogenesis, energy turns out to be just one (very important) parameter. One eats (and this is essential) and gets fatter: production and maintenance of complexity and organization requires energy. Typically, in allometric equations, so relevant in biology (see section 2.2), energy or mass appear as a parameter. Biology needs at least one different observable, in a different dimension, tentatively defined by K above, as phenotypic complexity – besides the correlated organization. Thus, in our approach, the key observable is complexity that is formed or renewed, as anti-entropy production. We can also see it as an operator, to be associated with time (the time for setting up complexity), as we shall argue, while we consider energy a parameter.

As explained at the beginning of this chapter, entropy is associated to all time irreversible processes, from energy flows to anti-entropy production. We call σ this

---

[7] In [Bailly & Longo, 2009] a brief introduction to Schrödinger's approach is given. In short, Schrödinger transforms an equation with the structure $E = \frac{p^2}{2m} + V(x)$, where $V(x)$ is a potential, by associating E and p to the differential *operators* $\partial/\partial t$ and $\partial/\partial x$, respectively.

key observable, global entropy production, which summarizes all time-irreversible phenomena. By its irreversibility, it is strongly linked to time — one can measure time by the formation of complexity, both in embryogenesis and evolution, which produces entropy, *per se*, as we said, and by all irreversible processes, that is by σ. In this, partly formal sense, (anti-)entropy production may be positively correlated to time and both may be seen as the "constitutive operators" of biological phenomena. We try by this, in a partly mathematized way, to stress that time has a different role on biology vs. physics: first, by its intrinsic non reversibility (in many physical theories time is reversible); second, it is the operator that constructs life. The equations somewhat justify this perspective.

In summary, we proposed to change the conceptual frame and the conceptual priorities with respect to physics: we associated the global entropy production σ to the differential operator given by time, $\partial/\partial t$ (Schrödinger does this for energy, which is conjugated to time, in quantum physics). Thus, our approach allowed to consider biological time as an "operator", both in this technical sense and in the global perspective of attributing to time a key constitutive role in evolution and in ontogenesis. But how to express this global observable?

9.4.1.1 A Balance Equation

In a footnote to [Schrödinger, 2000], Schrödinger proposes to analyze his notion of negative entropy as a form of Gibbs free energy G. We apply now this idea to proper anti-entropy $S^- = -kK$, where k is a positive dimensional constant and K is the phenotypic complexity above, which we also referred to as anti-entropy, by a small abuse of language, as they may be identified up to a constant.

Now, $G = H - TS$ is the system's enthalpy, where T is temperature, S is entropy and $H = U + PV$ (where U is the internal energy, P and V are pressure and volume).

By definition, *metabolism* R has the physical dimension of a power and corresponds to the difference between the fluxes of *generalized free energy* G, entering and exiting through the surface Σ:

$$R = \sum [J_G(x) - J_G(x+dx)] = -\sum dx (\text{Div } J_G) \quad (9.3)$$

Take the volume $\sum dx = 1$, then the conservation (or balance) equation is expressed in the general form:

$$R = -\text{Div } J_G = \frac{dG}{dt} + T\sigma \quad (9.4)$$

where σ represents the global production of entropy, that is σ is the entropy produced by *all* irreversible processes, including the production of biological complexity or anti-entropy. Thus, the global balance of metabolism for the "system of life" (the evolving biosphere) has the following form, where S^- and S^+ are anti-entropy and entropy, respectively:

$$R = \frac{dH}{dt} - T\left(\frac{dS^-}{dt} + \frac{dS^+}{dt}\right) + T\sigma \tag{9.5}$$

That is,

$$R = a\frac{dM}{dt} - T\left(\frac{dS^-}{dt} + \frac{dS^+}{dt}\right) + T\sigma \tag{9.6}$$

where $H \simeq aM$, for a mass M and a coefficient a, which has the magnitude of a speed squared — we relate energy and mass by dimensionality, as usual in physics.

$T\sigma$ is a crucial quantity: it contains our σ, modulo the temperature T, since R is a power. In the abstract, but effective, style of theoretical physics, we proceed by a " dimensional analysis". $T\sigma$ corresponds to the product of forces by fluxes (of matter, of energy — chemical energy, for instance — etc.). Now, a flux is proportional to a force, thus to a mass, and hence $T\sigma$ is proportional to a mass squared. It can then be written, up to a coefficient ζ_b and a constant term $T\sigma_0$ as:

$$T\sigma \approx \zeta_b M^2 + T\sigma_0 \tag{9.7}$$

ζ_b is a constant that depends only on the global nature of the biological system under study and it is 0 in absence of biological entities, as M is the bio-mass.

We may now use as "state function" for our analysis of *bio-mass diffusion* over time t and complexity K, a bio-mass density function $m(t,K)$, and use the operatorial approach relatively to equation 9.7. The full details of this approach are given in [Bailly & Longo, 2009]. In short, similarly to the construction of Schrödinger's diffusion equation from the equation $E = \frac{p^2}{2m} + V(x)$ in the footnote above (we associated E and p to the differential operators $\partial/\partial t$ and $\partial/\partial x$, respectively), we may transform equation 9.7 into a diffusion equation by associating $T\sigma$ and M to the differential operators $\partial/\partial t$ and $\partial/\partial K$, respectively. This further justifies 9.2 as derived now from $T\sigma$, which results form a global balance equation, and provides the source function $Q(t,K)$, under the form of a linear map $\alpha_b m$, due to the constant term $T\sigma_0$:

$$\frac{\partial m}{\partial t} = D_b \frac{\partial^2 m}{\partial K^2} + \alpha_b m \tag{9.8}$$

where D_b is a diffusion coefficient. Its solution

$$m(t,K) = \frac{A}{\sqrt{t}} \exp(at) \exp(-K^2/4Dt) \tag{9.9}$$

yields the diagram in figure 9.3.

In summary, while skipping all the technical details in [Bailly & Longo, 2009], we could derive, by mathematics and starting from Gould's informal hints, a general understanding as well as the behavior of the "evolution of complexity function" w. r. to time. And this fits data: at the beginning the linear source term gives an

9.4 (Anti-)Entropy in Evolution

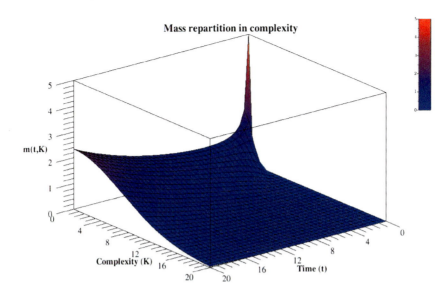

Fig. 9.3 *Time evolution of mass distribution over anti-entropy.* The initial condition is a finite mass at almost 0 anti-entropy, thus having the shape of a pulse. The biomass is latter driven by diffusion. The asymmetry of the distribution is given only by the propagation of the effects of the wall of minimum complexity.

exponential growth of free bacteria. Then, they complexify and compete, up to reaching the biological phenotype with lots of components.

- *Remark.* By our approach, we may provide a theoretical/mathematical justification of the ZFEL principle in [McShea & Brandon, 2010]:
  "ZFEL (Zero Force Evolutionary Law, general formulation): In any evolutionary system in which there is variation and heredity, there is a tendency for diversity and complexity to increase, one that is always present but may be opposed or augmented by natural selection, other forces, or constraints acting on diversity or complexity."
  In other words, ZFEL may be derived from our "asymmetric random diffusion principle", in a Darwinian context. That is, it follows from considering Darwin's two fundamental principles, *reproduction with modification* and *selection*, in presence of an asymmetric random diffusion. This, of course, further justifies ZFEL, yet not as a principle, but as a derived property of evolutionary systems.

Note that our three dimensional diagram 9.3, similarly to Gould's, is a global one: it only gives a qualitative, geometric, understanding of the process. It is like looking at life on Earth form Sirius and summarizing 4 billions years in 6 centimeters. Analogously to Gould's diagram, the "punctuated equilibria", say, and the major transitions and extinctions are not visible: the insight is from too far and too synthetic to appreciate them. It only theoretically justifies Gould's proposal and soundly extends it to time dependence, by mathematically deriving it from general principles: the dynamics of a diffusion by random paths, with an asymmetric origin. Its source is given by a "doubling" at each step (the reproduction of free bacteria), thus beginning by an exponential growth (due to the linear source function, in the differential equation). Life expansion is then bounded, canalized, selected in the interaction with the ever changing, co-constituted ecosystem. The core random complexification persists, while its "tail" exponentially decreases, see equation 9.9 and figure 9.3. In that tail, some neotenic big primates, with a huge neural network, turn out to be the random complexification of bacteria, a result of variability and of the immense massacres imposed by selection (it is generally believed that about 99% of the species that appeared on Earth are extinct). As we said, the major critical phase transitions are fully compatible with our qualitative diagram, similarly as Gould's diagram is compatible with the burst of diversity he claims in evolution (punctuated equilibria). We mentioned the formation of eukaryotes or of multicellular organisms, for example: they are critical changes along increasing evolutionary complexity.

Another important analogy can be made with Schrödinger's approach — his famous equation, not his book on life; this further justifies the reference to it for the analysis of this (rather ordinary) diffusion equation. Schrödinger dared to describe the deterministic evolution of the state function in Quantum Mechanics as the *dynamics of a law of probability*, or of a probability density or amplitude. This gives the intrinsic indetermination of the quantum system. We synthetically represented biological evolution as the *dynamics of a potential of variability*, the biomass density, under the left wall constraint. Again, this idea is essentially Gould's idea in his 1997 book: he sees evolution just as an asymmetric diffusion of random variability. We just made this point explicit and developed some computations as a consequences of the analogy with the equational determination in quantum mechanics and the operatorial approach used by Schrödinger. In particular, and in order to summarize:

- we looked at bio-mass abstractly, as a potential of variability, whose random diffusion over complexity leads to increasing complexity by propagating an original asymmetry,
- we proposed to see time as a fundamental biological operator and set the preliminary basis for defining a proper biological observable, phenotypic complexity.
- Gould's understanding of evolution stresses Darwin's two key principles, also at the core of our approach: descent with modification and selection. Moreover, some major critical transitions and punctuated "bursts" of change dramatically accelerate the global evolutionary modifications, while "exaptation" and related phenomena provide a further path towards modifications, see [Gould & Vrba, 1982]. All these principles rely on contingent, thus random, dynamics, without excluding, of course, the structural stability of organisms, species, ecosystem, which is also essential to life. It is stability via change and diversity, though, or, more exactly: randomness produces variability, which yields diversity and adaptivity (in individual, populations, species, ecosystems) which are essential components of biological stability.

- Along this line of thinking, evolutionary trajectories are based on a "non-conservation principle" for phenotypes: Darwin's descent with modification (to which selection applies), as a largely contingent result of biological activity and interactions (organism/ecosystem).

As for the last point above, the principle mentioned may seem in strong contrast with the main physical theories, which are largely based on conservation principles (energy and momentum conservation, typically), and the related theorems on symmetries in the equations, since Noether's, see chapter 5. Even far from equilibrium systems are based, as for most mathematical analysis, on flow or balance equations, which *in fine* refer to conservation (of energy or matter). Yet, these principles apply to new or proper biological observables, thus they are, a priori, compatible with and extend the underlying physical theories.

As a side remark, just note that Darwin dared to propose his "non-conservation principle" for phenotypes (descent with modification), about at a time where physics was proposing beautiful theories centered on conservation of energy (thermodynamics) and on the geodetic principles (Hamilton), which was also understood, later, in terms of symmetries and conservation. Of course, there doesn't need to be any theoretical incompatibility here, just different pertinent observables and parameters as well as distances from equilibrium. Yet, an autonomous biological thinking, such as Darwin's, is required before going towards a welcome unification of theories. And physics itself, by its method and ideas, may help to propose it, not only directly but also by its methods coupled to conceptual dualities and theoretical oppositions, as we are trying here.

9.5 Regeneration of Anti-entropy

Let us first recall why the biological notion of anti-entropy differs from other approaches, and in particular of the notion of negentropy. As we said above, negentropy simply opposes entropy and it is sometimes used to understand biological organization as a situation where, in spite of the presence of many energy production/consumption processes, an unusually low entropy (high organization) is maintained and made possible by the openness of the system. Note though that *negative* entropy, as entropy with a negative sign, has no physical meaning *per se* — as a physical observable. As a matter of fact, the third principle of thermodynamic states that the minimum entropy is zero, in the case of pure crystalline states at $0°$ K. From a statistical point of view, the number of microscopic configuration corresponding to a macrostate cannot be smaller than 1, so entropy cannot be negative. However, negative contributions in the expression of entropy can be found and interpreted, both in physics and biology — see next.

Our theoretical proposal leading to the term of anti-entropy is inspired by a conceptual symmetry between the relationships of matter and antimatter, on one side, and of physical entropy and "amount of biological complexity", on the other. We already recalled some basic properties this assumption leads to, and we will now present further considerations along this line. Notice first that the metabolic equation 9.6, by its formulation, can take into account the overall growth of anti-entropy occurring in development, but does not describe, even not *in abstracto*, how biological organization is *sustained* and more precisely *regenerated*. We will use our approach now for a closer analysis of the local interactions of entropy vs anti-entropy production, in comparison with existing theories and data.

In particle physics (relativistic quantum mechanics and quantum field theories), the collision of a particle and the corresponding antiparticle leads to the annihilation of both particles and the emission of photons, which are a radiative form of energy (corresponding in particular to momentum conservation). Reciprocally, energy leads to the spontaneous production of particle/anti-particle pairs (following the symmetries of the theory). This phenomenon spontaneously occurs in the vacuum because the latter is a state with no "real particles", but where energy is, nevertheless, not 0 — this can be viewed also through the time/energy uncertainty: their product never goes below Planck's h. More precisely, in quantum fields theories, the vacuum is generally understood as a extremely complex situation, described by virtual pairs of particles and antiparticles, which spontaneously appear and disappear on short time lengths[8]. Now, can these complex theoretical structures help us in order to obtain a better understanding of biological phenomena?

In the comparison between the physical and the biological concepts we are dealing with, we can highlight some common points and, also, some crucial differences. First, anti-entropy coexists with entropy over an extended period of time — the life of an organism, typically — while particles and anti-particles interact in a (point-wise) space-time singularity. This point is an aspect of our understanding of the *extended* "physical singularity" of life phenomena, that we treated in terms of extended criticality, see chapter 7. Moreover, particle/anti-particle pair can spontaneously be produced, whereas anti-entropy necessitates, for its growth, a preceding anti-entropy ("life from life")[9]. Observe now that, if anti-entropy is unlikely to reappear spontaneously, then some space-time extension of the coexistence of entropy and anti-entropy is needed in order to observe anti-entropy: an extended singularity — that is, our extended criticality, possibly, as implemented in organisms, continually producing it along phylogenesis.

---

[8] Notice that this situation usually leads to divergences in physical quantities, especially because the higher the level of energy of an experiment is, the larger the parts of this complex structure become physically relevant. However, finite differences between the coupling constants allow to understand the situation by the renormalization methods (the actual origin of these methods discussed in chapter 6). It is also interesting to note that this highly complex situation is due to the quantification of fields (called second quantification), which breaks the classical symmetries at quantum scales. Reciprocally, the classical fields are understood as the result of an infinite number of such interactions, by the renormalization of a linear combination of the possible interactions.

[9] There is at least one exception to the latter statement: the origin of life, a singularity we do not deal with, here. Even if we think that simple (proto-)organisms may appear spontaneously today, biologically *ex nihilo*, they should usually disappear very quickly, by getting consumed by phylogenetically older, more complex and organized organisms.

9.5 Regeneration of Anti-entropy

Last but not least, the key observables for particles and antiparticles are energy, momentum, charge, ..., the most crucial one being energy, generally parametrized over time, space .... As we often recalled, they all correspond to conserved quantities, whereas, in the context of entropy and anti-entropy, we are considering typically non conservative quantities. The lack of conservation is indeed expressed by the entropy production term σ, which is non-zero when irreversible processes occur in the system. Recall that σ is a fundamental quantity for us exactly because it is associated to all irreversible processes. Moreover, anti-entropy is not 0 — life is possible — only if it is permanently reconstructed, thus only if σ is not 0.

9.5.1 A Tentative Analysis of the Biological Dynamics of Entropy and Anti-entropy

We consider now a possible biological analog of the matter/anti-matter collisions, viewed in terms of the interplay entropy/anti-entropy. The basic idea is to further describe the coexistence of entropy and anti-entropy. This coexistence is not a static one, as it involves almost continual destruction and regeneration of anti-entropy. We propose to approach this in the formal terms of "collisions" between anti-entropy and entropy. This will allow us to relate anti-entropy to the metabolism required to sustain it.

In order to better understand the situation, we will first explain what may happen in the simplified situation, not sustainable for long times (biologically "instantaneous"), where the fluxes are null. In this case, equation 9.6 reads:

$$0 = -\frac{dS^-}{dt} - \frac{dS^+}{dt} + \sigma \tag{9.10}$$

Since $\sigma > 0$, the following may happen in an organism, at least in principle:

1. It can *sustain* its organization ($\frac{dS^-}{dt} = 0$) or even *increase* it ($-\frac{dS^-}{dt} > 0$) if the system can accommodate an accumulation of entropy ($\frac{dS^+}{dt} = -\frac{dS^-}{dt} + \sigma$). In order to do so, a reservoir of highly entropic matter is usually involved, which prevents this high entropic matter to interfere too much with biological organization (and affect or "disorganize" it). A typical example is the structure of the egg: whilst allantois collects high entropic liquids, the yolk sac contains low entropic reserves. Of course, in this case, isolation is not complete because of the gaseous exchanges, but we can nevertheless see such an organizational tendency, with both high $-\frac{dS^-}{dt} > 0$ and $\frac{dS^+}{dt} > 0$ (and $-\frac{dS^-}{dt}$ much higher than $\frac{dS^+}{dt}$, as organization quickly increases).

2. If the organism does not have the possibility to produce entropy and simultaneously maintain its organization, then it can use a part of its anti-entropy to reduce entropy. That is, it can "absorb" its own entropy production and then sustain it. It is crucial that this process implies a supplementary production of entropy associated with the transition from anti-entropy to negentropy, as lowering of entropy, since this transition is irreversible. Notice that such phenomenon occurs

in relatively common situations such as autophagy [Rabinowitz & White, 2010]. Besides the cases of starvation, it is thus a normal part in the process of organizational renewal. These processes may also contribute to (or be observed in) an actual decrease of organization (i.e. of anti-entropy), for example in degenerative diseases, [Pezard et al., 1998], or aging processes, as part, in our views, of the widely acknowledged entropic component of this latter fact of life, see above as well as [Aoki, 1994, Hayflick, 2007, Marineo & Marotta, 2005].

In order to better understand the opposing activities in the process above, we develop an analogy with colliding particles and anti-particles in physics. A "collision" may be described between a part of the anti-entropy δS^- and a part of the entropy δS^+. Consider a "biologically instantaneous" situation, that is a sufficiently short time so that we can disregard fluxes. Then one has $\delta S^+ = -\delta S^- + \sigma$. This assumption will allow us to propose a preliminary, simplified analysis. Further work is needed in order to have a more stable understanding of the situation.

As a consequence of the assumption on the very short biological time, the collision between entropy and anti-entropy can lead to both cases discussed above, provided that the existence of the *biological* time arrow is equivalent to $\sigma > 0$[10]. Case 1 is similar to the production of particle/anti-particle pairs, leading to an increase (in absolute value) of both entropy and anti-entropy. Case 2 is similar to the annihilation of a particle/antiparticle pair and it is a form of transformation of anti-entropy into negentropy, as a negative quantity of entropy, corresponding to a reduction of entropy. In both cases, the collisions involve a production of entropy along time, $\tau\sigma > 0$ (where the time interval τ is extremely small, or even a Dirac function).

The production of entropy limits the number of such collisions that can occur simultaneously, since we have to assume that the sum of all entropy produced at the same time are finite. As a result, since entropy production has a positive sign, only a finite number of collisions of the same nature (producing similar quantities of entropy) can occur in a finite amount of time. This situation is different from quantum field theory, where an infinite number of such collisions can occur and the resulting sum can remain finite, since the relevant quantities do not all have the same sign. Of course, this argument does not prevent the possibility of an infinite number of collisions in a finite duration, but the entropy production contributions has to be summable, which leads to a quantitative hierarchy of vanishing contributions — an infinite case that may thus make mathematical sense in physics, but not necessarily in biology.

---

[10] We stress once more that, since σ is associated to all irreversible processes, including the setting up and the maintenance of complexity and organization. Mathematically, $\sigma > 0$ represents our fundamental way to understand the strong irreversibility of biological time, which includes thermodynamical irreversibility, of course, but it also includes the properly biological formation and renewal of anti-entropy — in evolution, embryogenesis and ontogenesis. These are totally, deeply irreversible processes, for their proper phenomenology as "life organization constructors", well beyond thermodynamical irreversibility. As recalled above, in some cases, by using energy, one may reverse physical entropy (separate, say, mixed gases): no way to revert evolution, embryogenesis or ontogenesis.

9.5 Regeneration of Anti-entropy

The paradigmatic and simplest situation following this pattern of entropy/anti-entropy "collision" is the death of a cell in an organism. In this case, most of the negentropy obtained[11] is not stable as such, which means that the entropy will rapidly increase, or, in other words, that this negentropy will rapidly vanish. Moreover, this temporary decrease of entropy will in general lead, after some time, to a greater entropy than the initial entropy. Typically, the function of macrophages is to increase irreversibly the entropy of the remains of dead cells (or other objects), in a spatially constrained domain (namely in vacuoles) and to prevent by this further disorganization (loss of anti-entropy). After the collision (or sometimes before this event), the corresponding cell is replaced, which leads to a growth of anti-entropy (in absolute value) that compensate the loss of anti-entropy associated to the cell death. As we often stressed, this growth leads also to a certain amount of entropy production, by the slightly disordered nature of anti-entropy production. Notice, however, that this process is not necessarily as stationary as one might think. For example, significant scale free fluctuations in the cell numbers have been observed for blood cells of different categories, for time-scales of 1 to 200 days, see section 2.4 or [Perazzo et al., 2000]. Somewhat reciprocally, as we suggested, the replacing cell can be produced *before* the death of the old one, sometimes even leading to the lysis or the release of the preceding cell.

In order to analyze further the possible situations, we will now propose a graphical representation of the interactions between entropy and anti-entropy, very loosely inspired by Feynman's diagrams. This representation will allow us to schematize some singular events associated to the entropy/anti-entropy relations. We will split the representation space in two parts, one for entropy (top) and one for anti-entropy (down). The solid lines will represent the currently non-interacting quantities. Their distance to the central dotted axis corresponds to the corresponding values of these variables, whilst the winding curves will correspond to entropy production and the zigzags correspond to quantities involved in an interaction. The color corresponds to the sign of the involved quantities: positive is red and negative is blue. Figure 9.4 provides an elementary description of what happens in the case of an elementary disorganization.

When summing over diagrams of this kind, we find that the contribution of organizational renewal is *in fine* found in the entropy production, at the core of our mathematical analysis of Gould's diagram. More precisely, there are two typical contributions to entropy production: the entropy following from the destruction, *per se*, of a biological component and the entropy produced in its reconstruction, as mentioned above. Their close analysis is surely very complex and in particular needs to be extended over a substantial period of time. However, a crude macroscopic approximation of the situation can be of the form: $\sigma \simeq -aS^-/\tau_r$, where $0 < a < 1$ is the proportion of anti-entropy that is renewed. The introduction of a seems necessary. For example, even though there is a turnover for all cells in the lung of

---

[11] This negentropy corresponds to the low entropic inert matter, remaining as a trace of the former biological organization before the cell death.

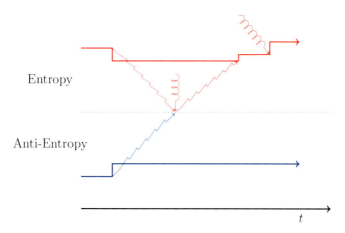

Fig. 9.4 *Diagram of an elementary loss of anti-entropy.* A small amount of anti-entropy $-\delta S^- < 0$ "collides" with a corresponding entropy $-\delta S^+ > 0$ and is transformed in an amount of entropy *produced* $-\delta S^- - \delta S^+ + \int \sigma > 0$ (the red, spring-like, winding-up curve), which adds to the entropy of the system (the climbing zigzag line that adds to entropy). Since this leads to an unstable and irreversible result (and since it is purely entropic), there is a subsequent entropy production (the second red winding-up curve; the time shift is meant to represent the fact that the collision lowers the entropy at first (the first descending step on the left), but usually leads to a higher entropy latter on).

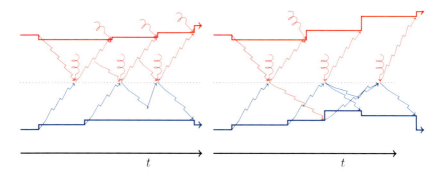

Fig. 9.5 *Other schema of entropy/anti-entropy interactions.* LEFT: the diagram represents, in temporal order, the lysis of a cell, the subsequent phagocytosis by a macrophage (involving a small increase of the anti-entropy of the macrophage, at least morphological), then the low entropic content interact with the anti-entropy of the macrophage. RIGHT: a cell is destroyed, but in this case we take into account its functional contribution. This is showed by the emission in the collision of a negative quantity of anti-entropy, which is then transformed into a quantity of anti-entropy in jeopardy. However, here a cell also divide (second collision) leading to a release of two cells (and an important amount of entropy). One of this cells replaces the destroyed cell and reestablishes the functional anti-entropy (third collision), however this process is also irreversible, leading to some reorganization and thus produces entropy. In this case the initial and final anti-entropy are the same.

9.5 Regeneration of Anti-entropy

mammals, there is no destruction/reconstruction of whole lungs for adults. See also the right diagram in 9.5 for a more subtle example. For adults at rest, we expect then that $R \simeq T\sigma \simeq -aTS^-/\tau_r$[12]. Notice than a more subtle approximation would be to assign specific renewal times to the different components of anti-entropy, some of them being infinite (which correspond to no renewal).

Interestingly, [McCarthy & Enquist, 2005] have roughly found the empirical relation $R = R_0 n^{0.58} M^{3/4}$ among different taxa, where n is the number of cell types. Assuming that $\tau_r \propto M^{1/4}$ (see [Lindstedt & Calder III, 1981, Savage et al., 2004]) and $N \propto M$ (which holds for part of the cell types, see [Savage et al., 2007]) we get:

$$R \simeq -aT\frac{S^-}{\tau_r} \propto -N\frac{S^-}{NM^{1/4}} \qquad (9.11)$$

$$\propto -M^{3/4}\frac{S^-}{N} \qquad (9.12)$$

The experimental results then leads to:

$$R \propto M^{3/4} n^{0.58} \qquad (9.13)$$

$$\text{thus} \qquad (9.14)$$

$$\frac{-S^-}{N} \propto n^{0.58} \qquad (9.15)$$

The latter equation differs strongly from the form of the combinatorial complexity we discussed, but this difference does not come as a surprise. Indeed, the latter result involves all anti-entropy renewal processes, and not only the renewal of the combinatorial component of anti-entropy. This empirical result then shows that anti-entropy and the number of cell types are correlated to a certain degree (beyond the combinatorial part). This results also corroborates our general approach to anti-entropy, which relates organizational complexity to metabolism.

Note that the general line of reasoning presented here should be further mathematically developed in the terms of a *time operator* formalism. As a matter of fact, our observables, anti-entropy, entropy and entropy production are difficult to evaluate directly in experiments, as their theoretical determination does not seem classical. On the contrary, their temporal structure is observable (at least in part), and it may be more easily associated to regularities, by loosing the physical specificity of trajectories that are parameterized over time.

---

[12] The metabolism of adults, normally, is not used to produce new organization. In equation (6) then, at constant mass M and with anti-entropy production (reconstruction) just enough to compensate entropy production, one is left with only $R \simeq T\sigma$, use then $\sigma \simeq -aS^-/\tau_r$ (see in [Bailly & Longo, 2009] for a closer analysis of the relative weight of the production of S^- and S^+ along individual life time).

9.6 Interpretation of Anti-entropy as a Measure of Symmetry Changes

In chapter 7, we proposed to understand biological phenomena, in comparison and contrast with physical theories, as a situation where the theoretical symmetries are "continually" changed: within structural stability and interval of viability, phylogenetic and ontogenetic trajectories are cascades of symmetry breaking and reconstruction. We will now show that such considerations allow to interpret anti-entropy in analogy to Boltzmann's approach of physical entropy. In section 9.4, following [Bailly & Longo, 2009], premises of these aspects are considered from a strictly combinatorial point of view, leading to a "constructive" definition of the three components of anti-entropy, $K = \alpha K_c + \beta K_m + \gamma K_f$.

The simplest way to understand how symmetries come into play is to look at these components of anti-entropy and exhibit the underlying symmetries that allow these definitions.

COMBINATORIAL COMPLEXITY, K_c: For a total number of cells N and for a number n_j of cells of cell type j, the combinatorial complexity is defined as:

$$K_c = \log\left(\frac{N!}{\prod_j n_j!}\right) \tag{9.16}$$

A classical combinatorial point of view consists in saying that this is the number of ways to classify N cells in j categories each of sizes n_j. More precisely, we recognize, inside the logarithm, the cardinal, $N!$, of the symmetry group S_N. This group is the group of transformations, called permutations, that exchange the labels of N elements. Similarly, $n_j!$ is the number of permutations among n_j units and has the biological meaning of permutations of cells within a cell type. In other words, permuting cells *within the same cell type* is a combinatorial invariant of the complexity of an organism. Thus, the group of permutations leaving the cell types invariants is the group $G_{type} = \prod S_{n_j}$, that is the group obtained as a direct product of the symmetries corresponding to permutations within each cell type. Formally, this group corresponds to the change of labels in each cell type, which can all be performed independently and conserve the classification by cell types. The cardinal of this group is $\prod_j n_j!$.

Then, the number of cell type configurations is the number of orbits generated by the right action of G_{type} on S_N. In other words, a cell type configuration is first given by a permutation of $[\![1,N]\!]$, which gives the random determination for N cells. Moreover, these transformations must be computed modulo any transformation of G_{type} that gives the same configuration (as we said, cells within each cell type are combinatorially equivalent — we will discuss this hypothesis below, in more biological terms). Lagrange theorem then gives the number of remaining transformations $\frac{N!}{\prod_j n_j!}$, which is the number of possible configurations. Clearly, if there is only one cell type, $K_c = \log \frac{N!}{N!} = 0$, thus an organism with just one cell type (typically, a unicellular being) has combinatorial complexity 0.

9.6 Interpretation of Anti-entropy as a Measure of Symmetry Changes

As a result, this measure of combinatorial complexity depends on the total number N of cells, but is actually *a measure of the symmetry breaking induced by the differentiation in cell types*.

Let's compare the situation with Boltzmann approach to entropy[13]. If one has a number of microscopic phase space states Ω having the same energy, the corresponding entropy is defined as $S = k_b \log(\Omega)$. In the case of gases, one considers that the particles are indiscernible (besides their positions in phase space, which are already taken into account). This means that one does not count twice situations which differ only by permuting particles. In other words, in thermodynamics, one formally understands the situation by saying that labels attached to particles are arbitrary. As a matter of fact, if Ω_0 were instead defined by fixing labels, one would have more states than needed. Thus, S is soundly defined by $S = k_b \log(\Omega_0/N!) = k_b \log(\Omega) - k_b \log(N!) > 0$. By considering this symmetry by permutation, one reduces the size of the microscopic possibility space, and, as a result, one obtains a value of entropy which is lower than the value computed without these symmetry considerations.

In our approach, we have $K_c = \sum_i \log(N!) - \log(n_i!)$ which is greater than 0, as soon as there is more than one cell type. The contribution to anti-entropy is given by $S_c^- = -kK_c$. Thus, the increase of the possibility space (the diversity or the differentiations) increases complexity (increases anti-entropy). More precisely, complexity, as absolute value of anti-entropy, is increased by the added symmetries, quantified by the term $\sum_i \log(n_i!)$. We understand then that anti-entropy can be analyzed, at least in this case, as an account of how many biological symmetries are broken by the cascade of differentiations. Formally, we can sum up the situation by saying that combinatorial complexity and its contribution to anti-entropy are based on a group of transformations, S_N, and a subgroup, G_{type}. The biologically relevant quantity is then the ratio of sizes of the groups S_N and G_{type}. This can be equivalently seen as the number of orbits of S_N under the right action of G_{type}.

MORPHOLOGICAL COMPLEXITY, K_m: This complexity is associated to the geometrical description of biologically relevant shapes. It is computed in particular by counting the number of connex areas. Note that this number corresponds to *space symmetry breakings* for motions covering this space — or ergodic motions. Then, one has to consider the number of shape singularities, in the mathematical sense, where singularities are invariants by action of diffeomorphisms. The fractal-like structures are particularly relevant since they correspond to an exponential increase of the number of geometrical singularities with the range of scales involved. Thus, fractal-like structures lead to a linear growth of anti-entropy with the order of magnitudes where fractality is observed (but also the variability in fractal shape should be involved, as scale symmetry changes).

FUNCTIONAL COMPLEXITY, K_f: This quantity is given by the number of possible graphs of interaction. As a result, the corresponding component of

---

[13] This account is based on "microcanonic ensembles", that is on the hypothesis of a symmetry for the probability distribution in the (microscopic) phase space: the states of equal energy have equal probabilities.

anti-entropy is given by the choice of one graph structure (with distinguished nodes) among the possible graphs. This involves the selection of the structure of possible graphs and, correspondingly, which resulting graphs are considered equivalent. In terms of symmetries, we first have a symmetry among the possible graphs which is reduced to a smaller symmetry, by the equivalence relation. For example, in [Bailly & Longo, 2009], the case is considered where the number of edges is fixed, so the considered symmetry group is engendered by the transformations which combine the deletion of an edge and the creation of another one. The orbits preserve the total number of edges, so that the orbit of a graph with $\langle k \rangle N$ edges are the graphs with this number of edges. The remaining symmetry was considered to be the identity.

We understand then that anti-entropy, or at least the decomposition we proposed here, is strictly correlated to the amount of symmetry changes. We will now look more closely at the case of combinatorial complexity since it involves only the groups of permutations and their subgroups, but at the same time will also allow us to express a crucial conceptual and mathematical point.

We indeed encounter a paradox in the case of combinatorial complexity. On one side, we have an assumption that cells of the same cell type are symmetric (interchangeable). On the other side, in section 9.2, we stressed that each cell division consists in a symmetry change. This apparent paradox depends on the scale we use to analyze the problem, as well as on the "plasticity" of the cells in a tissue or organ, as the possibility to be interchanged and/or to modify their individual organization. Typically, some cells, like probably liver cells function statistically (what matters is their average contribution to the function of the organ), while neurons may have strong specific activities, yet they may also deeply modify their structure (change number, forms and functionality of synaptic connections, for example). Thus, in the first case, type invariance or symmetry applies, while it does not apply in the second. We next consider the individual contribution of cells to the combinatorial complexity of an organism at different levels.

An organism with a large number of cells, N, with a proportion q_j of cells of cell type j (assuming also that there is a relatively large number of cells for each cell type), has two different quantities which yield the combinatorial complexities *per cell*, K_{c1} and K_{c2} (which are computed along the columns):

$$K_{c1} = \frac{\log(N!)}{N} \qquad K_{c2} = \frac{\log\left(\frac{N!}{\prod_j (q_j N)!}\right)}{N} \qquad (9.17)$$

$$\simeq \frac{\log\left((N/e)^N\right)}{N} \qquad \simeq \frac{\log\left(\frac{(N/e)^N}{\prod_j (q_j N/e)^{q_j N}}\right)}{N} \qquad (9.18)$$

$$\simeq \log(N) - 1 \qquad \simeq \log(N) - 1 - \sum_j q_j \left(\log(q_j) - 1 + \log(N)\right) \qquad (9.19)$$

$$\simeq \sum_j q_j \log(1/q_j) \qquad (9.20)$$

9.6 Interpretation of Anti-entropy as a Measure of Symmetry Changes

Now, both levels of cellular individuation are valid; but they have to be arranged in the right order. Cellular differentiation is the first and main aspect of the ability of cells to individuate in a metazoan, so we can assume that the main determinant of combinatorial complexity is K_{c2}. It is only after this contribution that the further process of cellular individuation occurs. The latter leads to a mean contribution to the complexity of the organism which is of $\sum_j a_j(q_j \log(q_j N) - 1)$ per cell, where a_j quantifies the ability of each cell type to change their organization and the relevance of this change. It seems reasonable to expect that the a_j are high in the cases, for example, of neurons or of cells of the immune system. On the contrary, the a_j should be especially low for red blood cells. The reason for this is not only their lack of DNA, but also because of their relatively simple and homogeneous cytoplasmic organization. Similarly, liver cells may have statistically irrelevant changes in their individual structure.

Thus, the contribution of cell types to anti-entropy derives first from the formation of new cell types, while the ability of cells to reproduce, with modification, within a cell type is a further important (numerically dominant) aspect of their individuation process. Note that this analysis does not suppose that a cell type for a cell is irreversibly determined, but it means that the contribution of cell type changes to anti-entropy are understood as changes of K_{c2}.

We can then provide a refined version of S_c^-, where a_{ct} is the "weight" accorded to the formation of different cell types:

$$\frac{S_c^-}{-Nk_b} = a_{ct} \sum_j q_j \log(1/q_j) + \sum_j a_j (q_j \log(q_j N) - 1) \quad (9.21)$$

$$= \sum_j [q_j(a_{ct} - a_j)\log(1/q_j) + a_j q_j(\log(N) - 1)] \quad (9.22)$$

$$= (a_{ct} - \langle a_j \rangle)\langle \log(1/q_j)\rangle + \langle(\langle a_j \rangle - a_j)\log(1/q_j)\rangle + \langle a_j \rangle(\log(N) - 1) \quad (9.23)$$

where $\langle x \rangle$ is the mean of x among all cells (so that the contribution of each cell type is proportional to its proportion in the organism). Both equations 9.21 and 9.23 are biologically meaningful. The terms in equation 9.21 correspond, by order of appearance, to the contribution of the categorization by cell types and to the contribution of individuation inside a cell type. In equation 9.23, we have obtained terms that can be assimilated to K_{c1} (last term) and to K_{c2} (first term), the latter being positive only if $a_{ct} - \langle a_j \rangle > 0$, meaning that the contribution associated to cell types is positive only if it is greater than the mean cellular individuation. This is logical since cell types make a positive contribution to the complexity only if the amount of cellular diversity they introduce is greater than the one that cellular individuation alone would introduce.

Last but not least, the second term has the sign of an anti-correlation between a_j and $\log(1/q_j)$, meaning that this term is positive when there are many low complexity cell types (given that this leads also to fewer cells per low complexity cell type or cell types with a very low complexity) and few high complexity cell types (with more cells or with a very high complexity). More precisely, using the

Cauchy-Schwartz equality case, we get that maximizing (and minimizing) this term (everything else being kept constant), leads to $\langle a_j \rangle - a_j \propto \log(1/q_j) - \langle \log(1/q_j) \rangle$. Then this optimization *a priori* leads to maximizing the second moment of information (in informational terms), at constant entropy (first moment), or in other terms, maximizing its *variance*. The situation gets quite complicated to solve analytically, but this point may be conceptually interesting[14].

In the above situation, the paradox resulted from looking with an increasing finer resolution at the individuation potential. However, the reciprocal situation can also occur. Let's consider the functional complexity, understood as the possibility of interactions between cells (here, the paradigmatic example will be neurons). Then, by assuming that there are N neurons with $\langle k \rangle$ average number of synapses for each neuron (where $\langle k \rangle$ is between 10^3 and 10^4 for humans), as presented in [Bailly & Longo, 2009], we get:

$$N_G = \binom{\binom{N}{2}}{\langle k \rangle N} \tag{9.24}$$

$$\frac{K_{f1}}{N} \simeq \langle k \rangle \log(N) \tag{9.25}$$

However, if we postulate that *any* graph of interaction is possible, then we get a total number of possible interactions which corresponds to a choice between interaction or no interaction for each entry of the interaction matrix (N^2 cells). However, the latter is symmetric; and we do not count the self-interactions (because they correspond to the complexity of the cell), so we obtain $N(N-1)/2$ binary choices, and then $2^{n(n-1)/2}$ possibilities:

$$\frac{K_{f2}}{N} \simeq \frac{N}{2} \tag{9.26}$$

There are two main lines of reasoning we can follow to understand this simple combinatorial result. The first is to look at the time structure of symmetry changes, since the symmetry changes occur as a temporal cascade. As a result, the temporal hierarchy of individuation is crucial. Here, we can refer to some phenomena concerning the graph of interaction of neurons. A crude description of the formation of neural networks is the following. First, a large number of "disordered" connections take place. Only after, the functional organization really increases by the decay of unused synapses (see for example [Luo & O'Leary, 2005]). Then, the "bigger" symmetry group involved in the description is of the form K_{f1}, with $\langle k \rangle$ mean number of connections; but then this symmetry group is reduced to obtain a smaller symmetry group with $\langle l \rangle$ mean number of connections. This operation can be seen as a change of symmetry group, from the transformations preserving the number of connections with $\langle k \rangle N$ connections to those preserving $\langle l \rangle N$ connections.

---

[14] Note that this situation is not very different from the statistical mechanics of hard spheres freeze by a correlation entropy term that is in competition with the configuration entropy term, which intuitively corresponds to disorder), see [Baus, 1987].

Of course there are many other possible components for a measure of biological complexity. This proposal, defined as anti-entropy, provides just a tentative backbone for transforming the informal notion of "biological organizational complexity" into a mathematical observable, that is into a real valued function defined over an organism. It should be clear that, once enriched well beyond this schematic definition and the further details given in [Bailly & Longo, 2009], phenotypic complexity is a proper (and fundamental) biological observable. It radically differs from the rarely quantified, largely informal, always discrete notion of "information" (informally understood as a map from topologically trivial structures to integer numbers), still dominating in particular in molecular biology, see [Longo et al., 2012a] for a critique of this latter notion. In [Longo et al., 2012c], we focus on the notion of levels of organization, also to be related to anti-entropy.

9.7 Theoretical Consequences of This Interpretation

In the section above, we proposed some technical consequences of the "microscopic" definition of anti-entropy. Using this method, we have seen that anti-entropy can mainly be understood in terms of symmetry changes. We will now consider the theoretical meaning of this situation in a more general way. As we exposed in chapter 7, we propose to understand biological systems as characterized by a cascade of symmetry changes. Note that by this we do not deny central role of inheritance and "structural stability" in the analysis of organisms, that is the "preservation" of forms and functions, along ontogenesis, reproduction and evolution. This peculiar form of stability, proper to biology, though, is very different from the mathematical stability understood in terms of invariants, thus symmetries, in physics. Note that also far from equilibrium systems are basically stable, either stationary (constant flow of energy and/or matter) or not. Since they are an organization of flows, each line, surface, structure etc. in them organizes flows as geodetics or alike. Moreover, flames, Bénard cells, micelles etc. are structurally identical, since the origin of Earth or more, they never changed.

Life structural stability, instead, is *also* and crucially based on changes: reproduction is always with variation, evolution is always based on phylogenetic modification, speciation and alike. No evolution, actually no life without changes. Moreover, diversity, a consequence of variability, thus of changes, is an essential component of structural stability, of an ecosystem, of a species, of a population, even of an individual (the diversity of cells, of organs and tissues — e.g. the internal variability of lungs, contribute to the adaptivity of an organism, thus to its structural stability in an ecosystem).

In our approach, we understood these changes as symmetry breakings and, by this, we could give a major role to symmetries along this book: there is no symmetry breaking, thus changes, without symmetries. Biological structural stability is not physico-mathematical invariance, but a complex blend of symmetries and their breakings/changes.

Now, our understanding of a "biological trajectory", a phylogenetic and ontogenetic path, as a cascade of symmetry changes yields a proper form of randomness to be associated to the construction and maintenance of biological organization. This perspective is particularly relevant for us, since it links the two main theoretical approaches to the living state of matter that we introduced here: extended criticality and anti-entropy.

More precisely, in phylogenesis, randomness is associated to the "choice" of different organizational forms, which occurs even when the biological objects are confronted with remarkably similar physical environment and physiological constraints. In classical non-linear dynamics, the symmetry breakings associated to bifurcations in trajectories is the geometric representation/counterpart of a random event: a minor fluctuation may lead to very different paths. In biology, the lungs of birds and mammals for example, have the same function in somewhat similar environments; they also are effectively compatible with a wide common range of body sizes, but they have phylogenetic histories which bifurcated long ago and, thus, extremely different architectures [Kay, 1998].

This example is particularly prone to lead to approximate common symmetries, since it relates to a vital function (respiration and therefore gas exchanges with the environment) shared by a wide class of organisms. It is noteworthy that numerous theoretical studies have analyzed lungs by optimality criteria [Horsfield, 1977, West et al., 1997, 1999, Gheorghiu et al., 2005]. However, the criteria for optimality are not the same among all these studies (minimum entropy production, maximum energetic efficiency, maximum surface/volume ratio, ...). Accordingly, even among mammals, structural variability remains high. For example, [Nelson et al., 1990] describes the differences in the geometrical scaling properties of human lungs on one side, and of rats, dogs and hamsters lungs on the other side. Moreover, [Mauroy et al., 2004] show that the criteria of energetic optimality and of robustness for the gas exchanges, with respect to geometric variations, are incompatible. More generally, optimization criteria are not particularly stable. In particular, robustness is essential but is nevertheless a relative notion: it depends on the property that we consider as robust as well as on the transformations with respect to which we expect the object of study to be robust [Lesne, 2008].

Similarly, the theoretical symmetries constituted in ontogenesis are the result of the interactions with the environment, on one side, and of the developmental trajectory already followed at a given time, on the other. In our perspective, this trajectory must then be understood as a history of symmetry changes. And, of course, the situation at a given moment does not "determine" the symmetry changes that the object will undergo. Biological systems are not "state determined systems" as their dynamics also depend on an history. The relative stability imposed by historical constraints is a crucial component of the peculiar randomness of the biological dynamics, as we consider that random events are associated to symmetry changes in a highly constraining historical context. These events are given by the interplay of the organism with its own physiology (and internal milieu) and with its environment, the latter being partially co-constituted by the theoretical symmetries of the

9.7 Theoretical Consequences of This Interpretation

organism, since many aspects of the environment depend also on the organism and its history.

In other terms, the conservation, in biology, is not entirely associated to the biological *proper observables*, the phenotype, and the same (physical) interface (e.g. energy exchange) with the environment may yield very different phenotypes; thus, there is no need to preserve a specific phenotype.

In short, the symmetry changes occurring in an organism can only be analyzed in terms of the previous theoretical symmetries (biology is, first, an historical science) and the differences in possible changes can be associated to different forms of randomness, in terms of symmetry breakings/changes:

- In the case of symmetry breakings, the symmetry change corresponds to the passage to a subgroup of the original symmetry group. As a result, the theoretical possibilities are predefinable (as the set of subgroups of the original group). This typically occurs in the case of physical phase transitions, and the result is then a macroscopic random event associated to the choice of how the symmetry gets broken, which is usually described by the direction of the order parameter (for example the sign or the direction of a global magnetization, which breaks the symmetry of a Hamiltonian). Typically, if an organism has an approximate rotational symmetry, we can say that this symmetry can be broken in a subgroup of this symmetry group, for example by providing a particular oriented direction. We then have a rotational symmetry along an axis, such as the origin of a polarity in embryogenesis. This can again be broken, for example into a discrete subgroup of order 5 (starfish). Another example is the breaking of metabolic allometry, corresponding to physiological changes in ontogeny [Glazier, 2005].
- Another form of symmetry change corresponds to the case where the symmetry changes are constituted on the basis of already determined theoretical symmetries (which can be altered in the process). This can be analyzed as the formation of additional observables (phenotypes) which are related to or the result of already existing ones. Then these symmetry changes are associated with already determined properties, but their specific form is nevertheless not predetermined. A typical example of this theoretical situation is the case of physically non-generic behaviors that can be found in the theoretical analysis of some biological models. In [Lesne & Victor, 2006], this kind of situation is argued to be widespread in biophysics and several examples are provided. From the point of view of the theoretical determination, this is a situation where there are predetermined possibilities of the phase space, not actually accessed yet (because of their non-genericity), prone to lead the biological system to develop its further organization on them, by a stabilization typically. The form of the biological response to these organizational opportunities of complexification is not, however, predetermined and then generates an original form of randomness. This theoretical account is close to the notion of "next adjacent niche", proposed in [Kauffman, 2002]; however, we emphasize here that the theoretical determination of these next organizational possibilities is only partially determined. For example imagine that a biological dynamic has approximately certain symmetries, which leads to a non-generic singular point, then it is possible (and maybe probable) that this point will be

stabilized in evolution, in an unknown way. Another example is the apparition of a new possible niche, but which we can identify as such. Example, are the apparition of swim bladder in some species of fish, following [Kauffman, 2002], see also section8.6, or the apparition of guts, but also external geological factors. In these cases, we know that the new potential niche may be colonized (a bacterium, says Kauffman, living only in the newly formed swim bladder), but we do not know the resulting organizational form of its inhabitants. Thus, the history as symmetry changes of the fish and of some bacteria superpose in an highly unpredictable way.

In such cases, the constitution of symmetry changes should be understood as having a peculiar random status, and there is no associated predictability. Gould's most quoted example of "exaptation", the formation of the bones of the internal hear from the double jaw of some tetrapods, some two hundred million years ago, is another example of this highly unpredictable evolutionary dynamics. Even more radically, the so-called Darwin's finches in the Galapagos Islands may be recalled, where hybridization lead to abrupt and unpredictable changes in beak shape, [Grant & Grant, 2002].

We have seen that the symmetry changes are related to randomness. Randomness, as symmetry changes, and its iterative accumulation are, however, the very fabric of biological organization. Therefore, we have a theoretical situation where order (biological organization) is a direct consequence of inheritance, structural stability and randomness, where randomness happens also to be a component of structural stability — by the stabilizing role of diversity, as we said. Its global analysis allowed us to give mathematical sense to Gould's evolutionary complexification along evolution, as a consequence of the random paths of an asymmetric diffusion (sections 9.3 and 9.4). A finer (or local) analysis suggested a way to understand also ontogenetic changes in these terms, that is as a random dynamics of symmetry changes.

This situation should be not confused with the cases of order by fluctuations or statistical stabilization — for example, by the central limit theorem. In our case, indeed, order is not the result of a statistical regularization of random dynamics into a stable form, which would transform them into a (mostly) deterministic frame. On the contrary, the random path of a cascade of symmetry changes yields the theoretical symmetries of the object (its specific phenotypes), which also contributes to its behavior. Moreover, while physical stability by statistical variability, e.g. the stability of an inert macroscopic object made out of quanta, is due to *very* large numbers of elementary components, the diversity contributing to the structural stability of a niche, a population, even an organ in an individual, is based on low numbers of individuals or cells.

In this context, the irreversibility of random components of the processes is taken into account by entropy production. The latter, or more precisely a part of the latter, is then associated to the ability of biological objects to generate variability, thus adaptability. In ontogenesis, this point confirms our analysis of the contribution of anti-entropy regeneration to entropy production, in association with variability, including cellular differentiation. This situation is also consistent with our analysis of anti-entropy as a measure of symmetry changes. Note that the symmetry changes,

9.7 Theoretical Consequences of This Interpretation

considered as relevant with respect to anti-entropy, may be taken into account, for example, in the coefficients corresponding to the individuation capacity of different cell types in our discussion above (see section 9.6).

In section 9.6, we used the notion of extended criticality and the associated cascades of symmetry changes in order to stress further the peculiar status of randomness in biology. Recall that, in all the main physical theories, random events take place in pregiven spaces of possibilities. One has only six possibilities for the result of dice tossing, two for coin flipping .... More generally, both in classical and quantum frames, the mathematical definition of a phase space (the space of observables and parameters) is the core theoretical step that precedes any form of knowledge construction. This space may be infinite or have increasing dimensions or infinite dimensions, like Hilbert or Fock spaces in some quantum analyses (e.g. to accommodate the creation of new particles), yet it is mathematically pregiven. These formal definitions are possible, by the invariant properties that characterize these spaces (mathematics is a science of invariants and invariant preserving transformations, thus of symmetries). All classical or quantum events, also highly unpredictable ones, thus random, take place in these predefined space of possible observable and parameters, would they be infinite (and of infinite dimensions) as we discussed in chapter 8.

In our understand, all through this book, in biology, the very space of observables phenotypes, that is the appropriate phase space for evolution (and ontogenesis) is highly unpredictable. The formation of new niches, Gould's exaptation and more, as recalled above, yield unpredictable phenotypes. "Hopeful Monsters" in ontogenesis provide the basis for phylogenetic variation — even though the ontogenetic constraints are immensely stronger. There is no way to predefine the space of future possibilities for life, as this is co-constituted by its own interactive dynamics, between the internal, organismal changing coherence and the ecosystem, which also changes. Cascades of symmetry breakings, in our approach in terms of extended criticality, continually change the "invariants", which are just biologically structurally stable, i.e. they include variability, adaptability and diversity as components. Thus, biological randomness adds on top of the many forms of physical randomness (classical and quantum randomness, of course also present in biology) and leads to the unpredictability of the very phase space. Mathematically, this is a form of "second order" randomness, yet to be analyzed — a tentative guideline for this is presented in [Buiatti & Longo, 2013].

Let's conclude this chapter by mentioning another theoretical, hopefully practical consequence of our approach. In the early paper on anti-entropy, [Bailly & Longo, 2009], there is no distinction between biological complexity and biological organization. Here, we more consistently used anti-entropy as a measure of "static" complexity (a sort of "anatomy" of a dead organism) and distinguished it from organization, as associated also to living functionalities, including autopoiesis. In section 9.6 we dared again to confuse them, as for the purposes of that section, we could use their covariance in the normal cases. That is, we assume that (increasing) organization requires (increasing) complexity and, reciprocally, that the latter yields an increase of organization, in general and normal cases. However, in discussions with

Carlos Sonnenschein and Ana Soto, well-known biologists of cancer, see [Sonnenschein & Soto, 1999, Soto & Sonnenschein, 2011, Baker, 2012], we came to notice that pathologies, cancer in particular, may lead to a decoupling of these two notions. That is, cancer may lead (or always leads?) to a greater complexity and a lower organization. Typically, mammary glands cancers increase the complexity of mammary ducts (the lumen is split in several parts, thus the topological complexity, our K_m increases), while organization, which includes functionality, decreases. Similarly the proliferation of villi, in intestinal cancer, increases the fractal dimension, again a component of our K_m, while functionality decreases. If this analysis is fully general, we could provide, also by a refinement of our K by more components, and a further insight into organization, a concrete measure in order to discriminate the pathological from the normal, at least in the case of cancer and related diseases. Of course, "normal" and "pathological" are notions that make no sense in physics. Yet, observe that we derived our analyses from a scientific methodology of which physical theorizing has been, historically, the major promoter, and even used some specific ideas from (quantum) physics.

Chapter 10
A Philosophical Survey on How We Moved from Physics to Biology

Abstract. In this book, the physical singularity of life phenomena has been analyzed by means of a permanent "constructive tension" with respect to the driving concepts and theories of the inert. In this chapter, we explicitly outline some key conceptual analogies, transferals of methodologies and of theoretical instruments between physics and biology, which have been at the core of our approach. By this, we stress significant differences and sometimes logical dualities used or to be further used to make biological phenomenalities intelligible. Our purpose in this chapter is to clarify how we applied, or at least how we tried to apply, a scientific method which has been at the core of the history of physics: the constructive objectivization of phenomena.

10.1 Introduction

Various physical theories (classical, relativistic, quantum, thermodynamic) make the inert intelligible in a remarkable way. Significant incompatibilities exist (the relativistic and quantum fields are not unified; they are in fact incompatible). However, some major principles of conceptual construction (see also [Bailly & Longo, 2011]) confer a great unity to contemporary theoretical physics. The geodesic principle and its accompaniment by "symmetries", see chapter 5 and [Weyl, 1983, Van Fraassen, 1989, Bailly & Longo, 2011], enable to grasp, under a conceptually unitary perspective, a wide area of knowledge regarding the inert. Biology, having to date been less "theorized" and mathematized, can also progress in the construction of its theoretical frameworks by means of analogies, extensions and differentiations regarding physical theories, even by means of conceptual dualities. Regarding dualities, we recall here one that is, we believe, fundamental and that has been extensively addressed in this book and in other writings, [Bailly & Longo, 2011, Frezza & Longo, 2010, Longo & Montévil, 2011a]): the *genericity* of physical objects (that is, their theoretical and experimental invariance) and the *specificity* of their trajectories (basically, their reconstruction by means of the geodesic principle or identification by mathematical techniques, by symmetries typically). In our perspective, this is inverted in biology, as it is transformed into the *specificity* (individuation and history) of the living object and the *genericity* of trajectories (evolutionary, ontogenetic: they are just "possibilities" within spaces — ecosystems — in co-constitution).

Let us now review the key concepts that we consider as relevant for the study of biological phenomena.

10.2 Physical Aspects

10.2.1 The Exclusively Physical

We exclude from our analyses those properties which come from physics (where they are often essential), but of which the transferal to biology is, from our point of view, misleading:

1. The GENERICITY of objects (the theoretical and experimental invariance of physical objects — or symmetry by replacement) does not apply to biology: the living object is historical and individuated; it is not "interchangeable", in general or with the generality of physics, not theoretically nor empirically.
2. The SPECIFICITY of trajectories (geodesics, in physics), because we exclude the prevalence of the geodesic principle (there is no "optimality") for ontogenetic and evolutive dynamics of "biological individualities" — cells, organisms, species (which we call, synthetically, "biolons"); in short, embryogenesis, development and evolution are not optimal trajectories, but *possible* ones, see chapter 7.
3. The STABILITY of the reference system as such. Besides classical physics, also in general relativity and in the energy/geometry relationships in space-time, the dimensions are set and do not vary during the phenomenal analysis. Instead, the space of observables in biology, of phenotypes for example, which can also be described by new "dimensions", is, itself, dynamically changing in an ecosystem. Using an informal analogy, we could say that the "*phase space*" (and the space of possibilities) of life phenomena is dynamically (co-)constituted, see chapter 8. As a matter of fact, in relativity theory, *space-time* is (co-)constituted by the energy/matter distribution, yet in stable dimensions and phase space — while in chapter 8 we claim that phase spaces change along biological processes, evolution in particular.

As discussed in chapter 5 and [Longo & Montévil, 2011a], the genericity of physical objects and the specificity of their trajectories depend on the theoretical symmetries which allows to constitute them. In biology, our hypothesis is that the properly biological theoretical symmetries are unstable. This leads to a change of the theoretical status of biological objects with respect to physical situations. We will come back to this point further below.

10.2.2 *Physical Properties of the "Transition" towards the Living State of Matter*

In the literature, we often find remarkable works concerning certain physical properties, sometimes transferred to the analyses of life phenomena, but which we will later consider in their exclusively biological form (i.e., that we only find in the living state; for example, critical transitions, which are pointwise in physics, are "extended" in our approach). In biology, we therefore *do not consider* them "as such", as they present themselves as components of the analysis of the inert, where they nevertheless provide a good starting point for reflections regarding life phenomena. For the moment, let's evoke them from a physical perspective ("as such") and stress that they partly pertain the biological theoretical vocabulary, but do not properly belong to it, in our view:

1. CRITICALITY as such (in physics, present in phase transitions, as a mathematical point with respect to the control parameter);
2. ORGANIZATIONAL CLOSURE as such (present in physical chemistry: micelles, vesicles — whose structure is entirely organized along geodetic principles, in contrast to living organisms);
3. PASSIVE PLASTICITY as such (present in changes of physical form or in phenomena of action/reaction/propagation in the manner of Turing, for example);
4. SCALING PROPERTIES as such (present in numerous physical phenomena and namely in critical transitions, anomalous diffusion, etc.);
5. GROWTH phenomena as such (present in the growth of crystals, for example);
6. CHIRALITY as such (present in the physics of particles or chemistry, for example);
7. Possibly negative variations of ENTROPY (present in the passing from disorder to order, in critical transition for example),
8. The DIMENSIONALITY of physical quantities (almost always present – in contrast, for example, to the pure numbers of biological rhythms, see section 2.2.3 and chapter 3);
9. The MEASUREMENT which is always understood as approximated, in classical frames;
10. The FRACTALITY as such of certain objects and dynamics (present in a number of physical phenomena, but also in *organs* of plants and animals as forced by their role in the exchange of energy and matter);
11. The chemistry of MACROMOLECULES and of *in vitro* physical chemistry.

10.3 Biological Aspects

The contingent materiality of life phenomena includes, typically, the physical chemistry specific to biology, our first group of properties:

10.3.0.1 A Few "Physical" Properties of Life Phenomena

1. The biological role of the CHIRALITY of molecules (amino acids, sugars) in the metabolism;
2. Various other physical INVARIANCES according to the level of organization (the chemical bases and geometric structure of DNA, relatively common to all living objects; the metabolic invariants, including the metabolism/mass/duration relationship).

In addition to the above physical properties, which specifically (and only) manifest themselves in life phenomena, the following are certainly part of biological *theorization*:

1. Analysis in terms of PHYSICO-CHEMICAL SUBSTRATES such as molecular cascades that may be found only in cells;
2. The MATHEMATICAL EXTENSION of certain physical laws including quantities that do not appear as such or in an operative way in physics (for example, our notion of anti-entropy in metabolic balances, recalled below, which extends well-known balance equations in thermodynamics by a new observable).

10.3.1 *The Maintenance of Biological Organization*

The setting of physiological activities (the functions of "orgons" — organelles, organs, populations, see [Bailly et al., 1993, Bailly & Longo, 2011]), is often accompanied by organizational closure which is accomplished by means of:

1. The METABOLISM and PHYSIOLOGICAL ACTIVITIES (essential to integration and to regulation) which interact and, in fact, superimpose one another;
2. The coupling between VARIOUS LEVELS OF ORGANIZATION, correlated in a causal manner, both "upwards" and "downwards", particularly by integration and regulation,
3. The FRACTALITY of orgons in their physiological functions (lung, vascular system, nervous system... intracellular structures);
4. The SCALING LAWS (allometry describes temporality and metabolism in function of the adult biological mass);
5. The importance of PURE NUMBERS (without physical dimensions) and of their RELATIVE INVARIANCE (total number of heartbeats, respirations... which are on average constant for mammals, and even among important groups of less studied species as for internal rhythms, see section 2.2.3).

We tried to conceptually frame these properties of the living state of matter by means of relatively new concepts, including that of *extended critical transition* in 1.4.2, as locus and framework for the phenomena, which we summarized above.

10.3.2 The Relationship to the Environment

To these functions, we must add the relationship to the environment that is not only dynamic, but adaptive *and* (or *because*) cognitive (as are protentional activities). Moreover, biological dynamic is also located at the level of the reference space (relevant parameters and observables), as, among other, an organism co-modifies its own environment:

1. ADAPTIVE PLASTICITY at all levels of organization, in the interaction with an environment;
2. The cognitive, present as soon as there is life, resides, in particular, in the CAPACITY TO DISCRIMINATE (the countable density of critical points within the zone of extended criticality mentioned below can represent this discriminatory capacity, by discontinuous passages (but without gaps) from one point to another);
3. The principle of COMPATIBILITY (tendency to achieve all possibilities compatible with the given constraints), which justifies the genericity of evolutive and ontogenetic trajectories;
4. The SPECIFICITY of the object and, as we were saying in section 10.2.1, the GENERICITY of trajectories (in opposition to physics);
5. The CHANGES IN REFERENCE SPACES, which induces and enables biological behavior, including in the number of relevant description dimensions (the "phase space" itself — relevant parameters and observables — changes over the course of the dynamics of life phenomena, as opposed to the physical frameworks, even quantum ones).

Again, most of these aspects are related to an instability of biological theoretical symmetries, associated here to the constitution by the biological object of the theoretically relevant environment.

10.3.3 Passage to Analyses of the Organism

CRITICAL TRANSITIONS are extensively discussed in the analysis of the passage, in particular, from the inert to the living, [Kauffman, 1993]. As such, they very well describe states of the inert that are interesting also for biology see [Binney et al., 1992, Mora & Bialek, 2011]. In physics, though, "coherent structures" appear over pointwise transitions, and normally in a reversible way. We are, however, facing a living state of matter when criticality is *irreversible* and *lasts* (till death). We deal with these issues by considering an organism as staying in a "continual" (ongoing) irreversible transition. Each mitosis, in a multicellular organism, yields an asymmetric bifurcation and the formation of a new coherence structure — A new tissular matrix ..., as components of a critical transition. In our approach, the interval of criticality is therefore extended in time and in all relevant control parameters (temperature, pressure...), see section 1.4.2. The key idea is that all the usual properties of critical phase transitions are preserved (the formation of coherence structures, diverging correlation lengths, symmetry changes ...). Yet, while, in physics, those only apply in a transition point (at least this is the mathematical representation, where the renormalization methods apply, see section 1.4.2), we consider the "transition" to be defined on a non-trivial interval. This occurs when rhythms (point 1 below and section 1.4.1), protentional activity (point 2 and 1.4.1) and organization, as anti-entropy (point 3 and 4, see also 1.4.4) jointly appear.

We may then conceive (but this discussion is not our aim, here) that, at the origin of the extended criticality of life, there may have been particular critical transitions of the inert matter, a global transition suddenly superposing all the ones we are dealing with. These may all be described as conceptual and material "bifurcations", with their organizational correlates: extension of criticality to an interval, by the formation of stabilizing membranes and of different levels of organization (as antientropy), bifurcation of the time dimension (autonomous rhythms). Yet, extended criticality is an ongoing phenomena for life, well beyond its origin. The five points below may be considered at the core of the synthesis in this book. We briefly propose to organize these "bifurcations", which mark the (conceptual) passage from a state of the inert to the living state, as a *constitution of*:

1. The second temporal dimension, the COMPACTIFIED time of biological rhythms;
2. The PROTENTION, as a "proactive gesture" in the interaction with the ecosystem, present even unicellular life;
3. ANTI-ENTROPY, as the establishment and maintenance of organization (which is opposed to disorganization — in particular to the entropy produced by all irreversible processes);
4. The distinction in SEVERAL LEVELS OF ORGANIZATION, at the core of the integration and regulation activity of any living unit (which may be conceptually unified as *orgons* — organelles, organs, populations — and *biolons* — cells, organisms, species).
5. An INSTABILITY OF THE THEORETICAL SYMMETRIES of the objects, which can be seen as a cascade of symmetry changes, over time, and leads to variability in the strong sense of changing theoretical symmetries.

In short, the intelligibility of life phenomena that we propose presupposes the existence ("somewhere", "at the origin of life") of correlated bifurcations whose understanding requires the addition of the new theoretical entities above. These are perfectly compatible with physical theorization, but they are not specific to it. In this sense, it is a matter of proposing compatible, but "strict" theoretical extensions of theories of the inert. Reduction may be a further step for the interested reductionist, who should prove that these theories are, first, conservative (in the sense of Logic), then only apparently "strict".

10.4 A Definition of Life?

Throughout the very old "physicalism / vitalism" debate, it has often been question of *defining* what is life. A small but remarkable book by Schrödinger [Schrödinger, 2000] contributed to reviving the debate in a way we find to be relevant, at least in

10.4 A Definition of Life?

its second part, and to which we refer in section 9. Did we provide, by this book, a "definition of life"? Did we, at least, work towards such a definition? Let's better specify how we see this question:

PRIMO. An "ideal" definition of life phenomena seems out of the question: there is no *Platonic idea* of life to be grasped in a definite manner or with the maximal conceptual stability and invariance specific to mathematical notions (as there is with the definition or *idea* of the triangle...). It is rather a question of defining a few *operational notions* enabling to draw out concepts with which to work for a systemic approach in biology. Moreover, physics does not define "matter" otherwise than by means of an operative duality or contraposition (with respect to the concept of energy or to that of vacuum or of anti-matter, for example). Yet another, very rigorous, "provable impossibility to define the object of study" is presented in the next section. Note that Darwin's approach to evolution does neither use nor need a definition of life, but needs to refer to organisms.

SEGUNDO. Any operational attempt, in our opinion, must be made with respect to the specific phenomenality of life phenomena: for example, it is possible that for any chosen finite list of "defining" properties of life, there would exist a sufficiently talented computer scientist able to create its virtual image to be rendered on a computer screen (it is quite simple to program an "autopoietic" system [Varela et al., 1974, Varela, 1989] or a formalized metabolic cycle in the manner of Rosen [Rosen, 2005] — see [Mossio et al., 2009], for example). However, not only any human being, but also the most simple-minded of animals would recognize it as a series of non-living "virtual images" (which are typically detectable through identical iteration, as indirectly suggested by Turing's imitation game, see [Longo, 2008]).

It is rather a question, thus, of proposing a possibly robust intelligibility of a phenomenality in its constitutive history, while keeping in mind the fact that *any constitution is contingent* — both the constitution (evolution) of life and of our historical understanding of it. That is, we stress the contingency of life and of our modest attempts to grasp its unfolding over a material evolution — better still: over one of the *possible* evolutions, taking place on *this* Earth, in *these* ecosystems and with *this* physical matter and history. Our point of view includes what biologists often express when they say that nothing can be understood in biology otherwise than in the light of evolution (Darwinian and in this world) and what historian claim to be the concrete historicity of science, as a non-arbitrary, but historical tool for constructing objectivity and objects of knowledge.

It should be clear that we do not discuss here how "life may have emerged from the inert", but rather we explore how to go from the current *theories* of inert to a sufficiently robust *theory* of the living. In particular, we hinted here to an analysis of the physical singularity and of the specificity of the living object, by looking first at the properties we would want to have (or *not*) in any theory of the "living state of matter". It is indeed an *incomplete* (see next) attempt at providing a conceptual framework guiding more specific analyses.

In the next methodological reflection, we will borrow from Mathematical Logic an understanding of the role of incompleteness in "our theoretical endeavors towards knowledge" (to put it in H. Weyl's words) and of its relation to conceptual or formal "definitions", of life in particular.

10.4.1 Interfaces of Incompleteness

Do we need to have a definition of life, in order to construct robust theories of the living state of matter?

Let's now answer to this question by an analogy with a frame where it may be dealt with the highest rigor: Mathematical Logic.

Is the concept of integer (thus "standard" or finite) number captured (defined, characterized) by the (formal) theory of numbers? Frege (1984) believed so, as the absolute concept of number was, in his view, fully characterized by Peano-Dedekind theory. In modern logical terms, we can say that, for Frege, Peano Arithmetic (*PA*) was "categorical". That is, *PA* was believed to have just one model, up to isomorphisms: the standard model of integers (the one which the reader learned about in elementary school, with 0, though, and formal induction). Thus, the theory was also meant to uniquely define "what a number is".

This turned out to be blatantly wrong. Löwenheim and Skolem (1915-20) proved that *PA* has infinitely many non-isomorphic models and, thus, that it is not categorical. Moreover, a simple theorem ("compactness") showed that no predicate, definable in *PA*, may isolate (define) all and exactly all the standard integers (see [Marker, 2002]). In short, any predicate valid on infinitely many standard integers, must hold also for (infinitely many) non-standard integers (which cannot be considered properly "finite") — this is known as the "overspill lemma". Gödel's incompleteness theorem reinforced these negative properties: *PA* is *incomplete* or, equivalently, it has lots of logically non-equivalent models, a much stronger property than *non-categoricity*.

A fortiori, there is no hope to characterize in a finitistic way the concept of standard integer number. One has to add an axiom of infinity (Set Theory) or proper second order quantification in order to do so, PA_2, and these are infinitary or impredicative formal frames (see [Longo, 1989]). Or, also, Set Theory with an axiom of infinity and PA_2 are not only strict extension, but they are *non-conservative* extension of *PA*: they prove propositions of *PA*, which are unprovable in PA[1].

---

[1] Whether the biological observables we focused on and their theories are strict extensions or not of the related physical theories is surely an interesting question. However, it would be much more interesting if one of our theories or their conjunction were shown to be non-conservative with respect to a (pertinent) theory of the inert. For example, Pasteur's famous example of statistically non-balanced chirality of some macromolecules is a property that can be stated in the language of physics, yet, as far as we know, it has not been derived from any physical theory. It would be fantastic if it could be justified within one of our frames, e.g. from a property of the phenotype at the cellular level ... extended criticality, say ....

In conclusion, in spite of its incompleteness, everybody soundly considers *PA* as the "natural" theory of numbers: it elegantly singles out the main relevant, and very robust, properties of numbers (0, successor, induction), even though it *cannot define what a number is*. In analogy to the impossibility of physics to define its own object of study, physical matter, as we mentioned at the beginning, we have here another example of sound theoretical frame, which cannot define, within itself, its own object of study, the natural number object. And we do not see a way to get out from the language of physics or of biology as Mathematical Logic can do, by using infinities: what would ever correspond to an axiom of infinity or to higher order quantification? Perhaps: ... "take the point of view (and the language) of God"?

We encourage thus the reader to pursue his/her theoretical work in biology without the anguishing search for a *definition* of life. And with the clear perspective of the intrinsic incompleteness of all our theoretical endeavors, [Longo, 2011a]: we can just hope to explicitly grasp and organize by theories some fragments of reality, whatever this word may mean. Let's try to do it towards the best of our knowledge, in a sufficiently broad and robust way, and in full theoretical and empirical freedom, without necessarily feeling stuck either to existing theories nor always searching for the "Ultimate (complete?) Theory" nor the "ultimate reduction". Molecular analyses are not useless, of course, nor wrong, a priori, they are just incomplete, in our opinion, as for describing phenotypes and their evolutionary or ontogenetic dynamics.

Similarly, the issue of the emergence of life from molecules is a very relevant one, but as long as we do not have a sufficiently robust, yet incomplete, theory of organisms, which "objects", with which properties, should ever been shown to emerge from inert matter?

10.5 Conclusion

Broadly speaking, and including the considerations in terms of extended criticality and symmetry changes, the principles that we propose, while addressing these particular observables and quantities, specific to life phenomena, constitute an *extension* of existing physical theories: they preserve the same formal mathematical structure and, if we set the value of the considered observables or parameters to 0, they lead us back to the case of the inert. That is, if there is no protention, no second temporal dimension, no extension of criticality, no value of anti-entropy, one returns to physical frames. Our theoretical propositions are thus compatible, although they may be irreducible as such, to "existing physical theories". That is, they are reducible to physics *as soon as* they are presented outside of the extended critical zone having its own temporality and its anti-entropy, or as soon as these specific quantities go to 0.

In short, the peculiar phenomenality of life deserves some new concepts and observables: we tried to propose such observables by the notions of extended critical transition, biological complexity and organization, proper time, ... The point is the pertinence, if any, of these treatments, "*per se*". Those who claim that all these concepts should be reduced to physical (existing?) theories are welcome to try: we would be very pleased and proud if the competent reductionists were able to rewrite them fully and faithfully (derive or embed them) in (existing) physical frames. But they should first look at the history of Physics itself, where novel theoretical frames are marked by the invention of new perspectives, new concepts as well as new irreducible observables. Their pertinence had to be judged "as such", within their domain of meaning, not on the grounds of their reducibility to existing, thus "safe", explanatory grounds. In any cases, should reduction or unification be performed, the first question is: *which theory* does one want to reduce to *which theory*? Reduction, as we learn from physics and logic, is a intertheoretical issue.

Note that, among our proposals, the concept of extended critical transition, in association with ubiquitous symmetry changes, leads to radical methodological changes, as associated to the specificity of objects and genericity of trajectories. It is probably the most radical change of perspective we propose. This proposal indeed alters, with respect to physics, the very theoretical nature of the scientific object, as for proper biological objects: organisms and phenotypes. As a result, physical notions like the space of theoretical determination (phase space) cannot have the same meaning and use. One of the main and maybe the main notion at the core of these changes is historicity. In evolution and in development, biological objects organize themselves and they do so in a never identical manner, as long as their organization allow them to survive. We want to emphasize again that the specificity of biological objects, associated to this historical determination and the unstable mathematical symmetries, calls for a change of perspective, with respect to physics, in the understanding of biological phenomena. Physical objects, even the most complex ones, are understood by their regularities (invariants and associated symmetries), while one of the most stable feature of biological objects is their variability, which engenders diversity and contributes by this to biological structural stability, at all levels of organization. It is the reason why we put variability, understood as symmetry changes, at the core of our approach to biological phenomena.

Appendix A
Mathematical Appendix

In this appendix, we provide some mathematical background on results discussed in this book. More precisely, we go back to scale symmetries, as discussed in chapter 2 and provide a proof of Noether's Theorem in addition to the presentation in chapter 5.

A.1 Scale Symmetries

Let us resume the discussion of section 2.1.1. We will consider the mathematical situation with more generality than in chapter 2, by recalling that symmetries correspond to the algebraic notion of group. We want to find the functions which have a scale symmetry. A strict scale symmetry is defined by the fact that to a dilatation of the parameter(s) corresponds a dilatation of the observed quantity. Then, we have the following result:

Theorem A.1 (Scaling). *Let us consider a function* $f : A \to \mathbb{R}$ *and a group* $G = \{\mathfrak{D}_\lambda | \lambda \in \mathbb{S}\}$ *acting on* A, *where* $\mathbb{S}$ *is a subgroup of* $(\mathbb{R}^{*+}, \times)$ *and* $\lambda \to \mathfrak{D}_\lambda$. *We suppose that* $\lambda \to f \circ \mathfrak{D}_\lambda$ *is continuous*[1] *and that there exist g such that* $\forall \lambda \in \mathbb{S}, f \circ \mathfrak{D}_\lambda = g(\lambda)f$.
Then, there exists α *such that* $\forall \lambda, g(\lambda) = \lambda^\alpha$.

Proof. If $f = 0$, then any α fulfill the conditions of the theorem. Let us consider now $f \neq 0$ and G according to the hypothesis of the theorem. We will proceed by considering sets of increasing size.

[1] $f \circ \mathfrak{D}_1 = g(1)f$ and $f \circ \mathfrak{D}_1 = f$ so $g(1) = 1$.
[N] Let us consider $a \in \mathbb{S}, a \neq 1$ and $n \in \mathbb{N}^*$. We have $f(\mathfrak{D}_{a^n}(x)) = g(a^n)f(x)$ and $f(\mathfrak{D}_{a^n}(x)) = f(\mathfrak{D}_a \circ \mathfrak{D}_{a^{n-1}}(x)) = g(a)f(\mathfrak{D}_{a^{n-1}}(x)) = g(a)g(a^{n-1})f(x)$ and then, $g(a^n) = g(a)g(a^{n-1})$. Finally, we obtain by recurrence $\forall n \in \mathbb{N}, g(a^n) = g(a)^n$.
[Z] Let us take $n \in \mathbb{N}$. We have $g(a^n a^{-n}) = g(1) = 1$ and $f(\mathfrak{D}_{a^n a^{-n}}(x)) = f(\mathfrak{D}_{a^n} \circ \mathfrak{D}_{a^{-n}}(x)) = g(a^n)f(\mathfrak{D}_{a^{-n}}(x)) = g(a^n)g(a^{-n})f(x)$, as a result $g(a^{-n}) = g(a)^{-n}$. We have shown that, $\forall n \in \mathbb{Z}, g(a^n) = g(a)^n$.

---
[1] This condition is always met if $\ln(\mathbb{S})$ is proportional to $\mathbb{Z}$.

If $\ln(\mathbb{S})$ is proportional to $\mathbb{Z}$, there exists a such that $\mathbb{S} = \{a^n \mid n \in \mathbb{Z}\}$. As a result for all $\lambda \in \mathbb{S}$, we have $\ln(\lambda)/\ln(a) \in \mathbb{Z}$. Thus $g(\lambda) = g(a)^{\ln(\lambda)/\ln(a)} = \lambda^{\ln(g(a))/\ln(a)}$, so $\alpha = \ln(g(a))/\ln(a)$ answers the question. In the following step, we will on the contrary assume that $\mathbb{S}$ is dense[2] in $\mathbb{R}$.

[S]   We will show that $\alpha = \ln(g(a))/\ln(a)$ fits the expected conditions.

Let us consider $\lambda \in \mathbb{S}$ and $p_k = \lambda^{n_k} a^{-m_k}$ where $\forall k$, $(n_k, m_k) \in \mathbb{N}^{*2}$ and (p_k) tends to 1. This last condition means that (m_k/n_k) tends to $\ln(\lambda)/\ln(a)$.

We have $f(\mathfrak{D}_{p_k}(x)) = g(p_k)f(x)$ which tends to $f(x)$ since $s \to f \circ \mathfrak{D}_s$ is continuous and (p_k) tends to 1, so $(g(p_k))$ tends to 1. Now, $g(p_k) = \frac{g(\lambda)^{n_k}}{g(a)^{m_k}}$ so $g(p_k)^{1/n_k} = \frac{g(\lambda)}{g(a)^{m_k/n_k}}$ tends to $\frac{g(\lambda)}{g(a)^{\ln(\lambda)/\ln(a)}}$ and to 1. As a result $g(\lambda) = g(a)^{\ln(\lambda)/\ln(a)} = \lambda^\alpha$.

In all cases, we thus have found α such that $g(\lambda) = \lambda^\alpha$.

Remark A.1. If $\ln(\mathbb{S})$ is proportional to $\mathbb{Q}$, we do not need the continuity hypothesis. Indeed, consider $p \in \mathbb{N}$ and $q \in \mathbb{N}^*$. We have $g(a^p) = g(a)^p$ and $g(a^p) = g((a^{p/q})^q) = g((a^{p/q}))^q$. As a result $g(a^{p/q}) = g(a)^{p/q}$.

Remark A.2. The continuity hypothesis is crucial in general. Let us for example decompose $\mathbb{R}$, as a $\mathbb{Q}$-vector space, as a linear sum of two non-trivial subspaces: $\mathbb{R} = A \oplus B$. We can then define $f : \mathbb{R}^+ \to \mathbb{R}^+$, so that for $x = \exp(a+b) \in \mathbb{R}$, with $a \in A$ and $b \in B$ we get $f(x) = \exp(2a + 3b)$. f is well defined because the decomposition is unique. For $\lambda = \exp(a_2 + b_2)$, we have $f(\lambda x) = f(\exp(a_2 + b_2 + a + b)) = \exp(2a + 2a_2 + 3b + 3b_2) = f(s)f(x)$. Thus the condition of scale symmetry is met, but not the continuity and neither is the existence of an α valid for all transformations.

Corollary A.1. *A function $f(x_1, \ldots, x_n) : \mathbb{R}^{*n}+$ which meet the criteria of theorem A.1 for the n groups, copies of $(\mathbb{R}^{*+}, \times)$ acting respectively on each x_i by the usual multiplication can be written as:*

$$f(x_1, \ldots, x_n) = f(1, \ldots, 1) \prod x_i^{\alpha_i} \tag{A.1}$$

A.2   Noether's Theorem

We will first formulate and prove Noether's theorem in the framework of classical Lagrangian mechanics. Then we provide this result in the (Lagrangian) field theoretic settings.

A.2.1   Classical Mechanics Version (Lagrangian)

In order to start from first principles, within Lagrangian mechanics, we will first derive the equations of motions from the variational principle, since these equations

---

[2] We recall that the subgroups of $(\mathbb{R}, +)$ are either proportional to $\mathbb{Z}$ or dense, which leads to the subgroups of $(\mathbb{R}^{*+}, \times)$ either having a logarithm proportional to $\mathbb{Z}$ or being dense.

A.2 Noether's Theorem

are used in the proof. The derivation of the equations exemplifies our discussion of the specificity of physical trajectories above and in chapter 7.

A.2.1.1 Variational Principle

Let us consider a classical system, governed by its Lagrangian $\mathscr{L}(t, q_1, \dot{q}_1, \ldots, q_n, \dot{q}_n)$, where $\dot{q}_i = \frac{dq_i}{dt}$. The state is then described as a $2n$ dimensional vector, we will write such a state as $\bar{q}(t)$. In order to simplify the notations, we will write in the following:

$$\mathscr{L}(t, q_1, \dot{q}_1, \ldots, q_n, \dot{q}_n) = \mathscr{L}(t, q_i, \dot{q}_i)$$

The variational principle, also called Hamilton's principle, states that the trajectory between two points in phase space is stationary with respect to the *action*:

$$\mathscr{S} = \int_{t_1}^{t_2} \mathscr{L}(t, q_i(t), \dot{q}_i(t)) dt \tag{A.2}$$

We can then derive the *Euler-Lagrange equations*, which are the equations of motion in a Lagrangian point of view. Let us consider a stationary trajectory in phase space $\bar{q}(t)$ and a small perturbation $\bar{\varepsilon}(t)$, with $\bar{\varepsilon}(t_1) = \bar{\varepsilon}(t_2) = 0$. We then have a corresponding change of the action:

$$\delta\mathscr{S} = \int_{t_1}^{t_2} \mathscr{L}(t, q_i(t) + \varepsilon_i(t), \dot{q}_i(t) + \dot{\varepsilon}_i(t)) - \mathscr{L}(t, q_i(t), \dot{q}_i(t)) \, dt \tag{A.3}$$

We will use from here on the Einstein summation convention, meaning that silent indexes that appear twice, one in an up position and one in a down position implies a summation over all possible values. Here, the concerned index is α which is thus associated to an implicit sum. The above gives at the first order:

$$\simeq \int_{t_1}^{t_2} \varepsilon_\alpha(t) \frac{\partial \mathscr{L}(t, q_i, \dot{q}_i)}{\partial q_\alpha}(t) + \dot{\varepsilon}_\alpha(t) \frac{\partial \mathscr{L}(t, q_i, \dot{q}_i)}{\partial \dot{q}_\alpha}(t) \, dt \tag{A.4}$$

With a partial integration, we obtain:

$$= \left[\varepsilon_\alpha(t) \frac{\partial \mathscr{L}(t, q_i, \dot{q}_i)}{\partial \dot{q}_\alpha}(t) \right]_{t_1}^{t_2} \\ + \int_{t_1}^{t_2} \varepsilon_\alpha(t) \frac{\partial \mathscr{L}(t, q_i, \dot{q}_i)}{\partial q_\alpha}(t) \; - \varepsilon_\alpha(t) \frac{d}{dt} \frac{\partial \mathscr{L}(t, q_i, \dot{q}_i)}{\partial \dot{q}_\alpha}(t) \, dt \tag{A.5}$$

On one side, the first term is null because $\bar{\varepsilon}(t_1) = \bar{\varepsilon}(t_2) = 0$, by definition. On the other side, $\delta\mathscr{S} = 0$ at the first order because $\bar{q}(t)$ is a stationary point of the Lagrangian. As a result, we obtain for every $\bar{\varepsilon}(t)$

$$0 = \int_{t_1}^{t_2} \varepsilon_\alpha(t) \frac{\partial \mathscr{L}(q_i, \dot{q}_i)}{\partial q_\alpha}(t) - \varepsilon_\alpha(t) \frac{d}{dt} \frac{\partial \mathscr{L}(q_i, \dot{q}_i)}{\partial \dot{q}_\alpha}(t) \, dt \tag{A.6}$$

And the latter proves the *Euler-Lagrange equations*:

$$0 = \frac{\partial \mathscr{L}(t, q_i, \dot{q}_i)}{\partial q_\alpha} - \frac{d}{dt} \frac{\partial \mathscr{L}(t, q_i, \dot{q}_i)}{\partial \dot{q}_\alpha} \tag{A.7}$$

This equations corresponds to the fundamental principle of dynamics with the momenta $p_\alpha = \frac{\partial \mathscr{L}(t, q_i, \dot{q}_i)}{\partial \dot{q}_\alpha}(t)$ and the forces $F_\alpha = \frac{\partial \mathscr{L}(t, q_i, \dot{q}_i)}{\partial q_\alpha}(t)$.

This short derivation shows that the Lagrangian formalism is equivalent to the Newtonian point of view. Another classical and powerful point of view is Hamiltonian mechanics. These different formalisms have different efficiency, depending on the problem considered.

A.2.1.2 Noether's Theorem

We will formulate Noether's theorem in this context and prove it.

Theorem A.2 (Noether, classical Lagrangian mechanics). *For the above Lagrangian, let us suppose that $\mathscr{S}$ is preserved under the action of a one parameter continuous group $\mathfrak{G}$ with infinitesimal generator[3] $v = \tau \frac{\partial}{\partial t} + \phi_\alpha \frac{\partial}{\partial q_\alpha} + \psi_\alpha \frac{\partial}{\partial \dot{q}_\alpha}$. Then, the quantity:*

$$C = \tau \mathscr{L} + \frac{\partial \mathscr{L}}{\partial \dot{q}_\alpha}(\phi_\alpha - \dot{q}_\alpha \tau) \tag{A.8}$$

is an invariant of the dynamic (meaning a quantity with a null derivative with respect to time).

Proof. In order to show that this quantity is conserved, we will consider its time derivative which we will prove to be 0.

EFFECT OF THE TRANSFORMATION Let us consider $g \in \mathfrak{G}$, we will note for every variable $g.x = \tilde{x}$.

$$\tilde{\mathscr{S}} = \int_{\tilde{t}_1}^{\tilde{t}_2} \mathscr{L}(\tilde{t}, \tilde{q}_i, \dot{\tilde{q}}_i) d\tilde{t} = \int_{t_1}^{t_2} \mathscr{L}(\tilde{t}, \tilde{q}_i, \dot{\tilde{q}}_i) \frac{d\tilde{t}}{dt} dt \tag{A.9}$$

Since, by hypothesis, the action is conserved, we have:

$$\tilde{\mathscr{S}} = \mathscr{S} \tag{A.10}$$

$$\int_{t_1}^{t_2} \mathscr{L}(\tilde{t}, \tilde{q}_i, \dot{\tilde{q}}_i) \frac{d\tilde{t}}{dt} dt = \int_{t_1}^{t_2} \mathscr{L}(t, q_i(t), \dot{q}_i(t)) dt \tag{A.11}$$

This equation is true for all t_1 and t_2 so:

---

[3] We cannot choose freely all the parameters of the generator, more precisely we will show in the proof that the ψ parameters is determined by the others (for example).

A.2 Noether's Theorem

$$\mathscr{L}(\tilde{t}, \tilde{q}_i, \dot{\tilde{q}}_i)\frac{d\tilde{t}}{dt} = \mathscr{L}(t, q_i(t), \dot{q}_i(t)) \tag{A.12}$$

Thus, at first order:

$$0 = \tau\frac{\partial\mathscr{L}}{\partial t} + \phi_\alpha\frac{\partial\mathscr{L}}{\partial q_\alpha} + \psi_\alpha\frac{\partial\mathscr{L}}{\partial \dot{q}_\alpha} + \mathscr{L}\frac{d\tau}{dt} \tag{A.13}$$

Notice that the last term is due to the changed scope of integration, the other terms come from the variation of the Lagrangian generated by the transformation.

RELATION BETWEEN THE PARAMETERS OF THE GROUP. At first order in ε, the following relation holds:

$$\dot{\tilde{q}}_\alpha(\tilde{t}) - \dot{q}_\alpha(t) = \psi_\alpha \varepsilon \tag{A.14}$$

$$= \frac{d}{d\tilde{t}}(q_\alpha(\tilde{t} - \varepsilon\tau) + \varepsilon\phi_\alpha) - \frac{d}{dt}q_\alpha(t) \tag{A.15}$$

$$= \varepsilon\left(\frac{d\phi_\alpha}{dt} - \frac{d\tau}{dt}\dot{q}_\alpha(t)\right) \tag{A.16}$$

This leads to:

$$\psi_\alpha = \frac{d\phi_\alpha}{dt} - \frac{d\tau}{dt}\dot{q}_\alpha(t) \tag{A.17}$$

REMAINING OF THE PROOF Let us start from equation A.13

$$0 = \tau\frac{\partial\mathscr{L}}{\partial t} + \phi_\alpha\frac{\partial\mathscr{L}}{\partial q_\alpha} + \psi_\alpha\frac{\partial\mathscr{L}}{\partial \dot{q}_\alpha} + \mathscr{L}\frac{d\tau}{dt} \tag{A.18}$$

$$= \frac{d}{dt}(\tau\mathscr{L}) - \tau\dot{q}_\alpha\frac{\partial\mathscr{L}}{\partial q_\alpha} - \tau\ddot{q}_\alpha\frac{\partial\mathscr{L}}{\partial \dot{q}_\alpha} + \phi_\alpha\frac{\partial\mathscr{L}}{\partial q_\alpha} + \psi_\alpha\frac{\partial\mathscr{L}}{\partial \dot{q}_\alpha} \tag{A.19}$$

We use equation A.17

$$= \frac{d}{dt}(\tau\mathscr{L}) - \tau\dot{q}_\alpha\frac{\partial\mathscr{L}}{\partial q_\alpha} - \tau\ddot{q}_\alpha\frac{\partial\mathscr{L}}{\partial \dot{q}_\alpha} + \phi_\alpha\frac{\partial\mathscr{L}}{\partial q_\alpha} + \left(\frac{d\phi_\alpha}{dt} - \frac{d\tau}{dt}\dot{q}_\alpha(t)\right)\frac{\partial\mathscr{L}}{\partial \dot{q}_\alpha} \tag{A.20}$$

$$= \frac{d}{dt}(\tau\mathscr{L}) + \frac{\partial\mathscr{L}}{\partial q_\alpha}(\phi_\alpha - \tau\dot{q}_\alpha) + \frac{\partial\mathscr{L}}{\partial \dot{q}_\alpha}\frac{d}{dt}(\phi_\alpha - \tau\dot{q}_\alpha) \tag{A.21}$$

We remark that:

$$\frac{d}{dt}\left(\frac{\partial \mathscr{L}}{\partial \dot{q}_\alpha}(\phi_\alpha - \tau \dot{q}_\alpha)\right) = \frac{\partial \mathscr{L}}{\partial \dot{q}_\alpha}\frac{d}{dt}(\phi_\alpha - \tau \dot{q}_\alpha) + \frac{d}{dt}\left(\frac{\partial \mathscr{L}}{\partial \dot{q}_\alpha}\right)(\phi_\alpha - \tau \dot{q}_\alpha) \quad (A.22)$$

$$\frac{\partial \mathscr{L}}{\partial \dot{q}_\alpha}\frac{d}{dt}(\phi_\alpha - \tau \dot{q}_\alpha) = \frac{d}{dt}\left(\frac{\partial \mathscr{L}}{\partial \dot{q}_\alpha}(\phi_\alpha - \tau \dot{q}_\alpha)\right) - \frac{d}{dt}\left(\frac{\partial \mathscr{L}}{\partial \dot{q}_\alpha}\right)(\phi_\alpha - \tau \dot{q}_\alpha) \quad (A.23)$$

We thus obtain, combining equations A.21 and A.23:

$$0 = \frac{d}{dt}(\tau \mathscr{L}) + \frac{\partial \mathscr{L}}{\partial q_\alpha}(\phi_\alpha - \tau \dot{q}_\alpha) + \frac{d}{dt}\frac{\partial \mathscr{L}}{\partial \dot{q}_\alpha}(\phi_\alpha - \tau \dot{q}_\alpha) - \frac{d}{dt}\frac{\partial \mathscr{L}}{\partial \dot{q}_\alpha}(\phi_\alpha - \tau \dot{q}_\alpha) \quad (A.24)$$

$$= \left(\frac{\partial \mathscr{L}}{\partial q_\alpha} - \frac{d}{dt}\frac{\partial \mathscr{L}}{\partial \dot{q}_\alpha}\right)(\phi_\alpha - \tau \dot{q}_\alpha) + \frac{d}{dt}(\tau \mathscr{L}) + \frac{d}{dt}\left(\frac{\partial \mathscr{L}}{\partial \dot{q}_\alpha}(\phi_\alpha - \tau \dot{q}_\alpha)\right) \quad (A.25)$$

We recognize the Euler-Lagrange equations (A.7) in the first term, which is thus null, so

$$0 = \frac{d}{dt}(\tau \mathscr{L}) + \frac{d}{dt}\left(\frac{\partial \mathscr{L}}{\partial \dot{q}_\alpha}(\phi_\alpha - \tau \dot{q}_\alpha)\right) = \frac{dC}{dt} \quad (A.26)$$

Thus C is an invariant of the dynamic.

We provide examples of the application of this theorem and more qualitative discussion in setion 5.3 that this mathematical appendix aims to complete.

A.2.2 Field Theoretic Point of View

We will consider now a relativistic field theoretic version. The interesting aspects involved is the notion of *current* that the theorem allows to define and the shift to a space-time point of view instead of the specific role played by time in the classical point of view. See for example for a general introduction to field theories [Altland & Simons, 2006].

Theorem A.3 (Noether, Field theory). *For continuous and differentiable fields Φ_i on space-time (whose coordinates are noted X^μ, $\mu = 0, \ldots, 3$) consider the action:*

$$\mathscr{S} = \int_\Omega \mathscr{L}\left(\Phi_i, \frac{\partial \Phi_i}{\partial X_\mu}, X^\mu\right) d^4 X \quad (A.27)$$

Let us suppose that $\mathscr{S}$ is preserved under the action of a continuous group $\mathfrak{G}$ with a finite number of parameters (index by r). Let us note its infinitesimal

A.2 Noether's Theorem

generators[4] as $v = x_{\mu,r}\frac{\partial}{\partial X_\mu} + \phi_{\alpha,r}\frac{\partial}{\partial \Phi_\alpha} + \psi_{\alpha,\mu,r}\frac{\partial}{\partial\left(\frac{\partial \Phi_\alpha}{\partial X_\mu}\right)}$. Then, we define the Noether *current densities:*

$$j_{\mu,r} = -\frac{\partial \mathscr{L}}{\partial\left(\frac{\partial \Phi_\beta}{\partial X_\mu}\right)}\phi_{\alpha,r} + \left[\frac{\partial \mathscr{L}}{\partial\left(\frac{\partial \Phi_\alpha}{\partial X_\mu}\right)}\frac{\partial \Phi_\alpha}{\partial X_\nu} - \mathscr{L}\delta_{\mu,\nu}\right]x_{\nu,r} \tag{A.28}$$

(where $\delta_{\mu,\nu}$ is the Kronecker symbol: 1 iff its parameter are equal, else 0). These currents verify:

$$\text{Div } j_r = \frac{\partial j_{\mu,r}}{\partial X_\mu} = 0 \tag{A.29}$$

This equality means, through the Gauss–Ostrogradsky theorem[5], that the quantity associated to j_r (its integral over a space-time volume Ω) equals its flow through the boundaries of Ω. As a result, the currents can be seen as currents of quantities, called charge, which are conserved[6].

Remark A.3. Here we have considered a 4-dimensional space-time, but this choice is mathematically arbitrary and only motivated by the usual physical applications.

Remark A.4. Noether's theorem, in its classical mechanics version, is a special case of this field version. Indeed, if we consider spacial coordinates as a 3-dimensional field over a 1-dimensional space-time (only a time dimension), then we fall back to the version provided by Lagrangian mechanics.

---

[4] As in the classical case, the ψ parameters are defined by the other parameters.
[5] This theorem, also known as divergence theorem, states that for a sufficiently regular volume and field the integral of the field on the volume is equal to its fluxes across the boundaries. Intuitively:
  Sources − Sink = fluxes in − fluxes out
  As a result, a null divergence leads to: fluxes in = fluxes out.
[6] Note that the definition of Noether's current is up to a solenoidal vector field (aka, a field with null divergence).

References

[Agutter & Wheatley, 2004] Agutter, P., Wheatley, D.: Metabolic scaling: Consensus or controversy? Theoretical Biology and Medical Modelling 1, 13 (2004)

[Allin, 1975] Allin, E.: Evolution of the mammalian middle ear. J. Morphol. 147(4), 403–437 (1975)

[Almaas et al., 2007] Almaas, E., Vázquez, A., Barabási, A.-L.: Scale-Free Networks in Biology. World Scientific Pub. Co. Inc. (2007)

[Als-Nielsen & Birgeneau, 1977] Als-Nielsen, J., Birgeneau, R.J.: Mean field theory, the ginzburg criterion, and marginal dimensionality of phase transitions. American Journal of Physics 45(6), 554–560 (1977)

[Altland & Simons, 2006] Altland, A., Simons, B.: Condensed matter field theory. Cambridge U. P. (2006)

[Amzallag, 2002] Amzallag, G.N.: La raison malmenée. De l'origine des idées reçues en biologie moderne. CNRS edn. (2002)

[Aoki, 1994] Aoki, I.: Entropy production in human life span: A thermodynamical measure for aging. Age 17, 29–31 (1994)

[Aon et al., 2003] Aon, M., Cortassa, S., Marbán, E., O'Rourke, B.: Synchronized whole cell oscillations in mitochondrial metabolism triggered by a local release of reactive oxygen species in cardiac myocytes. Journal of Biological Chemistry 278(45), 44735–44744 (2003)

[Aon et al., 2004a] Aon, M., Cortassa, S., O'Rourke, B.: Percolation and criticality in a mitochondrial network. Proceedings of the National Academy of Sciences 101(13), 4447 (2004a)

[Aon et al., 2004b] Aon, M., O'Rourke, B., Cortassa, S.: The fractal architecture of cytoplasmic organization: scaling, kinetics and emergence in metabolic networks. Molecular and Cellular Biochemistry 256(1), 169–184 (2004b)

[Aon et al., 2008] Aon, M., Roussel, M., Cortassa, S., O'Rourke, B., Murray, D., Beckmann, M., Lloyd, D.: The scale-free dynamics of eukaryotic cells. PloS One 3(11), 3624 (2008)

[Aoyagi et al., 2000] Aoyagi, N., Ohashi, K., Tomono, S., Yamamoto, Y.: Temporal contribution of body movement to very long-term heart rate variability in humans. Am. J. Physiol. Heart Circ. Physiol. 278(4), H1035–H1041 (2000)

[Arjun & van Oudenaarden, 2008] Arjun, R., van Oudenaarden, R.: Stochastic gene expression and its consequences. Cell 135(2), 216–226 (2008)

[Aspect et al., 1982] Aspect, A., Grangier, P., Roger, G.: Experimental realization of Einstein-Podolsky-Rosen-Bohm Gedankenexperiment: A new violation of Bell's inequalities. Physical Review Letters 49(2), 91–94 (1982)

[Avnir et al., 1998] Avnir, D., Biham, O., Lidar, D., Malcai, O.: Is the geometry of nature fractal? Science 279(5347), 39–40 (1998)

[Bailly, 1991] Bailly, F.: L'anneau des disciplines. Revue Internationale de Systémique 5(3) (1991)

[Bailly et al., 1988] Bailly, F., Gaill, F., Mosseri, R.: Fonctions biologiques, niveaux d'organisation et dimensions fractales. Revue Internationale de Systémique 2 (1988)

[Bailly et al., 1993] Bailly, F., Gaill, F., Mosseri, R.: Orgons and biolons in theoretical biology: Phenomenological analysis and quantum analogies. Acta Biotheoretica 41(1-2), 3–11 (1993)

[Bailly & Longo, 2003] Bailly, F., Longo, G.: Objective and epistemic complexity in biology. In: Invited Lecture, International Conference on Theoretical Neurobiology, New Delhi (2003)

[Bailly & Longo, 2007] Bailly, F., Longo, G.: Randomness and determinism in the interplay between the continuum and the discrete. Mathematical Structures in Computer Science 17(02), 289–305 (2007)

[Bailly & Longo, 2008] Bailly, F., Longo, G.: Extended critical situations: the physical singularity of life phenomena. Journal of Biological Systems 16(2), 309 (2008)

[Bailly & Longo, 2009] Bailly, F., Longo, G.: Biological organization and anti-entropy. Journal of Biological Systems 17(1), 63–96 (2009)

[Bailly & Longo, 2011] Bailly, F., Longo, G.: Mathematics and the natural sciences; The Physical Singularity of Life. Imperial College Press, London (2011); Preliminary version in French: Hermann, Vision des sciences (2006)

[Bailly et al., 2011] Bailly, F., Longo, G., Montévil, M.: A 2-dimensional geometry for biological time. Progress in Biophysics and Molecular Biology 106(3), 474–484 (2011)

[Baish & Jain, 2000] Baish, J.W., Jain, R.K.: Fractals and cancer. Cancer Res. 60(14), 3683–3688 (2000)

[Bak et al., 1988] Bak, P., Tang, C., Wiesenfeld, K.: Self-organized criticality. Physical Review A 38(1), 364–374 (1988)

[Baker, 2012] Baker, S.G.: Paradoxes in carcinogenesis should spur new avenues of research: an historical perspective. Disruptive Science and Technology 1(2), 100–107 (2012)

[Balakrishnan & Ashok, 2010] Balakrishnan, J., Ashok, B.: The role of hopf bifurcation dynamics in sensory processes. Journal of Theoretical Biology (2010)

[Balleza et al., 2008] Balleza, E., Alvarez-Buylla, E.R., Chaos, A., Kauffman, S., Shmulevich, I., Aldana, M.: Critical dynamics in genetic regulatory networks: Examples from four kingdoms. PLoS One 3(6), e2456 (2008)

[Bancaud et al., 2009] Bancaud, A., Huet, S., Daigle, N., Mozziconacci, J., Beaudouin, J., Ellenberg, J.: Molecular crowding affects diffusion and binding of nuclear proteins in heterochromatin and reveals the fractal organization of chromatin. The EMBO Journal (2009)

[Bassingthwaighte et al., 1994] Bassingthwaighte, J., Liebovitch, L., West, B.: Fractal physiology. American Physiological Society, New York (1994)

[Baus, 1987] Baus, M.: Statistical mechanical theories of freezing: An overview. Journal of Statistical Physics 48, 1129–1146 (1987)

[Belitz et al., 2005] Belitz, D., Kirkpatrick, T.R., Vojta, T.: How generic scale invariance influences quantum and classical phase transitions. Rev. Mod. Phys. 77(2), 579–632 (2005)

[Bell, 1998] Bell, J.: A Primer in Infinitesimal Analysis. Cambridge U.P. (1998)

References

[Berry & Chaté, 2011] Berry, H., Chaté, H.: Anomalous subdiffusion due to obstacles: A critical survey. arXiv:1103.2206v1, 1–33 (2011)

[Berthoz, 2002] Berthoz, A.: The Brain's Sense of Movement. Harvard U.P. (2002)

[Binney et al., 1992] Binney, J., Dowrick, N., Fisher, A., Newman, M.: The Theory of Critical Phenomena: An Introduction to the Renormalization Group. Oxford U. P. (1992)

[Bishop, 1999] Bishop, C.: The maximum oxygen consumption and aerobic scope of birds and mammals: Getting to the heart of the matter. Proceedings of the Royal Society - Biological Sciences 266(1435), 2275–2281 (1999)

[Bokma, 2004] Bokma, F.: Evidence against universal metabolic allometry. Functional Ecology 18(2), 184–187 (2004)

[Boser et al., 2005] Boser, S., Park, H., Perry, S., Menache, M., Green, F.: Fractal geometry of airway remodeling in human asthma. Am. J. Respir. Crit. Care Med. 172(7), 817–823 (2005)

[Botzung et al., 2008] Botzung, A., Denkova, E., Manning, L.: Experiencing past and future personal events: Functional neuroimaging evidence on the neural bases of mental time travel. Brain and Cognition 66(2), 202–212 (2008)

[Bourgine & Stewart, 2004] Bourgine, P., Stewart, J.: Autopoiesis and cognition. Artificial Life 10, 327 (2004)

[Boxenbaum & DiLea, 1995] Boxenbaum, H., DiLea, C.: First-time-in-human dose selection: allometric thoughts and perspectives. The Journal of Clinical Pharmacology 35(10), 957–966 (1995)

[Brillouin, 1956] Brillouin, L.: Science and Information Theory. Academic Press, New York (1956)

[Bros & Iagolnitzer, 1973] Bros, J., Iagolnitzer, D.: Causality and local mathematical analyticity: study. Ann. Inst. Henri Poincaré. 18(2) (1973)

[Brouzés & Farge, 2004] Brouzés, E., Farge, E.: Interplay of mechanical deformation and patterned gene expression in developing embryos. Current Opinion in Genetics & Development 14(4), 367–374 (2004)

[Buiatti & Longo, 2013] Buiatti, M., Longo, G.: Randomness and multi-level interactions in biology. Theory of Biosciences. TIBI-D-12-00030R1 (to appear, 2013)

[Bunde et al., 2000] Bunde, A., Havlin, S., Kantelhardt, J., Penzel, T., Peter, J., Voigt, K.: Correlated and uncorrelated regions in heart-rate fluctuations during sleep. Physical Review Letters 85(17), 3736–3739 (2000)

[Byers, 1999] Byers, N.: E. Noether's discovery of the deep connection between symmetries and conservation laws. In: The Heritage of Emmy Noether in Algebra, Geometry, and Physics, vol. 12, pp. 67–81. Israel Mathematical Conference, Tel Aviv (1999)

[Camalet et al., 2000] Camalet, S., Duke, T., Julicher, F., Prost, J.: Auditory sensitivity provided by self-tuned critical oscillations of hair cells. In: Proceedings of the National Academy of Sciences, pp. 3183–3188 (2000)

[Canals et al., 2000] Canals, M., Olivares, R., Labra, F., Novoa, F.: Ontogenetic changes in the fractal geometry of the bronchial tree in Rattus norvegicus. Biological Research 33, 31–35 (2000)

[Canguilhem, 1972] Canguilhem, G.: Le normal et le pathologique. Presses universitaires de France (1972)

[Carter et al., 2003] Carter, J., Banister, E., Blaber, A.: Effect of endurance exercise on autonomic control of heart rate. Sports Medicine 33(1), 33–46 (2003)

[Catoni et al., 2008] Catoni, F., Boccaletti, D., Cannata, R.: Mathematics of Minkowski Space. Birkhäuser Verlag Basel (2008)

[Cavagna et al., 2010] Cavagna, A., Cimarelli, A., Giardina, I., Parisi, G., Santagati, R., Stefanini, F., Viale, M.: Scale-free correlations in starling flocks. Proceedings of the National Academy of Sciences 107(26), 11865–11870 (2010)

[Ceron-Carrasco et al., 2009] Ceron-Carrasco, J., Requena, A., Perpete, E., Michaux, C., Jacquemin, D.: Double proton transfer mechanism in the adenine-uracil base pair and spontaneous mutation in rna duplex. Chemical Physics Letters 484, 64–68 (2009)

[Chaline, 1999] Chaline, J.: Les horloges du vivant: un nouveau stade de la théorie de l'évolution? Hachette Littératures (1999)

[Collini et al., 2010] Collini, E., Wong, C., Wilk, K., Curmi, P., Brurner, P., Scholes, G.D.: Coherently wired light harvesting in photosynthetic marine algae at ambient temperature. Nature 463, 644–648 (2010)

[Connes, 1994] Connes, A.: Non-commutative Geometry. Academic Press (1994)

[Cranford, 1983] Cranford, J.: Body temperature, heart rate and oxygen consumption of normothermic and heterothermic western jumping mice (Zapus princeps). Comparative Biochemistry and Physiology - A 74(3), 595 (1983)

[d'Alessio, 2004] d'Alessio, P.: Aging and the endothelium. Experimental Gerontology 39(2), 165–171 (2004)

[De Vico et al., 2009] De Vico, G., Peretti, V., Losa, G.: Fractal organization of feline oocyte cytoplasm. European Journal of Histochemistry 49(2), 151 (2009)

[Dehaene, 1997] Dehaene, S.: The number sense. Oxford University Press (1997)

[Delamotte, 2004] Delamotte, B.: A hint of renormalization. American Journal of Physics 72, 170 (2004)

[Delides et al., 2005] Delides, A., Panayiotides, I., Alegakis, A., Kyroudi, A., Banis, C., Pavlaki, A., Helidonis, E., Kittas, C.: Fractal dimension as a prognostic factor for laryngeal carcinoma. Anticancer Res. 25(3B), 2141 (2005)

[Demetrius, 2004] Demetrius, L.: Caloric restriction, metabolic rate, and entropy. The Journals of Gerontology Series A: Biological Sciences and Medical Sciences 59(9), B902–B915 (2004)

[Descartes, 1724] Descartes, R.: Les méditations métaphysiques. chez Robert-Marc d'Espilly (1724)

[Dietrich, 2003] Dietrich, M.: Richard goldschmidt: hopeful monsters and other 'heresies'. Nature Reviews Genetics 4, 68–74 (2003); Historical Article, Journal Article, Portraits

[Dirac, 1928] Dirac, P.A.M.: The quantum theory of the electron. Proceedings of the Royal Society of London. Series A 117(778), 610–624 (1928)

[Dmi'el, 1970] Dmi'el, R.: Growth and metabolism in snake embryos. Journal of Embryology and Experimental Morphology 23(3), 761–772 (1970)

[Drake et al., 2007] Drake, J., Fuller, M., Zimmermann, C., Gamarra, J.: From energetics to ecosystems: the dynamics and structure of ecological systems. Springer (2007)

[Duffin & Zener, 1969] Duffin, R., Zener, C.: Geometric programming, chemical equilibrium, and the anti-entropy function. Proceedings of the National Academy of Sciences 63(3), 629 (1969)

[Edelman & Tononi, 2000] Edelman, G., Tononi, G.: Reentry and the dynamic core: neural correlates of conscious experience, pp. 139–151. MIT Press (2000)

[Edelman & Tononi, 2001] Edelman, G., Tononi, G.: A universe of consciousness: How matter becomes imagination. Basic Books (2001)

[Edgar, 1993] Edgar, G.: Classics on Fractals. Basic Books (1993)

[Einstein et al., 1935] Einstein, A., Podolsky, B., Rosen, N.: Can quantum-mechanical description of physical reality be considered complete? Physical Review 47(10), 777–780 (1935)

References

[Eke et al., 2002] Eke, A., Herman, P., Kocsis, L., Kozak, L.R.: Fractal characterization of complexity in temporal physiological signals. Physiological Measurement 23(1), R1 (2002)

[Eldredge & Gould, 1972] Eldredge, N., Gould, S.J.: Punctuated equilibria: An alternative to phyletic gradualism. In: Schopf, T.J.M. (ed.) Models in Paleobiology, pp. 82–115. Freeman, Cooper and Co., San Francisco (1972)

[Engel et al., 2007] Engel, G., Calhoun, T., Read, E., Ahn, T., Mançal, T., Yuan-Chung, C., Blankenship, R., Fleming, G.: Evidence for wavelike energy transfer through quantum coherence in ptotosynthetic systems. Nature 447, 782–786 (2007)

[Esen et al., 2001] Esen, F., Özbeit, F., Esen, H.: Fractal scaling of heart rate variability in young habitual smokers. Turkish Journal of Medical Sciences 4, 317–322 (2001)

[Falconer & Wiley, 2003] Falconer, K., Wiley, J.: Fractal geometry: Mathematical foundations and applications. Wiley, New York (2003)

[Feynman & Gleick, 1967] Feynman, R., Gleick, J.: The character of physical law. MIT Press (1967)

[Field, 1987] Field, D.: Relations between the statistics of natural images and the response properties of cortical cells. Journal of the Optical Society of America A 4(12), 2379–2394 (1987)

[Field et al., 1993] Field, D., Hayes, A., Hess, R.: Contour integration by the human visual system: Evidence for a local association field. Vision Research 33(2), 173–193 (1993)

[Fisher, 1998] Fisher, M.: Renormalization group theory: Its basis and formulation in statistical physics. Reviews of Modern Physics 70(2), 653–681 (1998)

[Fisher, 1930] Fisher, R.: The Genetical Theory of Natural Selection. Clarendon (1930)

[Fleury, 2000] Fleury, V.: Branching morphogenesis in a reaction-diffusion model. Phys. Rev. E 61, 4156–4160 (2000)

[Fleury, 2009] Fleury, V.: Clarifying tetrapod embryogenesis, a physicist's point of view. The European Physical Journal Applied Physics 45 (2009)

[Fox Keller, 1995] Fox Keller, E.: Refiguring Life: Metaphors of Twentieth-Century Biology. Columbia University Press (1995)

[Fox Keller, 2002] Fox Keller, E.: The century of the gene. Harvard University Press (2002)

[Fox Keller, 2005] Fox Keller, E.: Revisiting "scale-free" networks. BioEssays 27(10), 1060–1068 (2005)

[Frezza & Longo, 2010] Frezza, G., Longo, G.: Variations on the theme of invariants: conceptual and mathematical dualities in physics vs biology. Human Evolution 25(3-4), 167–172 (2010)

[Gabrys et al., 2005] Gabrys, E., Rybaczuk, M., Kedzia, A.: Fractal models of circulatory system. symmetrical and asymmetrical approach comparison. Chaos, Solitons & Fractals 24(3), 707–715 (2005)

[Galileo, 1638] Galileo, G.: Discorsi e dimostrazioni matematiche, intorno à due nuove scienze. Elsevier (1638)

[Gerkema, 2002] Gerkema, M.: Ultradian Rhythms, p. 207. Springer (2002)

[Gheorghiu et al., 2005] Gheorghiu, S., Kjelstrup, S., Pfeifer, P., Coppens, M.-O.: Is the lung an optimal gas exchanger? In: Losa, G.A., Merlini, D., Nonnenmacher, T.F., Weibel, E.R. (eds.) Fractals in Biology and Medicine. Mathematics and Biosciences in Interaction, pp. 31–42. Birkhäuser, Basel (2005)

[Giguere et al., 1988] Giguere, L., Cote, B., St-Pierre, J.: Metabolic rates scale isometrically in larval fishes. Marine Ecology Progress Series. Oldendorf 50(1), 13–19 (1988)

[Gillooly et al., 2006] Gillooly, J., Allen, A., Savage, V., Charnov, E., West, G., Brown, J.: Response to Clarke and Fraser: effects of temperature on metabolic rate. Ecology 20(2), 400–404 (2006)

[Gillooly et al., 2005] Gillooly, J., Allen, A., West, G., Brown, J.: The rate of DNA evolution: Effects of body size and temperature on the molecular clock. Proceedings of the National Academy of Sciences 102(1), 140–145 (2005)

[Gillooly et al., 2001] Gillooly, J., Brown, J., West, G., Savage, V., Charnov, E.: Effects of size and temperature on metabolic rate. Science 293(5538), 2248–2251 (2001)

[Glazier, 2005] Glazier, D.: Beyond the: variation in the intra- and interspecific scaling of metabolic rate in animals. Biological Reviews 80(04), 611–662 (2005)

[Glazier, 2006] Glazier, D.: The 3/4-power law is not universal: Evolution of isometric, ontogenetic metabolic scaling in pelagic animals. Bioscience 56(4), 325–332 (2006)

[Glazier, 2009] Glazier, D.: Activity affects intraspecific body-size scaling of metabolic rate in ectothermic animals. Journal of Comparative Physiology B: Biochemical, Systemic, and Environmental Physiology 179(7), 821–828 (2009)

[Glenny et al., 2007] Glenny, R., Bernard, S., Neradilek, B., Polissar, N.: Quantifying the genetic influence on mammalian vascular tree structure. Proceedings of the National Academy of Sciences 104(16), 6858–6863 (2007)

[Goldberger et al., 2000] Goldberger, A., Amaral, L., Glass, L., Hausdorff, J., Ivanov, P., Mark, R., Mietus, J., Moody, G., Peng, C.-K., Stanley, H.: PhysioBank, PhysioToolkit, and PhysioNet: Components of a new research resource for complex physiologic signals. Circulation 101(23), e215–e220 (2000)

[Goldberger et al., 2002a] Goldberger, A., Amaral, L., Hausdorff, J., Ivanov, P., Peng, C.-K., Stanley, H.: Fractal dynamics in physiology: alterations with disease and aging. Proceedings of the National Academy of Sciences 99(suppl. 1), 2466 (2002a)

[Goldberger et al., 2002b] Goldberger, A., Peng, C.-K., Lipsitz, L.: What is physiologic complexity and how does it change with aging and disease? Neurobiology of Aging 23(1), 23–26 (2002b)

[Golding & Cox, 2006] Golding, I., Cox, E.: Physical nature of bacterial cytoplasm. Physical Review Letters 96(9), 98102 (2006)

[Gould, 1977] Gould, S.: The return of hopeful monsters. Natural History 86, 22–30 (1977)

[Gould, 1989] Gould, S.: Wonderful life. Norton (1989)

[Gould, 1997] Gould, S.: Full house: The spread of excellence from Plato to Darwin. Three Rivers Pr. (1997)

[Gould & Vrba, 1982] Gould, S.J., Vrba, E.S.: Exaptation — a missing term in the science of form. Paleobiology 8(1), 4–15 (1982)

[Grant & Grant, 2002] Grant, P.R., Grant, B.R.: Unpredictable evolution in a 30-year study of darwin's finches. Science 296(5568), 707–711 (2002)

[Gray & Winkler, 2003] Gray, H., Winkler, J.: Electron tunneling through proteins. Q. Rev. Biophys. 36, 341–372 (2003)

[Green et al., 1988] Green, M.B., Schwarz, J.H., Witten, E.: Superstring theory. Cambridge University Press (1988)

[Grigolini et al., 2009] Grigolini, P., Aquino, G., Bologna, M., Luković, M., West, B.J.: A theory of noise in human cognition. Physica A: Statistical Mechanics and its Applications 388(19), 4192–4204 (2009)

[Günther & Morgado, 2005] Günther, B., Morgado, E.: Allometric scaling of biological rhythms in mammals. Biological Research 38, 207–212 (2005)

[Hardy, 1934] Hardy, J.D.: The radiation of heat from the human body iii. Journal of Clinical Investigation 13(4), 615–620 (1934)

References

[Hayflick, 2007] Hayflick, L.: Entropy explains aging, genetic determinism explains longevity, and undefined terminology explains misunderstanding both. PLoS Genet. 3(12), e220 (2007)

[Heams, 2013] Heams, T.: Randomness in biology. Structures in Comp. Sci. (to appear, 2013)

[Hellbrugge et al., 1964] Hellbrugge, T., Lange, J., Rutenfranz, J., Stehr, K.: Circadian periodicity of physiological functions in different stages of infancy and childhood. Annals of the New York Academy of Sciences 117(1), 361–373 (1964); Photo-Neuro-Endocrine Effects in Ciradian Systems, with Particular Reference to the Eye

[Hempleman et al., 2005] Hempleman, S.C., Kilgore, D.L., Colby, C., Bavis, R.W., Powell, F.L.: Spike firing allometry in avian intrapulmonary chemoreceptors: matching neural code to body size. Journal of Experimental Biology 208(16), 3065–3073 (2005)

[Holstein, 2000] Holstein, B.: A brief introduction to chiral perturbation theory. Czechoslovak Journal of Physics 50, 9–23 (2000)

[Horsfield, 1977] Horsfield, K.: Morphology of branching trees related to entropy. Respiration Physiology 29(2), 179 (1977)

[Hubbell, 2001] Hubbell, S.: The unified neutral theory of biodiversity and biogeography. In: Monographs in Population Biology. Princeton University Press, Princeton (2001)

[Husserl, 1970] Husserl, E.: Origin of geometry. In: The Crisis of European Sciences and Transcendental Phenomenology: An Introduction to Phenomenological Philosophy. Northwestern University Press, Evanston (1970)

[Ivanov et al., 2001] Ivanov, P., Amaral, L., Goldberger, A., Havlin, S., Rosenblum, M., Stanley, H., Struzik, Z.: From 1/f noise to multifractal cascades in heartbeat dynamics. Chaos 11(3), 641–652 (2001)

[Iyengar et al., 1996] Iyengar, N., Peng, C.-K., Morin, R., Goldberger, A., Lipsitz, L.: Age-related alterations in the fractal scaling of cardiac interbeat interval dynamics. American Journal of Physiology- Regulatory, Integrative and Comparative Physiology 271(4), 1078 (1996)

[Jean, 1994] Jean, R.: Phyllotaxis: A Systemic Study in Plant Morphogenesis. Cambridge Studies in Mathematics (1994)

[Jensen, 1998] Jensen, H.: Self-Organized Criticality, Emergent Complex Behavior in Physical and Biological Systems. Cambridge lectures in Physics (1998)

[Jensen & Sibani, 2007] Jensen, H.J., Sibani, P.: Glassy dynamics. Scholarpedia 2(6), 2030 (2007)

[Jeong et al., 2000] Jeong, H., Tombor, B., Albert, R., Oltvai, Z., Barabási, A.: The large-scale organization of metabolic networks. Nature 407(6804), 651–654 (2000)

[Kant, 1781] Kant, I.: Critique of Pure Reason. transl. Kemp Smith, N. Palgrave Macmillan (1781); this edition published (1929)

[Karsenti, 2008] Karsenti, E.: Self-organization in cell biology: a brief history. Nature Reviews Molecular Cell Biology 9(3), 255–262 (2008)

[Kauffman, 1993] Kauffman, S.: The origins of order. Oxford U.P. (1993)

[Kauffman, 2001] Kauffman, S.: Prolegomenon to a general biology. Annals of the New York Academy of Sciences 935, 18–36 (2001)

[Kauffman, 2002] Kauffman, S.: Investigations. Oxford University Press, USA (2002)

[Kauffman, 2012] Kauffman, S.: The end of a physics worldview (2012), http://www.mylab.fi/en/statement/the_end_of_a_physics_worldview/

[Kay, 1998] Kay, I.: Introduction to animal physiology. BIOS Scientific (1998)

[Khanin & Wit, 2006] Khanin, R., Wit, E.: How scale-free are biological networks. Journal of Computational Biology 13(3), 810–818 (2006)

[Kim et al., 2002] Kim, S.-H., Kaminker, P., Campisi, J.: Telomeres, aging and cancer: In search of a happy ending. Oncogene 21, 503–511 (2002)

[Kirkwood, 2008] Kirkwood, T.: Gerontology: Healthy old age. Nature 455(7214), 739–740 (2008)

[Kirman et al., 2003] Kirman, C.R., Sweeney, L.M., Meek, M.E., Gargas, M.L.: Assessing the dose-dependency of allometric scaling performance using physiologically based pharmacokinetic modeling. Regulatory Toxicology and Pharmacology 38(3), 345–367 (2003)

[Kiyono et al., 2004] Kiyono, K., Struzik, Z., Aoyagi, N., Sakata, S., Hayano, J., Yamamoto, Y.: Critical scale invariance in a healthy human heart rate. Physical Review Letters 93(17), 178103 (2004)

[Kleiber, 1932] Kleiber, M.: Body size and animal metabolism. Hilgardia 6, 315–353 (1932)

[Kleiber, 1961] Kleiber, M.: The fire of life. An introduction to animal energetics. John Wiley & Sons, Inc., New York (1961)

[Kochen & Specker, 1967] Kochen, S., Specker, E.: The problem of hidden variables in quantum mechanics. Journal of Mathematics and Mechanics 17, 59–87 (1967)

[Kucharski et al., 2008] Kucharski, R., Maleszka, R., Foret, S., Maleszka, A.: Nutritional control of reproductive status in honeybees via dna methylation. Science 319, 1827–1830 (2008)

[Kupiec, 1983] Kupiec, J.: A probabilistic theory of cell differentiation, embryonic mortality and dna c-value paradox. Specul. Sci. Techno. 6, 471–478 (1983)

[Labra et al., 2007] Labra, F., Marquet, P., Bozinovic, F.: Scaling metabolic rate fluctuations. Proc. Natl. Acad. Sci. USA 104(26), 10900–10903 (2007)

[Laguës & Lesne, 2003] Laguës, M., Lesne, A.: Invariance d'éhelle, Belin, Paris (2003)

[Lambert, 2007] Lambert, F.: Entropy and the second law of thermodynamics (2007), http://www.entropysite.com

[Laskar J., 1994] Laskar, J.: Large scale chaos in the solar system. Astron. Astrophys. 287 (1994)

[Lazebnik, 2002] Lazebnik, Y.: Can a biologist fix a radio? or, what i learned while studying apoptosis. Cancer Cell 2(3), 179–182 (2002)

[Le Méhauté et al., 1998] Le Méhauté, A., Nigmatullin, R., Nivanen, L.: Flèches du temps et géométrie fractales, Hermès (1998)

[Le Van Quyen, 2003] Le Van Quyen, M.: Disentangling the dynamic core: a research program for a neurodynamics at the large-scale. Biological Research 36, 67–88 (2003)

[Lecointre et al., 2001] Lecointre, G., Le Guyader, H., Visset, D.: Classification phylogénétique du vivant, Belin (2001)

[Lesne, 2003] Lesne, A.: Approches multi-échelles en physique et en biologie. Thèse d'habilitation à diriger des recherches (2003)

[Lesne, 2006] Lesne, A.: Complex networks: from graph theory to biology. Letters in Mathematical Physics 78(3), 235–262 (2006)

[Lesne, 2008] Lesne, A.: Robustness: Confronting lessons from physics and biology. Biol. Rev. Camb. Philos. Soc. 83(4), 509–532 (2008)

[Lesne & Victor, 2006] Lesne, A., Victor, J.-M.: Chromatin fiber functional organization: Some plausible models. Eur. Phys. J. E Soft Matter 19(3), 279–290 (2006)

[Lewis & Rees, 1985] Lewis, M., Rees, D.: Fractal surfaces of proteins. Science 230(4730), 1163–1165 (1985)

[Lindner et al., 2008] Lindner, A., Madden, R., Demarez, A., Stewart, E., Taddei, F.: Asymmetric segregation of protein aggregates is associated with cellular aging and rejuvenation. Proceedings of the National Academy of Sciences 105(8), 3076–3081 (2008)

[Lindstedt & Calder III, 1981] Lindstedt, S., Calder III, W.: Body size, physiological time, and longevity of homeothermic animals. Quarterly Review of Biology, 1–16 (1981)

[Liu & Ochman, 2007] Liu, R., Ochman, H.: Stepwise formation of the bacterial flagellar system. Proceedings of the National Academy of Sciences 104(17), 7116–7121 (2007)

[Longo, 1989] Longo, G.: Some aspects of impredicativity: Weyl's philosophy of mathematics and today's type theory. In: E., et al. (eds.) Logic Colloquium 87 (European Summer Meeting of the A.S.L.), North-Holland. Invited lecture (1989)

[Longo, 2008] Longo, G.: Laplace, turing and the "imitation game" impossible geometry: randomness, determinism and programs in turing's test. In: Epstein, R., Roberts, G., Beber, G. (eds.) Parsing the Turing Test, pp. 377–411. Springer (2008)

[Longo, 2009] Longo, G.: Critique of computational reason in the natural sciences. In: Gelenbe, E., Kahane, J.-P. (eds.) Fundamental Concepts in Computer Science. Imperial College Press/World Scientific (2009)

[Longo, 2011a] Longo, G.: Interfaces of incompleteness (2011a) (to appear); Originale in italiano per "La Matematica" 4, Einaudi (2010)

[Longo, 2011b] Longo, G.: Mathematical infinity "in prospettiva" and the spaces of possibilities. "Visible", a Semiotics Journal 9 (2011b)

[Longo, 2012] Longo, G.: On the relevance of negative results. In: Conference on Negation, Duality, Polarity, Marseille (2008) (Proceedings Ininfluxus, Electronic Journal) (2012), http://www.influxus.eu/article474.html

[Longo et al., 2012a] Longo, G., Miquel, P.-A., Sonnenschein, C., Soto, A.M.: Is information a proper observable for biological organization? Progress in Biophysics and Molecular Biology 109(3), 108–114 (2012a)

[Longo & Montévil, 2011a] Longo, G., Montévil, M.: From physics to biology by extending criticality and symmetry breakings. Progress in Biophysics and Molecular Biology 106(2), 340–347 (2011a); Systems Biology and Cancer

[Longo & Montévil, 2011b] Longo, G., Montévil, M.: Protention and retention in biological systems. Theory in Biosciences 130, 107–117 (2011b)

[Longo & Montévil, 2012] Longo, G., Montévil, M.: Randomness increases order in biological evolution. In: Dinneen, M.J., Khoussainov, B., Nies, A. (eds.) WTCS 2012 (Calude Festschrift). LNCS, vol. 7160, pp. 289–308. Springer, Heidelberg (2012)

[Longo & Montévil, 2013] Longo, G., Montévil, M.: Extended criticality, phase spaces and enablement in biology. Chaos, Solitons & Fractals, Invited Paper, Special Issue (2013)

[Longo et al., 2012b] Longo, G., Montévil, M., Kauffman, S.: No entailing laws, but enablement in the evolution of the biosphere. In: Genetic and Evolutionary Computation Conference, GECCO 2012, Philadelphia, PA, USA. ACM (2012b) (invited paper)

[Longo et al., 2012c] Longo, G., Montévil, M., Pocheville, A.: From bottom-up approaches to levels of organization and extended critical transitions. Frontiers in Physiology 3(232) (2012c) (invited paper)

[Longo et al., 2013] Longo, G., Montévil, M., Sonnenschein, C., Soto, A.M.: Biology's theoretical principles and default states (to appear, 2013)

[Longo et al., 2010] Longo, G., Palamidessi, C., Paul, T.: Some bridging results and challenges in classical, quantum and computational randomness. In: Zenil, H. (ed.) Randomness through Computation. World Scientific (2010)

[Longo & Perret, 2013] Longo, G., Perret, N.: Anticipation, protection and biological inertia (to appear, 2013)

[Longo & Viarouge, 2010] Longo, G., Viarouge, A.: Mathematical intuition and the cognitive roots of mathematical concepts. Topoi. 29(1), 15–27 (2010); Horsten, L., Starikova, I., (eds.) Special issue on Mathematical knowledge: Intuition, visualization, and understanding

[Lopes & Betrouni, 2009] Lopes, R., Betrouni, N.: Fractal and multifractal analysis: A review. Medical Image Analysis 13(4), 634–649 (2009)

[Lorthois & Cassot, 2010] Lorthois, S., Cassot, F.: Fractal analysis of vascular networks: Insights from morphogenesis. Journal of Theoretical Biology 262(4), 614–633 (2010)

[Losa, 2006] Losa, G.: Do complex cell structures share a fractal-like organization? Microscopie 5, 53–56 (2006)

[Losa et al., 1992] Losa, G., Baumann, G., Nonnenmacher, T.: Fractal dimension of pericellular membranes in human lymphocytes and lymphoblastic leukemia cells. Pathology, Research and Practice 188(4-5), 680 (1992)

[Lovecchio et al., 2012] Lovecchio, E., Allegrini, P., Geneston, E., West, B.J., Grigolini, P.: From self-organized to extended criticality. Frontiers in Physiology 3(98) (2012)

[Luo & O'Leary, 2005] Luo, L., O'Leary, D.: Axon retraction and degeneration in development and disease. Annual Review of Neuroscience 28, 127–156 (2005)

[Lévy-Leblond, 2007] Lévy-Leblond, J.M.: Deux discours sur la taille de l'enfer, by Galilée, chapter Postface: Galilée, de l'Enfer de Dante au purgatoire de la science (2007)

[Machta et al., 2011] Machta, B.B., Papanikolaou, S., Sethna, J.P., Veatch, S.L.: A minimal model of plasma membrane heterogeneity requires coupling cortical actin to criticality. Biophysical Journal 100(7), 1668–1677 (2011)

[Maffini et al., 2004] Maffini, M.V., Soto, A.M., Calabro, J.M., Ucci, A.A., Sonnenschein, C.: The stroma as a crucial target in rat mammary gland carcinogenesis. Journal of Cell Science 117(8), 1495–1502 (2004)

[Maina & van Gils, 2001] Maina, J., van Gils, P.: Morphometric characterization of the airway and vascular systems of the lung of the domestic pig, Sus scrofa: Comparison of the airway, arterial and venous systems. Comparative Biochemistry and Physiology A 130(4), 781–798 (2001)

[Makowiec et al., 2006] Makowiec, D., Galažska, R., Dudkowska, A., Rynkiewicz, A., Zwierz, M.: Long-range dependencies in heart rate signals–revisited. Physica A: Statistical Mechanics and its Applications 369(2), 632–644 (2006)

[Mandelbrot, 1983] Mandelbrot, B.: The fractal geometry of nature. W.H. Freeman (1983)

[Marbà et al., 2007] Marbà, N., Duarte, C., Agustí, S.: Allometric scaling of plant life history. Proceedings of the National Academy of Sciences 104(40), 15777–15780 (2007)

[Marineo & Marotta, 2005] Marineo, G., Marotta, F.: Biophysics of aging and therapeutic interventions by entropy-variation systems. Biogerontology 6, 77–79 (2005)

[Marker, 2002] Marker, D.: Model theory: an introduction. Springer (2002)

[Marratto, 2012] Marratto, S.: The Intercorporeal Self. Merleau-Ponty on Subjectivity. State University of New York Press, Albany (2012)

[Massin et al., 2000] Massin, M., Maeyns, K., Withofs, N., Ravet, F., Gerard, P.: Circadian rhythm of heart rate and heart rate variability. Archives of Disease in Childhood 83(2), 179–182 (2000)

[Masters, 2004] Masters, B.: Fractal analysis of the vascular tree in the human retina. Biomedical Engineering 6(1), 427 (2004)

[Mauroy et al., 2004] Mauroy, B., Filoche, M., Weibel, E., Sapoval, B.: An optimal bronchial tree be dangerous. Nature 427, 633–636 (2004)

[Maynard-Smith & Szathmary, 1997] Maynard-Smith, J., Szathmary, E.: The Major Transitions in Evolution. Oxford U.P. (1997)

[McCarthy & Enquist, 2005] McCarthy, M., Enquist, B.: Organismal size, metabolism and the evolution of complexity in metazoans. Evolutionary Ecology Research 7(5), 681 (2005)

[McShea & Brandon, 2010] McShea, D., Brandon, R.: Biology's first law: the tendency for diversity and complexity to increase in evolutionary systems. University of Chicago Press (2010)

[Michl et al., 2006] Michl, M., Ouyang, N., Fraek, M.-L., Beck, F.-X., Neuhofer, W.: Expression and regulation of $\alpha\beta$-crystallin in the kidney in vivo and in vitro. Pflügers Archiv. 452, 387–395 (2006)

[Misslin, 2003] Misslin, R.: Une vie de cellule. Revue de Synthèse 124(1), 205–221 (2003)

[Montévil, 2013] Montévil, M.: Biological measurement and extended critical transitions (to be submitted, 2013)

[Mora & Bialek, 2011] Mora, T., Bialek, W.: Are biological systems poised at criticality? Journal of Statistical Physics 144, 268–302 (2011), doi:10.1007/s10955-011-0229-4

[Moran & Wells, 2007] Moran, D., Wells, R.: Ontogenetic scaling of fish metabolism in the mouse-to-elephant mass magnitude range. Comparative Biochemistry and Physiology - A 148(3), 611–620 (2007)

[Moreno & Mossio, 2013] Moreno, A., Mossio, M.: Biological autonomy. A Philosophical and Theoretical Enquiry. Springer, Dordrecht (2013)

[Mortola & Lanthier, 2004] Mortola, J., Lanthier, C.: Scaling the amplitudes of the circadian pattern of resting oxygen consumption, body temperature and heart rate in mammals. Comparative Biochemistry and Physiology - A 139(1), 83–95 (2004)

[Mossio et al., 2009] Mossio, M., Longo, G., Stewart, J.: A computable expression of closure to efficient causation. Journal of Theoretical Biology 257(3), 489–498 (2009)

[Mossio & Moreno, 2010] Mossio, M., Moreno, A.: Organisational closure in biological organisms. History and Philosophy of the Life Sciences 32(2-3), 269–288 (2010)

[Nagy, 2005] Nagy, K.: Field metabolic rate and body size. Journal of Experimental Biology 208(9), 1621–1625 (2005)

[Nelson et al., 1990] Nelson, T., West, B., Goldberger, A.: The fractal lung: Universal and species-related scaling patterns. Cellular and Molecular Life Sciences 46, 251–254 (1990)

[Nicolas, 2006] Nicolas, F.: Quelle unité pour l'œvre musicale? In: Les mathématiques, les idées et le réel physique. Vrin, Paris (2006)

[Nicolis & Prigogine, 1977] Nicolis, G., Prigogine, I.: Self-organization in non-equilibrium systems. Wiley, New York (1977)

[Nietzsche, 1886] Nietzsche, F.: Jenseits von Gut und Böse (1886)

[Niklas & Enquist, 2001] Niklas, K., Enquist, B.: Invariant scaling relationships for interspecific plant biomass production rates and body size. Proceedings of the National Academy of Sciences 98(5), 2922–2927 (2001)

[Noble, 2006] Noble, D.: The music of life. Oxford U. P., Oxford (2006)

[Noble, 2008] Noble, D.: Claude bernard, the first systems biologist, and the future of physiology. Experimental Physiology 93(1), 16–26 (2008)

[Noble, 2009] Noble, D.: Could there be a synthesis between western and oriental medicine, and with sasang constitutional medicine in particular? Evidence-Based Complementary and Alternative Medicine 6(S1), 5–10 (2009)

[Noble, 2010] Noble, D.: Biophysics and systems biology. Philosophical Transactions of the Royal Society A: Mathematical, Physical and Engineering Sciences 368(1914), 1125 (2010)

[Noble, 2011] Noble, D.: The music of life: sourcebook, Oxford (2011)

[Noether, 1918] Noether, E.: Invariante Variationsprobleme. Nachrichten von der Gesellschaft der Wissenschaften zu Göttingen, 235–257 (1918)

[Nottale, 1993] Nottale, L.: Fractal space-time and microphysics: towards a theory of scale relativity. World Scientific (1993)

[Nykter et al., 2008a] Nykter, M., Price, N., Aldana, M., Ramsey, S., Kauffman, S., Hood, L., Yli-Harja, O., Shmulevich, I.: Gene expression dynamics in the macrophage exhibit criticality. Proceedings of the National Academy of Sciences 105(6), 1897 (2008a)

[Nykter et al., 2008b] Nykter, M., Price, N., Larjo, A., Aho, T., Kauffman, S., Yli-Harja, O., Shmulevich, I.: Critical networks exhibit maximal information diversity in structure-dynamics relationships. Physical Review Letters 100(5), 58702 (2008b)

[Nyström, 2007] Nyström, T.: A bacterial kind of aging. PLoS Genet. 3(12), e224 (2007)

[Ochs et al., 2004] Ochs, M., Nyengaard, J., Jung, A., Knudsen, L., Voigt, M., Wahlers, T., Richter, J., Gundersen, H.: The number of alveoli in the human lung. Am. J. Respir. Crit. Care Med. 169(1), 120–124 (2004)

[Olshansky & Rattan, 2005] Olshansky, S., Rattan, S.: At the heart of aging: Is it metabolic rate or stability? Biogerontology 6, 291–295 (2005)

[Overduin & Wesson, 1997] Overduin, J., Wesson, P.: Kaluza-Klein gravity. Physics Reports 283, 303–380 (1997)

[Pagni, 2012] Pagni, E.: Corpo Vivente Mondo. Aristotele e Merleau-Ponty a confronto. Firenze University Press, Firenze (2012)

[Paldi, 2003] Paldi, A.: Stochastic gene expression during cell differentiation: order from disorder? Cell Mol. Life Sci. 60, 1775–1779 (2003)

[Paumgartner et al., 1981] Paumgartner, D., Losa, G., Weibel, E.: Resolution effect on the stereological estimation of surface and volume and its interpretation in terms of fractal dimensions. Journal of Microscopy 121(Pt. 1), 51 (1981)

[Peng et al., 1995] Peng, C., Havlin, S., Stanley, H., Goldberger, A.: Quantification of scaling exponents and crossover phenomena in nonstationary heartbeat time series. Chaos: An Interdisciplinary Journal of Nonlinear Science 5(1), 82–87 (1995)

[Peng et al., 2002] Peng, C., Mietus, J., Liu, Y., Lee, C., Hausdorff, J., Stanley, H.E., Goldberger, A., Lipsitz, L.: Quantifying fractal dynamics of human respiration: Age and gender effects. Ann. Biomed. Eng. 30(5), 683–692 (2002)

[Peng et al., 1993] Peng, C.-K., Mietus, J., Hausdorff, J., Havlin, S., Stanley, H., Goldberger, A.: Long-range anticorrelations and non-gaussian behavior of the heartbeat. Phys. Rev. Lett. 70(9), 1343–1346 (1993)

[Perazzo et al., 2000] Perazzo, C., Fernandez, E., Chialvo, D., Willshaw, P.: Large scale-invariant fluctuations in normal blood cell counts: A sign of criticality? Fractals 8(3), 279–283 (2000)

[Perfetti & Goldman, 1976] Perfetti, C., Goldman, S.: Discourse memory and reading comprehension skill. Journal of Verbal Learning and Verbal Behavior 15(1), 33–42 (1976)

[Perret, 2013] Perret, N.: Épistémologie constitutive pour les sciences du vivant: sur la catégorie de causalité en biologie (2013)

[Perutz, 1987] Perutz, M.: Schrödinger's what is Life? and molecular biology. Cambridge University Press, Cambridge (1987)

[Peters, 1986] Peters, R.: The ecological implications of body size. Cambridge U. P. (1986)

[Petitot, 2003] Petitot, J.: The neurogeometry of pinwheels as a sub-riemannian contact structure. Journal of Physiology, Paris 97(2-3), 265–309 (2003)

[Petitot, 2008] Petitot, J.: Neurogéométrie de la vision. Ed de l'École Polytechnique (2008)

[Pezard et al., 1998] Pezard, L., Martinerie, J., Varela, F., Bouchet, F., Guez, D., Derouesné, C., Renault, B.: Entropy maps characterize drug effects on brain dynamics in alzheimer's disease. Neuroscience Letters 253(1), 5–8 (1998)

References

[Phillips, 2009a] Phillips, J.: Scaling and self-organized criticality in proteins I. Proceedings of the National Academy of Sciences 106(9), 3107–3112 (2009a)

[Phillips, 2009b] Phillips, J.: Scaling and self-organized criticality in proteins II. Proceedings of the National Academy of Sciences 106(9), 3113–3118 (2009b)

[Pikkujamsa et al., 1999] Pikkujamsa, S., Makikallio, T., Sourander, L., Raiha, I., Puukka, P., Skytta, J., Peng, C.-K., Goldberger, A., Huikuri, H.: Cardiac interbeat interval dynamics from childhood to senescence: comparison of conventional and new measures based on fractals and chaos theory. Circulation 100(4), 393 (1999)

[Pilgram & Kaplan, 1999] Pilgram, B., Kaplan, D.T.: Nonstationarity and 1/f noise characteristics in heart rate. Am. J. Physiol. 276(1Pt. 2), R1–R9 (1999)

[Rabinowitz & White, 2010] Rabinowitz, J.D., White, E.: Autophagy and metabolism. Science 330(6009), 1344–1348 (2010)

[Renner et al., 2005] Renner, U., Schütz, G., Vojta, G.: Diffusion on fractals. In: Heitjans, P., Käger, J. (eds.) Diffusion in Condensed Matter, pp. 793–811. Springer, Heidelberg (2005)

[Ribeiro et al., 2010] Ribeiro, T., Copelli, M., Caixeta, F., Belchior, H., Chialvo, D., Nicolelis, M., Ribeiro, S., Sporns, O.: Spike avalanches exhibit universal dynamics across the sleep-wake cycle. PloS One 5(11), 11167–11177 (2010)

[Risser et al., 2006] Risser, L., Plouraboué, F., Steyer, A., Cloetens, P., Le Duc, G., Fonta, C.: From homogeneous to fractal normal and tumorous microvascular networks in the brain. Journal of Cerebral Blood Flow & Metabolism 27(2), 293–303 (2006)

[Rosen, 2005] Rosen, R.: Life itself: a comprehensive inquiry into the nature, origin, and fabrication of life. Columbia U. P. (2005)

[Rossberg et al., 2011] Rossberg, A.G., Farnsworth, K.D., Satoh, K., Pinnegar, J.K.: Universal power-law diet partitioning by marine fish and squid with surprising stability-diversity implications. Proceedings of the Royal Society B: Biological Sciences 278(1712), 1617–1625 (2011)

[Rovelli, 1996] Rovelli, C.: Black hole entropy from loop quantum gravity. Phys. Rev. Lett. 77, 3288–3291 (1996)

[Rubner, 1883] Rubner, M.: über den Einfluss der Körpergrösse auf Stoff- und Kraftwechsel. Zeitschrift für Biologie 19, 535–562 (1883)

[Ruud, 1954] Ruud, J.: Vertebrates without erythrocytes and blood pigment. Nature 173(4410), 848 (1954)

[Saigusa et al., 2008] Saigusa, T., Tero, A., Nakagaki, T., Kuramoto, Y.: Amoebae anticipate periodic events. Physical Review Letters 100(1), 018101 (2008)

[Savage et al., 2007] Savage, V., Allen, A., Brown, J., Gillooly, J., Herman, A., Woodruff, W., West, G.: Scaling of number, size, and metabolic rate of cells with body size in mammals. Proceedings of the National Academy of Sciences 104(11), 4718 (2007)

[Savage et al., 2004] Savage, V., Gilloly, J., Woodruff, W., West, G., Allen, A., Enquist, B., Brown, J.: The predominance of quarter-power scaling in biology. Ecology 18, 257–282 (2004)

[Scheffer et al., 2009] Scheffer, M., Bascompte, J., Brock, W., Brovkin, V., Carpenter, S., Dakos, V., Held, H., Van Nes, E., Rietkerk, M., Sugihara, G.: Early-warning signals for critical transitions. Nature 461(7260), 53–59 (2009)

[Schmidt-Nielsen, 1984] Schmidt-Nielsen, K.: Scaling. Cambridge U. P (1984)

[Schrödinger, 2000] Schrödinger, E.: What Is Life? Cambridge U.P. (2000)

[Schulte-Frohlinde & Kleinert, 2001] Schulte-Frohlinde, V., Kleinert, H.: Critical properties of phi4-theories. World Scientific (2001)

[Sen, 2010] Sen, S.: Symmetry, symmetry breaking and topology. Symmetry 2(3), 1401–1422 (2010)

[Sethna, 2006] Sethna, J.P.: Statistical mechanics: Entropy, order parameters, and complexity. Oxford University Press, New York (2006)

[Shinde et al., 2011] Shinde, D.P., Mehta, A., Mishra, R.K.: Searching and fixating: Scale-invariance vs. characteristic timescales in attentional processes. Europhysics Letters 94(6), 68001 (2011)

[Shmulevich et al., 2005] Shmulevich, I., Kauffman, S., Aldana, M.: Eukaryotic cells are dynamically ordered or critical but not chaotic. Proceedings of the National Academy of Sciences 102(38), 13439–13444 (2005)

[Silva & Annamalai, 2008] Silva, C., Annamalai, K.: Entropy generation and human aging: Lifespan entropy and effect of physical activity level. Entropy 10(2), 100–123 (2008)

[Silverman, 2008] Silverman, M.: Quantum Superposition. Springer (2008)

[Sinclair et al., 2000] Sinclair, S., McKinney, S., Glenny, R., Bernard, S., Hlastala, M.: Exercise alters fractal dimension and spatial correlation of pulmonary blood flow in the horse. Journal of Applied Physiology: Respiratory, Environmental and Exercise Physiology 88(6), 2269–2278 (2000)

[Singer et al., 1995] Singer, D., Schunck, O., Bach, F., Kuhn, H.: Size effects on metabolic rate in cell, tissue, and body calorimetry. Thermochimica Acta 251, 227–240 (1995)

[Smith et al., 1999] Smith, P., Morrison, I., Wilson, K., Fernández, N., Cherry, R.: Anomalous diffusion of major histocompatibility complex class i molecules on hela cells determined by single particle tracking. Biophys. J. 76(6), 3331–3344 (1999)

[Smith et al., 1996] Smith, T., Lange, G., Marks, W.: Fractal methods and results in cellular morphology–dimensions, lacunarity and multifractals. Journal of Neuroscience Methods 69(2), 123–136 (1996)

[Sohal & Weindruch, 1996] Sohal, R., Weindruch, R.: Oxidative stress, caloric restriction, and aging. Science 273(5271), 59–63 (1996)

[Song & Lehrer, 2003] Song, H., Lehrer, P.: The effects of specific respiratory rates on heart rate and heart rate variability. Applied Psychophysiology and Biofeedback 28(1), 13–23 (2003)

[Sonnenschein & Soto, 1999] Sonnenschein, C., Soto, A.: The society of cells: cancer and control of cell proliferation. Springer, New York (1999)

[Sonnenschein & Soto, 2000] Sonnenschein, C., Soto, A.: Somatic mutation theory of carcinogenesis: why it should be dropped and replaced. Molecular Carcinogenesis 29(4), 205–211 (2000)

[Sornette et al., 1995] Sornette, D., Johansen, A., Dornic, I.: Mapping self-organized criticality onto criticality. J. Phys. I France 5(3), 325–335 (1995)

[Soto et al., 2008] Soto, A., Sonnenschein, C., Miquel, P.-A.: On physicalism and downward causation in developmental and cancer biology. Acta Biotheoretica 56(4), 257–274 (2008)

[Soto & Sonnenschein, 2011] Soto, A.M., Sonnenschein, C.: The tissue organization field theory of cancer: a testable replacement for the somatic mutation theory. Bioessays 33(5), 332–340 (2011)

[Stern et al., 2009] Stern, G., Beel, J., Suki, B., Silverman, M., Westaway, J., Cernelc, M., Baldwin, D., Frey, U.: Long-range correlations in rectal temperature fluctuations of healthy infants during maturation. PLoS One 4(7), e6431 (2009)

[Stewart, 2004] Stewart, J.: La vie existe-t-elle? Editions Vuibert (2004)

[Stiegler, 2001] Stiegler, B.: Nietzsche et la biologie. Puf, Paris (2001)

[Strocchi, 2005] Strocchi, F.: Symmetry breaking. Lecture Notes in Physics, vol. 732. Springer, Heidelberg (2005)

[Suki et al., 1994] Suki, B., Barabási, A., Hantos, Z., Peták, F., Stanley, H.: Avalanches and power-law behaviour in lung inflation. Nature 368(6472), 615–618 (1994)

References

[Szpunar et al., 2007] Szpunar, K., Watson, J., McDermott, K.: Neural substrates of envisioning the future. Proceedings of the National Academy of Sciences 104(2), 642 (2007)

[Taniguchi et al., 2010] Taniguchi, Y., Choi, P.J., Li, G.-W., Chen, H., Babu, M., Hearn, J., Emili, A., Xie, X.S.: Quantifying e. coli proteome and transcriptome with single-molecule sensitivity in single cells. Science 329(5991), 533–538 (2010)

[Thamrin et al., 2010] Thamrin, C., Stern, G., Frey, U.: Fractals for physicians. Paediatric Respiratory Reviews 11(2), 123–131 (2010)

[Thompson, 2007] Thompson, E.: Mind in Life. Biology, Phenomenology and the Sciences of Mind. Harvard University Press, Cambridge (2007)

[Tofani, 1999] Tofani, S.: Physics help chemistry to improve medicine: a possible mechanism for anticancer activity of static and elf magnetic fields. Physica Medica 4, 291–294 (1999)

[Tolić-Nørrelykke et al., 2004] Tolić-Nørrelykke, I., Munteanu, E.-L., Thon, G., Oddershede, L., Berg-Sørensen, K.: Anomalous diffusion in living yeast cells. Phys. Rev. Lett. 93(7), 078102 (2004)

[Toll, 1956] Toll, J.: Causality and the dispersion relation: Logical foundations. Phys. Rev. 104(6), 1760–1770 (1956)

[Toulouse et al., 1977] Toulouse, G., Pfeuty, P., Barton, G.: Introduction to the renormalization group and to critical phenomena. Wiley, London (1977)

[Trieb et al., 1976] Trieb, G., Pappritz, G., Lützen, L.: Allometric analysis of organ weights. I. Rats. Toxicology and Applied Pharmacology 35(3), 531–542 (1976)

[Tsallis & Tirnakli, 2010] Tsallis, C., Tirnakli, U.: Nonadditive entropy and nonextensive statistical mechanics–some central concepts and recent applications. Journal of Physics: Conference Series 201, 012001 (2010)

[Van Fraassen, 1989] Van Fraassen, B.: Laws and symmetry. Oxford University Press, USA (1989)

[Van Gelder, 1999] Van Gelder, T.: Wooden iron? Husserlian phenomenology meets cognitive science, pp. 245–265. Stanford U. P. (1999)

[van Saarloos & Kurtze, 1984] van Saarloos, W., Kurtze, D.A.: Location of zeros in the complex temperature plane: Absence of lee-yang theorem. Journal of Physics A: Mathematical and General 17(6), 1301 (1984)

[Varela, 1979] Varela, F.: Principles of biological autonomy. North Holland, New York (1979)

[Varela, 1989] Varela, F.: Autonomie et connaissance. Seuil, Paris (1989)

[Varela, 1997] Varela, F.: Patterns of life: Intertwining identity and cognition. Brain and Cognition 34(1), 72–87 (1997)

[Varela, 1999] Varela, F.: The specious present: A neurophenomenology of time consciousness, pp. 266–314. Stanford U. P. (1999)

[Varela et al., 1974] Varela, F., Maturana, H., Uribe, R.: Autopoiesis: The organization of living systems, its characterization and a model. Biosystems 5(4), 187–196 (1974)

[Varela et al., 2003] Varela, M., Jimenez, L., Fariña, R.: Complexity analysis of the temperature curve: New information from body temperature. European Journal of Applied Physiology 89, 230–237 (2003)

[Vaz & Varela, 1978] Vaz, N., Varela, F.: Self and non-sense: an organism-centered approach to immunology. Medical Hypotheses 4(3), 231–267 (1978)

[Vilar & Rubí, 2001] Vilar, J.M.G., Rubí, J.M.: Thermodynamics "beyond" local equilibrium. Proceedings of the National Academy of Sciences 98(20), 11081–11084 (2001)

[Waddington, 1977] Waddington, G.: Stabilisation in systems: Chreods and epigenetic landscapes. Futures 9(2), 139–146 (1977)

[Wagner et al., 1996] Wagner, C., Nafz, B., Persson, P.: Chaos in blood pressure control. Cardiovascular Research 31(3), 380 (1996)

[Wang et al., 2001] Wang, Z., O'Connor, T., Heshka, S., Heymsfield, S.: The reconstruction of Kleiber's law at the organ-tissue level. Journal of Nutrition 131(11), 2967–2970 (2001)

[Weibel, 1994] Weibel, E.: Design of biological organisms and fractal geometry. In: Fractals in Biology and Medicine, vol. 1. Springer (1994)

[Weibel et al., 2004] Weibel, E., Bacigalupe, L., Schmitt, B., Hoppeler, H.: Allometric scaling of maximal metabolic rate in mammals: Muscle aerobic capacity as determinant factor. Respiratory Physiology and Neurobiology 140(2), 115–132 (2004)

[Weinberg, 1995] Weinberg, S.: The Quantum Theory of Fields. Cambridge University Press (1995)

[Weiss et al., 2004] Weiss, M., Elsner, M., Kartberg, F., Nilsson, T.: Anomalous subdiffusion is a measure for cytoplasmic crowding in living cells. Biophys. J. 87(5), 3518–3524 (2004)

[Weiss, 1992] Weiss, V.: Major genes of general intelligence. Personality and Individual Differences 13(10), 1115–1134 (1992)

[Werner, 2007] Werner, G.: Metastability, criticality and phase transitions in brain and its models. Biosystems 90(2), 496–508 (2007)

[Werner, 2010] Werner, G.: Fractals in the nervous system: conceptual implications for theoretical neuroscience. Frontiers in Physiology 1(0) (2010)

[West, 2006] West, B.: Where medicine went wrong: Rediscovering the path to complexity. Studies of nonlinear phenomena in life sciences, vol. 11. World Scientific, Teaneck (2006)

[West, 2010] West, B.: Fractal physiology and the fractional calculus: a perspective. Frontiers in Physiology 1(0) (2010)

[West et al., 1986] West, B., Bhargava, V., Goldberger, A.: Beyond the principle of similitude: renormalization in the bronchial tree. Journal of Applied Physiology 60(3), 1089–1097 (1986)

[West & Brown, 2005] West, G., Brown, J.: The origin of allometric scaling laws in biology from genomes to ecosystems: Towards a quantitative unifying theory of biological structure and organization. Journal of Experimental Biology 208(9), 1575–1592 (2005)

[West et al., 1997] West, G., Brown, J., Enquist, B.: A general model for the origin of allometric scaling laws in biology. Science 276(5309), 122–126 (1997)

[West et al., 1999] West, G., Brown, J., Enquist, B.: The fourth dimension of life: Fractal geometry and allometric scaling of organisms. Science 284(5420), 1677–1679 (1999)

[Weyl, 1983] Weyl, H.: Symmetry. Princeton Univ. Pr. (1983)

[White & Seymour, 2003] White, C., Seymour, R.: Mammalian basal metabolic rate is proportional to body mass 2/3. Proceedings of the National Academy of Sciences of the United States of America 100(7), 4046–4049 (2003)

[Wildman & Kling, 1978] Wildman, D., Kling, M.: Semantic, syntactic, and spatial anticipation in reading. Reading Research Quarterly 14(2), 128–164 (1978)

[Wilson et al., 2009] Wilson, B., Hart, G., Parcell, A.: Cardiac inter-beat interval complexity is influenced by physical activity. Medical Physiology Online 2 (2009)

[Winfree, 2001] Winfree, A.: The geometry of biological time, 2nd edn. Springer (2001)

[Winkler et al., 2005] Winkler, J.R., Gray, H.B., Prytkova, T.R., Kurnikov, I.V., Beratan, D.N.: Electron Transfer through Proteins, pp. 15–33. Wiley-VCH Verlag GmbH and Co. KGaA (2005)

[Witten et al., 1997] Witten, M., Tinajero, J., Sobonya, R., Lantz, R., Quan, S., Lemen, R.: Human alveolar fractal dimension in normal and chronic obstructive pulmonary disease subjects. Research Communications in Molecular Pathology and Pharmacology 98(2), 221 (1997)

References

[Yadav et al., 2010] Yadav, C., Verma, M., Ghosh, S.: Statistical evidence for power law temporal correlations in exploratory behaviour of rats. Biosystems 102(2-3), 77–81 (2010)

[Yan et al., 2008] Yan, R., Yan, G., Zhang, W., Wang, L.: Long-range scaling behaviours of human colonic pressure activities. Communications in Nonlinear Science and Numerical Simulation 13(9), 1888–1895 (2008)

[Zamir, 2001] Zamir, M.: Fractal dimensions and multifractility in vascular branching. Journal of Theoretical Biology 212(2), 183–190 (2001)

[Zinn-Justin, 2007] Zinn-Justin, J.: Phase transitions and renormalization group. Oxford University Press, New York (2007)